T0295171

Materials for Civil Engineering: Testing and Applications

Materials for Civil Engineering: Testing and Applications

Edited by Cayson Carter

www.statesacademicpress.com

States Academic Press,
109 South 5th Street,
Brooklyn, NY 11249, USA

Visit us on the World Wide Web at:
www.statesacademicpress.com

ISBN: 978-1-63989-757-5

Trademark Notice: Registered trademark of products or corporate names are used only for explanation and identification without intent to infringe.

Cataloging-in-Publication Data

Materials for civil engineering : testing and applications / edited by Cayson Carter.
    p. cm.
Includes bibliographical references and index.
ISBN 978-1-63989-757-5
1. Civil engineering. 2. Building materials. 3. Testing. I. Carter, Cayson.
TA153 .M38 2023
624--dc23

Table of Contents

Preface

Civil engineering refers to the application of scientific and physical principles for the design, development and maintenance of both the man-made and naturally built environments. This comprises infrastructure including buildings, highways, airports, sewage systems, bridges, pipelines, railways, canals, power plants and dams. The basic materials utilized in civil engineering applications are bitumen and bituminous materials, reinforced and structural steels, cement and concrete, structural clay, and concrete units. The success of civil engineering projects heavily relies on the quality of the materials used, therefore it is crucial to have testing tools that can deliver precise and repeatable results to customers. The testing of materials in civil engineering can be done through destructive and non-destructive methods. Non-destructive testing (NDT) refers to a type of method used to evaluate the long-term viability and strength of existing concrete structures using a set of non-invasive inspection procedures. This book aims to shed light on the various materials used in civil engineering as well as their testing and applications. Researchers and students in the field of civil engineering will be assisted by it.

After months of intensive research and writing, this book is the end result of all who devoted their time and efforts in the initiation and progress of this book. It will surely be a source of reference in enhancing the required knowledge of the new developments in the area. During the course of developing this book, certain measures such as accuracy, authenticity and research focused analytical studies were given preference in order to produce a comprehensive book in the area of study.

This book would not have been possible without the efforts of the authors and the publisher. I extend my sincere thanks to them. Secondly, I express my gratitude to my family and well-wishers. And most importantly, I thank my students for constantly expressing their willingness and curiosity in enhancing their knowledge in the field, which encourages me to take up further research projects for the advancement of the area.

Editor

1

Tensile and Shear Testing of Basalt Fiber Reinforced Polymer (BFRP) and Hybrid Basalt/Carbon Fiber Reinforced Polymer (HFRP) Bars

Kostiantyn Protchenko *⬤, Fares Zayoud ⬤, Marek Urbański and Elżbieta Szmigiera

Faculty of Civil Engineering, Warsaw University of Technology, 00-637 Warsaw, Poland;
fares.zayoud.stud@pw.edu.pl (F.Z.); m.urbanski@il.pw.edu.pl (M.U.); e.szmigiera@il.pw.edu.pl (E.S.)
* Correspondence: k.protchenko@il.pw.edu.pl

Abstract: The use of sustainable materials is a challenging issue for the construction industry; thus, Fiber Reinforced Polymers (FRP) is of interest to civil and structural engineers for their lightweight and high-strength properties. The paper describes the results of tensile and shear strength testing of Basalt FRP (BFRP) and Hybrid FRP (HFRP) bars. The combination of carbon fibers and basalt fibers leads to a more cost-efficient alternative to Carbon FRP (CFRP) and a more sustainable alternative to BFRP. The bars were subjected to both tensile and shear strength testing in order to investigate their structural behavior and find a correlation between the results. The results of the tests done on BFRP and HFRP bars showed that the mechanical properties of BFRP bars were lower than for HFRP bars. The maximum tensile strength obtained for a BFRP bar with a diameter of 10 mm was equal to approximately 1150 MPa, whereas for HFRP bars with a diameter of 8 mm, it was higher, approximately 1280 MPa. Additionally, better results were obtained for HFRP bars during shear testing; the average maximum shear stress was equal to 214 MPa, which was approximately 22% higher than the average maximum shear stress obtained for BFRP bars. However, HFRP bars exhibited the lowest shear strain of 57% that of BFRP bars. This confirms the effectiveness of using HFRP bars as a replacement for less rigid BFRP bars. It is worth mentioning that after obtaining these results, shear testing can be performed instead of tensile testing for future studies, which is less complicated and takes less time to prepare than tensile testing.

Keywords: Fibers Reinforced Polymer (FRP); Hybrid Fibers Reinforced Polymer (HFRP); Basalt Fibers Reinforced Polymer (BFRP); tensile strength testing; shear strength testing

1. Introduction

Concrete structures built during the previous century will be in need of modernization and renovation, which will lead to an increase in resources and costs associated with these buildings. Hence, it is necessary to develop inexpensive and efficient modernizing techniques in order to prevent structural damage. Throughout the years, researchers have experimented with different materials in various contexts, and Fiber Reinforced Polymers (FRP) have proved to be an applicable material for strengthening and reinforcing concrete structures.

For decades, FRP has been widely used in the construction industry as a substitute for steel due to its lightweight, increased corrosion resistance, and improved strength. FRP can be used in a wide range of applications in construction, such as reinforcement of concrete, formwork, modular structures, bridge decks, and external reinforcement for strengthening structures.

FRP reinforcement exhibits a linear elastic behavior prior to failure but fails in a brittle manner. The weakness of FRP reinforcement in terms of its mechanical behavior is observed in the transverse direction, which can be seen in shear resistance testing, where the strength characteristics of tested samples are significantly reduced. Moreover, the FRP reinforcement lacks plastic behavior. However, FRP reinforcement has several advantages, such as high tensile strength in the longitudinal direction, long durability, high corrosion resistance, and easy application.

Tensile tests have been conducted by several researchers over the past decades on different FRP bars to determine their mechanical and physical properties [1–7]. Conclusions drawn from these earlier studies [4–10] indicate that the mechanical characteristics of FRP bars are mainly dependent on fiber volume and type. Studies have also shown that FRP bars exhibit elastic linear behavior up to failure, with a modulus of elasticity lower than that of steel.

In the studies of Walsh, Chang, and Wallenberger et al. [8–10], extensive research can be found on the most common fibers used to manufacture FRP rebars, such as carbon, aramid, and glass. Hollaway et al. [11] found that fibers exhibited a linear elastic behavior under tensile loading up to failure without showing any yield.

Nanni et al. [12] investigated the tensile properties of hybrid rods for concrete reinforcement. The tested specimens were made of 1.0 m long rods, including the anchorage system. The results of uniaxial tensile testing, conducted in order to determine stress–strain relationships and tensile properties, showed that the stress–strain diagrams displayed a bilinear behavior.

Zhishen Wu et al. found that the tensile modulus of the fibers influenced the fatigue failure mode, where the Glass Fiber Reinforced Polymers (GFRP) and Basalt Fiber Reinforced Polymers (BFRP) composites exhibited transverse cracks that indicated a stress transfer to the adhesive layer with a modulus of about 90 GPa, whereas the Carbon Fiber Reinforced Polymers (CFRP) and Poly(p-phenylene benzobisoxazole) (PBO) composites showed longitudinal cracks along the test coupons with a modulus of approximately 250 GPa [13].

Marta Kosior-Kazberuk performed an investigation on the application of Basalt FRP (BFRP) bars and Hybrid FRP (HFRP) bars as reinforcement for full-scale concrete beams. The results showed that the FRP bars used in the tensile zone worked in a similar way to steel reinforcement. Rapid destruction in the tensile zone with crushing of concrete in the compression zone, deformation of compressed bars, stirrups breaking within cracks, and failure of tensile reinforcement were observed [14].

Shi et al. [15] investigated the tensile behavior of FRP and Hybrid FRP sheets in Freeze–Thaw (FT) cycling environments. Additional degradation of the tensile strength and rupture elongation of FRP sheets were observed as a result of sustained loading during FT cycling, while the elastic modulus of the FRP sheets was not influenced by the sustained loading during FT cycling. Hence, it is necessary to consider service loads during durability tests to reflect the actual conditions faced by FRP composites in harsh civil engineering environments.

Many researchers have studied the shear strength behavior of different FRP bars. The limited application of BFRP in the past decades led to the limited investigation of its shear behavior. However, Benmokrane et al. [16] proved that the type of the resin plays a significant role in the inter-laminar shear strength of BFRP bars, as BFRP bars with epoxy resin showed higher durability than BFRP bars with vinyl ester resin. In addition, the durability of GFRP bars was lower than for BFRP bars with epoxy resin.

Alam et al. [17] studied the shear strength of FRP reinforced members without transverse reinforcement and reported that the normalized shear strength increased linearly with the cube root of the axial stiffness of the reinforcing bars. However, there were similar results for the shear strength of FRP reinforced elements where it was proportional to the axial stiffness of the longitudinal reinforcement obtained by Alam et al. [17], El-Sayed et al. [18,19], and Razaqpur et al. [20]. Hence, the conclusion that the shear strength was proportional to the amount of longitudinal reinforcement was reached by Alkhrdaji et al. [21], and the effect of the reinforcement stiffness and amount of reinforcement on the shear strength of FRP elements, despite no significant influence of the longitudinal reinforcement ratio on the shear strength, was observed by Yost et al. [22].

Khalifa et al. [23], Täljsten [24], and Triantafillou [25] developed a theory that aims at describing FRP stress distribution along a shear crack with closed-form equations, as opposed to the regression-based formula that Triantafillou et al. [26] introduced. FRP contribution to the resisting shear and the FRP resultant across the crack can be computed when this formula is correctly defined.

Based on the literature, axial tensile testing is most commonly used for defining the mechanical properties of composite bars, and some of the researchers have studied the shear properties of the bars through tensile testing. Current research is concentrated on the defining properties of BFRP and HFRP bars through axial-tensile testing and shear testing, as well as finding a correlation between these two testing methods.

The variables in this study were types of FRP bars and the diameter of bars. The outcomes were compared, and a correlation between the results was found. Establishing this correlation allowed the identification of mechanical properties of FRP bars basing on shear testing, which is easier to prepare and takes less time than tensile testing.

Purpose and Novelty of the Work

RC bending elements are the basic structural components that cover the operational load as well as the load from partition walls and transfer them further to vertical elements (such as columns and load-bearing walls). Reinforcement of these elements should be characterized by high strength values and appropriate corrosion resistance. In order to determine these properties, strength tests should be carried out. On the one hand, the tensile strength test for FRP bars is complex, relatively labor-intensive, and expensive. On the other hand, testing the shear strength of FRP bars is simple, low-cost, and fast. By carrying out a quick and simple shear test, it is possible to determine the tensile strength and stiffness (modulus of elasticity) of both homogeneous BFRP bars and HFRP bars on the basis of the method presented in the article. Additionally, the method allows optimizing the tensile strength and stiffness based on the determined correlation between shear strength and tensile strength.

2. The Concept of Hybrid FRP Bars

In spite of the fact that FRP has numerous prevalent material properties, such as high corrosion resistance, high specific stiffness, high specific strength, and durability, the brittle nature and high cost of FRP limits its extensive usage in many industries. To overcome these obstacles, combinations of FRP and ordinary materials are being explored by several researchers [27]. Appropriate properties can be obtained by utilizing a combination of FRP and steel [28,29] or by combining different sorts of FRP materials in one structure [30]. The advantages of hybrid structural systems include cost-effectiveness and the ability to optimize the cross-section based on the material properties of each constituent material [31].

HFRP bars can be an optimal solution since hybrid bars have a better combination of mechanical properties and are less expensive than FRP bars with homogeneous fibers. On the other hand, they are corrosion resistant and have significantly better strength characteristics than conventional steel reinforcements [32]. Hybridization between different constituents focuses on picking out the advantages of each constituent, while the disadvantages can be improved [33].

Due to the high cost of carbon, the ideal solution is the usage of HFRP bars, where the part of low-cost basalt fibers was substituted by high-cost carbon fibers. Hence, the hybridization of carbon/basalt fibers is less expensive than CFRP; concurrently, this combination is characterized by better mechanical properties in comparison with BFRP [34–36].

Carbon fibers are chosen due to their high properties in the longitudinal direction as well as their strong anisotropy. However, basalt fibers were chosen for their environmentally friendly producing process and low-cost. In addition, basalt fibers are significantly less brittle when used in composites [37]. An additional reason for such selection was similar strain parameters for both types of fibers. Mechanical properties of the constituents utilized for preparing HFRP bars are represented in Table 1.

Table 1. Properties of constituents utilized for preparing Hybrid Fiber Reinforced Polymers (HFRP) rebars [34].

Properties	Units	Carbon Fibers LS	Basalt Fibers	Epoxy Resin
Density	g/cm³	1.90–2.10	2.60–2.80	1.16
Diameter	μm	7.00–11.00	11.20–13.40	–
E_{11}	GPa	232.00	89.00	3.45
E_{22}	GPa	15.00	89.00	3.45
v_{12}	–	0.28	0.26	0.35
v_{23}	–	0.49	0.26	0.35
G_{12}	GPa	24.00	21.70	1.28
G_{23}	GPa	5.03	21.70	1.28
σ_{11}	MPa	2500–3500	1153–2100	55–130

E_{ii} is the modulus of elasticity along axis i, v_{ij} is the Poisson ratio that corresponds to a contraction in direction j when an extension is applied in direction i, G_{ij} is the shear modulus in direction j on the plane whose normal is in direction i, and σ_{ii} is the tensile strength in the direction i [34].

Composite materials properties can be calculated according to the Rule of Mixtures (ROM) (axial loading–Voigt model) based on the literature [34–36]. Based on the Rule of Mixtures (ROM) equation, Young's modulus and other parameters were determined for the Hybrid Carbon/Basalt FRP (HC/BFRP) and for different combinations depending on carbon to basalt fiber volume fractions. Different levels of fibers substitution (carbon-to-basalt C/B) were proposed: 1:1, 1:2, 1:3, 1:4, and 1:9, which is represented in Figure 1 [36].

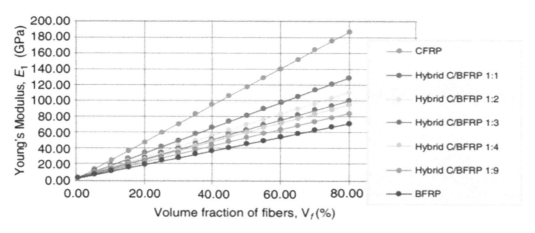

Figure 1. Theoretical relationship of fiber volume fraction and Young's modulus (obtained from Voigt's Model) for Hybrid Carbon/Basalt FRP [36,37].

The Voigt model does not consider the location of the fibers; therefore, numerical analysis was made for two configurations: (i) carbon fibers in the core region and basalt fibers in the near-surface region, (ii) carbon fibers in the near-surface region and basalt fibers in the core region of the bar section. The numerical simulation of the tensile strength test for HFRP bars was performed by Finite Element Methods (FEM). The bars consisted of two parts, the core region (cylindrical form) and the surface region (tube form), which were perfectly interconnected. The constant pressure was applied on both sides along the longitudinal axis. The obtained results from numerical modeling were compared with analytical considerations and were found to be in a good convergence with each other.

Numerical results indicated that the arrangement of the fibers was not of high importance for the final mechanical properties of HFRP bars with different combinations of fibers. More on the analytical and numerical investigation of Hybrid FRP bars can be found in companion papers [36,37]. Two different HFRP bars' configurations were produced by the manufacturing company. For the first configuration, carbon fibers were placed in the core region, while the second configuration was when carbon fibers were in the near-surface region. However, some technological issues were observed while placing carbon near-surface, increasing heterogeneity in fiber distribution, and local scorching of carbon fibers caused by rising temperature. Finally, carbon fibers were placed mostly in the core region of the bar.

HFRP bars utilized for this work were composed of epoxy matrix and basalt fibers, with a volume ratio of 1:4 (i.e., 16% carbon fibers, 64% basalt fibers, and 20% matrix). Due to this substitution, HFRP bars are characterized by a much higher stiffness, which enables more efficient application as reinforcement for concrete elements subjected to bending and compression [37–39]. More about the hybridization of FRP bars used in work can be found in these companion papers [38,39].

In the tensile and shear strength tests conducted, two types of FRP were used: BFRP and HFRP. Twisting FRP braids were done in order to improve adhesion with concrete (the equivalent of ribbing for steel bars) [36].

3. Tensile Strength Test

3.1. Experimental Procedure

The tensile strength testing of different BFRP and HFRP bars was carried out according to the ACI 440.3R and ASTM D7205/D7205M methodologies [40].

In the tensile strength test, anisotropic FRP materials of different diameters were tested according to the methodologies. The two different types of FRP bars had different diameters, specifically Ø6, Ø8, Ø10, Ø12, Ø14, and Ø18 mm. Hence, for each type of BFRP and HFRP bar, five samples of each diameter were tested. In total, tensile tests were carried out on sixty bars.

Basalt fiber strength in the transverse direction is very low compared to its extremely high longitudinal strength. Hence, the use of appropriate anchorage at both ends of the bar subjected to the tensile test on the testing apparatus is a requirement. To meet this requirement, two steel pipes with a length of 400 mm, an external diameter of 40 mm, and a thickness of 5 mm each were designed. The steel caps at the end of the steel pipes were designed with an opening in the center for the bar.

The free space between the bar and the pipe was filled with a special adhesive layer. The material filling the anchoring pipes was determined on the basis of previous tests done at the Warsaw University of Technology [39,40]. Figure 2a shows the BFRP bars prepared for tensile strength testing; Figure 2b shows the HFRP bars in the tensile testing machine.

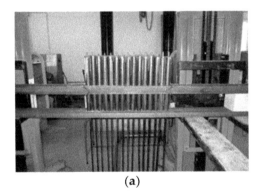
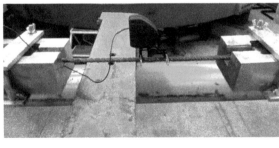

(a)   (b)

Figure 2. (**a**) Basalt Fiber Reinforced Polymers (BFRP) bars prepared for testing, (**b**) Hybrid Fiber Reinforced Polymers (HFRP) bars in the tensile testing machine.

The tensile strength test was carried out in accordance with [40] standard for pultruded FRP bars. However, the tensile strength of the specimens can be calculated by dividing the measured load by the transverse cross-sectional area of FRP bars of the corresponding type, $A_{i,FRP}$. To experimentally determine the modulus of elasticity of FRP bars, Equation (1) can be used.

$$E_{11,iFRP} = \frac{(P_1 - P_2)}{(\varepsilon_1 - \varepsilon_2) \cdot A_{i,FRP}}$$
(1)

where:

- $E_{11,iFRP}$ —modulus of elasticity of corresponding FRP bars along the longitudinal axis,
- P_1 and P_2—applied loads corresponding to 50% and 25% of the ultimate load, respectively,
- ε_1 and ε_2—strains corresponding to 50% and 25% of the ultimate strains, respectively,
  $A_{i,FRP}$—cross-sectional area for BFRP and HFRP bars, respectively.

3.2. Calculations and Results

The deformation measurement was performed up to 70% of the breaking load of the bars due to the possibility of strain gauge damage.

The average values obtained for BFRP and HFRP bars of tensile strength, tensile strain, and tensile modulus for six different diameters of each type are presented in Tables 2–4, respectively.

Table 2. Tensile strength results of FRP bars.

Bar	Parameter	Ø6	Ø8	Ø10	Ø12	Ø14	Ø18
BFRP	f_t, MPa	1148.81	1103.33	1152.54	1111.00	1101.94	877.98
	SD *, MPa	41.18	22.87	35.27	40.23	28.54	11.17
	COV **, %	3.58	2.07	3.06	3.62	2.59	1.27
HFRP	f_t, MPa	1083.78	1277.92	1138.65	1008.52	1160.06	809.44
	SD *, MPa	59.64	55.41	34.89	36.51	44.75	19.03
	COV **, %	4.72	4.34	3.06	3.62	3.86	2.35

* SD–Standard Deviation; ** COV–Coefficient of Variation.

Table 3. Modulus of elasticity of FRP bars.

Bar	Parameter	Ø6	Ø8	Ø10	Ø12	Ø14	Ø18
BFRP	E_f, GPa	46.47	43.87	44.27	47.63	46.02	42.20
	SD, GPa	2.58	0.86	1.71	0.89	0.23	1.58
	COV, %	5.55	1.95	3.56	1.87	0.49	3.74
HFRP	E_f, GPa	72.71	73.89	64.77	53.35	72.12	52.43
	SD, GPa	1.67	3.07	3.50	3.93	2.20	2.46
	COV, %	2.29	4.15	4.76	5.78	3.05	4.70

Table 4. Strains at FRP bars rupture.

Bar	Parameter	Ø6	Ø8	Ø10	Ø12	Ø14	Ø18
BFRP	ε_t, %	2.48	2.52	2.40	2.33	2.39	2.08
	SD, GPa	0.15	0.05	0.15	0.11	0.06	0.10
	COV, %	5.92	2.09	6.09	4.60	2.71	4.84
HFRP	ε_t, %	1.74	1.73	1.55	1.49	1.61	1.55
	SD, GPa	0.08	0.07	0.10	0.10	0.06	0.04
	COV, %	4.38	4.33	6.65	6.70	3.73	2.59

Figure 3 displays the stress–strain relationships of tension FRP bars tested with mean values for diameters Ø6, Ø8, Ø14, and Ø18 mm.

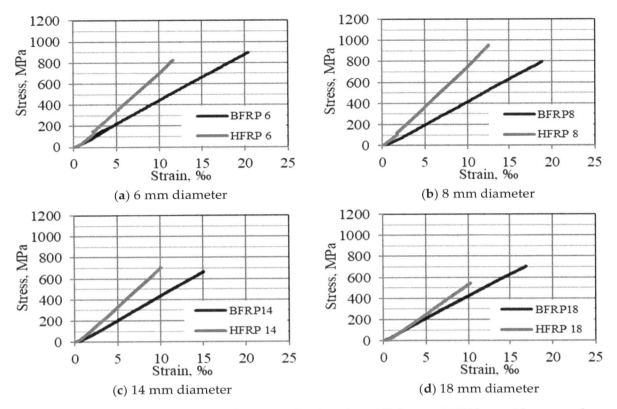

Figure 3. Stress–strain relationships of tension Fiber Reinforced Polymers (FRP) bars with mean values for diameters: (**a**) Ø6, (**b**) Ø8, (**c**) Ø14, (**d**) Ø18 mm.

In the tensile property values obtained for BFRP and HFRP, the deformation values obtained for HFRP were lower than the analogous deformation values for BFRP at rupture.

Additionally, comparing the (Coefficient of Variation, COV) values of the two types of FRPs used in the test, much higher (COV) values were obtained for HFRP bars compared to the tested BFRP bars. However, this phenomenon might occur due to technical issues indicated in the previous chapter, which might occur during the manufacturing of bars composed of several FRP roving. Hence, the (COV) value depends on the diameter of the bars. The average (COV) values for BFRP bars were twice as small as the corresponding COV values for HFRP. The changes in mechanical properties for the different diameters and FRP types are represented in Figure 4.

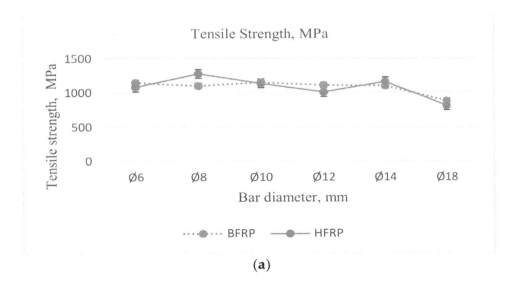

(**a**)

Figure 4. *Cont.*

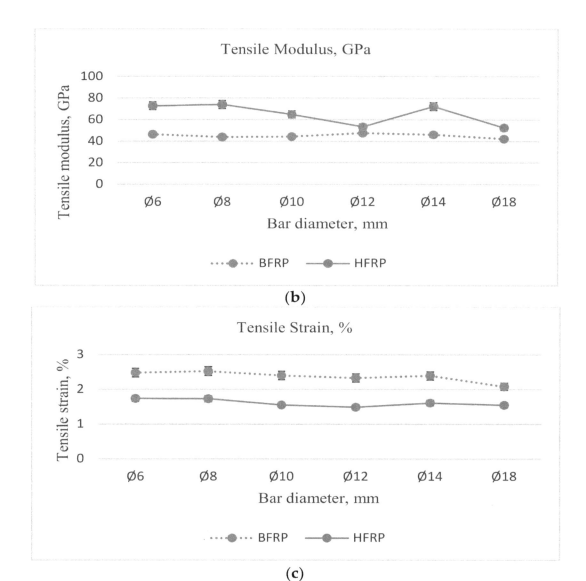

Figure 4. Mechanical properties of FRP bars obtained from tensile testing: (**a**) Tensile strength, (**b**) Modulus of elasticity, (**c**) Tensile strain.

Comparing the tensile strength of BFRP and HFRP bars with diameters of Ø8 and Ø14 mm, the values for BFRP bars of diameter Ø8 mm were smaller than HFRP bars of the same diameter by 15% and slightly smaller for bars of diameter Ø14 mm by approximately 5%. However, the tensile strength values for BFRP bars with diameters Ø6, Ø10, Ø12, and Ø18 mm were slightly higher than HFRP bars with the same diameters. Values for BFRP bars with diameters Ø6, Ø12, and Ø18 mm were slightly higher than HFRP bars by approximately 5%, and for diameter Ø10 mm by approximately 2%.

The rupture strain for all HFRP bar diameters is about 1/3 lower than the rupture strain of the BFRP bars. This is due to the smaller ultimate tensile strains of the carbon fiber (compared to the strain of basalt fiber) contained in the HFRP bars.

Observations made during the study confirm theoretical forecasts. The ultimate strain of hybrid bars corresponds to the strain of its fibers, which are characterized by the lowest ultimate strain among the fibers contained in the HFRP bar. The values obtained for the tensile modulus of elasticity for HFRP bars for all diameters were greater than the values obtained for the tensile modulus of elasticity for BFRP bars for all diameters.

The results obtained during this test confirm the viability of using HFRP bars as a replacement for less rigid BFRP bars.

4. Shear Strength Testing

4.1. Experimental Procedure

The shear strength test of FRP bars was carried out in accordance with the methodology described in ACI 440.3R-04 and ASTM D7617/D7617M–11 [40].

The shear strength testing was carried out on BFRP and HFRP bars with diameters Ø6, Ø8, Ø10, Ø12, Ø14, and Ø18 mm. For the shear strength test, in order to ensure that the bar sample shears in two planes simultaneously, a steel device was constructed with blades converging along a surface perpendicular to the sample axis. The device consists of one upper blade, two lower blades, and a sample holder. The sample holder has a V-shaped groove for placing FRP samples and a rectangular notch for holding the lower and upper blades in the center of the upper part of the device. The sample holder has 100 mm width, 110 mm height, and 230 mm length. It is attached to the instrument stand with screws that stabilize it in order to eliminate horizontal movement of the bar. Figure 5 represents the components of the shear testing device and the shear failure of the FRP bar. A total of 60 specimens of BFRP and sixty specimens of HFRP bar were tested, where five test specimens were produced for each diameter of every FRP bar type used during the shear strength test.

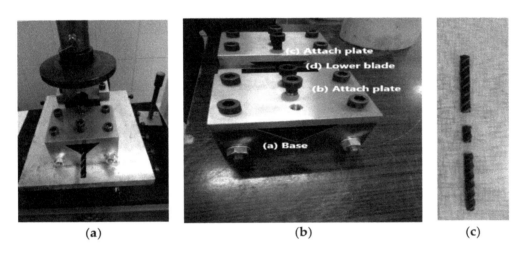

| (a) | (b) | (c) |

Figure 5. (**a**) Double shear testing device, Instron 3382, (**b**) Description of Instron 3382 testing machine, (**c**) Shear failure of the FRP bar.

All FRP bars that were tested for shear strength were composed of a spiral braid of basalt fiber. All the FRP bars were tested at speeds up to 50 MPa/min for a range of time between 3.5 min and 4.5 min, with the test ending when a sharp decrease in shear stress was observed.

According to ACI 440.6-08 and ASTM A615/A615M [40] methodologies, shear stress fs was calculated according to Equation (2) using an equivalent cross-section of bars.

$$f_s = \frac{F_u}{2 \cdot A_{i,FRP}} \qquad (2)$$

where:

- F_u —the maximum shear force, N.
- $A_{i,FRP}$ —the equivalent cross-section of the tested bar, mm^2.

4.2. Calculations and Results

Failure of the tested bars due to shearing could occur in two ways; two failure modes are shown in the Figure 6. The blue "B" line (straight line), which occurred most often, was typical for simultaneous shearing of a bar in two planes. In contrast, the red "R" (dot line) graph showed the situation when shearing occurred in one and then slightly later in the other cross-section of the tested bar.

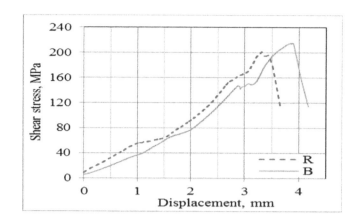

Figure 6. Stress-displacement diagram of transverse shear test for two HFRP bars with a mean of 14 mm.

Figure 7 shows a relationship of shear stress to strain for three types of BFRP and HFRP bars with nominal diameters of 6, 8, 10, 12, 14, 18 mm.

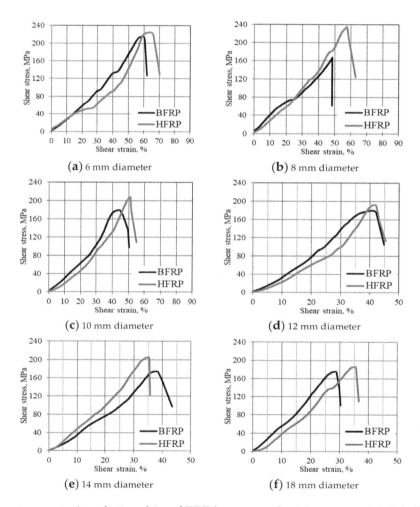

Figure 7. Shear stress–strain relationship of FRP bars tested with means. (a) Ø6, (b) Ø8, (c) Ø10, (d) Ø12, (e) Ø14, (f) Ø18 mm.

It was impossible to clearly determine the values of the shear modulus due to the non-uniform slope of the shear curves, in contrast to the stress–strain relationship. Comparing BFRP bars to HFRP bars, it was noticed that BFRP bars had greater shear deformation and less shear strength compared to HFRP bars. Table 5 shows shear strength values obtained for different FRP bars.

Table 5. Shear strength testing results of FRP bars for different diameters.

Bar	Parameter	Unit	Ø6	Ø8	Ø10	Ø12	Ø14	Ø18
BFRP	f_s	MPa	210.66	205.37	175.33	185.46	172.85	172.25
	SD	MPa	3.84	11.57	6.46	4.70	8.59	6.90
	COV	%	1.82	5.63	3.68	2.54	4.97	4.01
HFRP	f_s	MPa	202.04	229.44	198.31	187.09	196.84	176.35
	SD	MPa	17.51	4.06	6.17	5.99	12.06	9.75
	COV	%	8.67	1.77	3.11	3.20	6.13	5.53

The type of FRP bar affects the COV variation for shear strength. The quite significant spread of FRP bar test results confirms that they are heterogeneous materials consisting of a matrix and fibers. Studies show that the average lateral shear strength of BFRP bars ranges from 170 to 210 MPa. While for HFRP bars, it is up to 20% higher due to the participation of carbon fibers.

The transverse shear strength displays a slight downward trend as the bar diameter increases, most evidently in HFRP bars. However, minor variation is observed in the shear strength data, with the COV greater than in the rebar tension tests. Figure 8 shows the comparison of shear strength of FRP bars for different diameters.

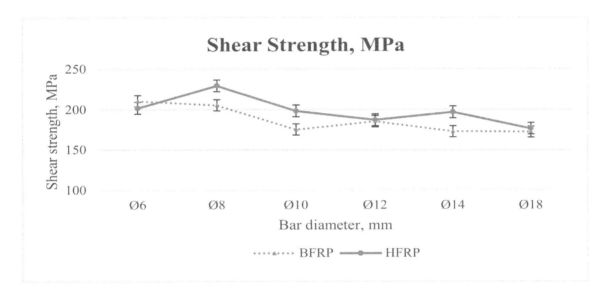

Figure 8. Comparison of average shear strength testing results of different FRP bars for all diameters.

Comparing the average shear strength of BFRP bars to HFRP bars, the values for BFRP bars were lower than the average shear strength values of HFRP bars except for diameter Ø6 mm, where it was slightly higher by approximately 1%. However, regarding the other diameters, the average shear strength of BFRP bars was lower for diameters Ø8, Ø10, and Ø14 mm by approximately 19% and for diameters Ø12 and Ø18 mm by approximately 1%. This might suggest that the shear strength of BFRP bars was lower than for HFRP.

5. Results and Discussions

From the results obtained from the tensile and shear strength testing of BFRP and HFRP with different diameters, a correlation was made between the obtained values in order to find a method for calculating tensile strength from the data obtained from shear strength experimental testing.

In Table 6, the correlation between shear strength and tensile strength of different types of FRP bars with different diameters is shown.

Table 6. Shear strength to tensile strength ratio.

Bar Type	Parameter	Ø6	Ø8	Ø10	Ø12	Ø14	Ø18
BFRP	f_s/f_t	0.183	0.186	0.152	0.167	0.157	0.196
HFRP	f_s/f_t	0.186	0.180	0.174	0.186	0.170	0.218

From Table 6, the average ratio (k_d) according to bar diameter can be calculated from Equation (3):

$$k_d = \frac{\sum \frac{f_s}{f_t}}{n_t} \tag{3}$$

where n_t–number of tested bar types ($n_t = 2$).

Based on Equation (3), the coefficient of accuracy corresponding to a given diameter can be identified. The values are presented in Table 7. Tables 8 and 9 represent comparison between the tensile strength values of experimental and calculated results.

Table 7. Coefficient of accuracy (k_d) of FRP bars according to diameter.

FRP Bars	Ø6	Ø8	Ø10	Ø12	Ø14	Ø18
Coefficient of Accuracy, k_d	0.185	0.183	0.163	0.176	0.163	0.207

Table 8. Tensile strength calculation according to diameter ratios ($f_t = f_s/k_d$).

Bar Diameter	k_d	BFRP	HFRP
mm	–	MPa	MPa
Ø6	0.185	1138.70	1092.11
Ø8	0.183	1122.24	1253.77
Ø12	0.163	1075.64	1216.63
Ø14	0.176	1053.75	1063.01
Ø16	0.163	1060.43	1207.61
Ø18	0.207	832.12	851.93

Table 9. Comparison between the tensile strength values of experimental and calculated results.

Bar type	Parameter	Units	Ø6	Ø8	Ø10	Ø12	Ø14	Ø18
BFRP	$f_t(exp)$	MPa	1148.81	1103.33	1152.54	1111.00	1101.94	877.98
	$f_t(\varnothing)$	MPa	1138.70	1122.24	1075.64	1053.75	1060.43	832.12
HFRP	$f_t(exp)$	MPa	1083.78	1277.92	1138.65	1008.52	1160.06	809.44
	$f_t(\varnothing)$	MPa	1092.11	1253.77	1216.63	1063.01	1207.61	851.93
BFRP Accuracy	$f_t(ex)/f_t(\varnothing)$	%	0.88%	−1.71%	6.67%	5.15%	3.77%	5.22%
HFRP Accuracy	$f_t(ex)/f_t(\varnothing)$	%	−0.77%	1.89%	−6.85%	−5.40%	−4.10%	−5.25%

Accuracy represented in Table 9 was calculated according to Equation (4):

$$FRP_{accuracy} = \frac{f_t(exp) - f_t(\varnothing)}{f_t(exp)} \cdot 100 \tag{4}$$

where:

- $f_t(exp)$ —Tensile strength obtained experimentally,
- $f_t(\varnothing)$ —Tensile strength obtained theoretically.

Figure 9 presents results obtained from Equation (3) according to the diameter of FRP bars proposed in order to calculate the tensile strength of FRP bars from the shear strength obtained experimentally; the accuracy of the values obtained theoretically using coefficient (k_d) varied between −7% and +7% in comparison with the tensile strength experimental values, using the Equation (5) for BFRP and HFRP:

$$f_{tensile} = \frac{f_{shear}}{k_d} \tag{5}$$

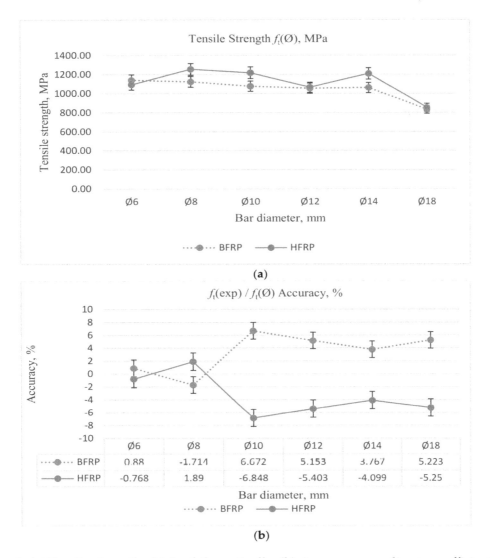

(a)

(b)

Figure 9. (a) Tensile strength obtained theoretically. (b) Accuracy according to coefficient (k_d).

Thus, it is possible to obtain the tensile strength of FRP bars by applying shear test strength only using Equation (5), taking into consideration the additional number of samples to be tested in order to eliminate errors and obtaining better accuracies. The coefficient (k_d) can be identified for any specific type of FRP, which will allow us to determine tensile strength based on shear testing.

The modulus of elasticity for the tested bar diameters depends mostly on the volume fraction of carbon and basalt fibers. The differences in volume fraction can be caused due to a various number of rovings used in the pultrusion process, which do not correspond to the bar diameter. In the case of

BFRP bars with diameters of 10, 12, and 14 mm, the volume fraction of basalt fibers was greater than the average value by about 10%, while for the diameters of 18 mm, it was lower by over 10%. On the other hand, for the HFRP bars with diameters of 14 mm, the volume fraction of fibers was higher by more than 10% than the average value, while for the diameters of 18 mm, it was 20% lower.

The tensile strength, in turn, is influenced by the shear lag effect, which causes that as the diameter of the bar increases, there is an increasing disproportion of the stress of the fibers in the cross-section of the bar.

Summarizing, the values of modulus of elasticity and tensile strength, as well as shear strength, depend on a combination of two factors, specifically the volume fraction of fibers and the shear lag effect.

6. Conclusions

The tests conducted in this study provided insight into the mechanical behavior of the types of FRP bars analyzed. Considering the results observed during tests done on BFRP bars for all diameters, it can be seen that the modulus of elasticity was quite low compared to HFRP bars, lower, on average, by approximately 18%. However, the main objective of this research was to check the maximum strength of HFRP and how it is compared to BFRP, as well as to find the stress–strain correlation. The influence of material hybridization was explored in order to examine the possibility of using less complicated shear testing as an alternative to tensile testing.

Based on experimental tests and numerical and analytical considerations for BFRP and HFRP bars, the following conclusive remarks can be drawn:

1. The tensile stress–strain correlation of BFRP bars was tested for given diameters. The average maximum stress obtained equaled to approximately 700 MPa, with an average strain approximately equal to 17%. However, the HFRP bars tested exhibited average maximum stress approximately equal to 725 MPa with the average strain approximately equal to 11%. This confirmed the viability of using HFRP bars as a replacement for less rigid BFRP bars.

2. Based on the tests performed, HFRP bars exhibited better results in terms of their tensile stress–strain relationship. However, comparing the obtained results for HFRP to BFRP bars, the tensile strength and modulus of elasticity of HFRP bars were higher by approximately 68% and 16%, respectively. The 6% lower elongation of HFRP bars could be explained by the superior extensibility properties of carbon fiber.

3. Additionally, according to the shear stress–strain relationship, better results were obtained for HFRP bars with average maximum shear stress equal to 214 MPa, which was approximately 22% higher than the average maximum shear stress obtained for BFRP bars. However, HFRP bars obtained a lower shear strain, 57% compared to BFRP bars.

4. Comparing experimental to analytical/numerical calculations done by the authors, the predicted analytical/numerical results were much better than those obtained experimentally. The difference in mechanical properties of FRP bars could be explained by the exact chemistry of the polymer matrix and the strength of polymer/BFRP or polymer/HFRP interactions.

5. Comparing the tensile stress–strain relationship to shear stress–strain for all bars and diameters, it was observed that BFRP and HFRP bars could withstand more tensile force and had a lower accompanying strain in tensile tests than shear tests.

The preparation for shear testing is less complicated than for tensile testing; therefore, on the basis of the results and observations, shear testing can substitute tensile testing for this class of composite materials.

Author Contributions: Conceptualization, K.P. and E.S.; Data curation, M.U.; Formal analysis, M.U.; Investigation, K.P. and F.Z.; Methodology, M.U. and E.S.; Supervision, M.U.; Validation, E.S.; Writing—original draft, K.P. and F.Z.; Writing—review & editing, E.S. All authors have read and agreed to the published version of the manuscript.

Acknowledgments: This research was funded by the National Centre for Research and Development (NCRD), grant number NCBR: PBS3/A2/20/2015, and the APC was funded by the Warsaw University of Technology. Special thanks to Andrzej Garbacz for useful private communication.

References

1.   Pleimann, L. Strength, modulus of elasticity, and bond of deformed FRP rods. In *Advanced Composites Materials in Civil Engineering Structures: Proceedings of the Specialty Conference*; ASCE: New York, NY, USA, 1991; pp. 99–110.

2.   Chaallal, O.; Benmokrane, B. Glass-fiber reinforcing rod: Characterization and application to concrete structures and grouted anchors. In *Materials: Performance and Prevention of Deficiencies and Failures: Proceedings of* the Materials Engineering Congress; ASCE: New York, NY, USA, 1992; pp. 606–617.

3.   Mukae, K.; Kamagai, S.; Nakai, H.; Asai, H. Characteristics of aramid FRP rods. In *Proceedings of the Symposium on Fiber Reinforced Polymer Reinforcement for Concrete Structures*; American Concrete Institute: Detroit, MI, USA, 1993.

4.   Yamasaki, Y.; Masuda, Y.; Tanano, H.; Shimizu, A. Fundamental properties of continuous fiber bars. In *Proceedings of the Symposium on Fiber Reinforced Polymer Reinforcement for Concrete Structures; ACI SP 138*; American Concrete Institute: Detroit, MI, USA, 1993.

5.   Nanni, A.; Okamoto, T.; Tanigaki, M.; Osakada, S. Tensile properties of braided FRP rods for concrete reinforcement. *Cem. Concr. Compos.* **1993**, *15*, 121–129.

6.   Porter, M.; Barnes, B. Tensile testing of glass fiber composite rod. In *Advanced Composites Materials in Civil Engineering Structures: Proceedings of the Specialty Conference*; ASCE: New York, NY, USA, 1991; pp. 123–131.

7.   Uomoto, T.; Hodhod, H. Properties of fiber reinforced plastic rods for prestressing tendons. In *Proceedings of the Symposium on Fiber Reinforced Polymer Reinforcement for Concrete Structures; ACI SP 138*; American Concrete Institute: Detroit, MI, USA, 1993.

8.   Walsh, P.J. Carbon Fibres. In *ASM Handbook Volume 21-Composites*; ASM International: Material Park, OH, USA, 2001; pp. 35–40.

9.   Chang, K.K. Aramid Fibers. In *ASM Handbook, Volume 21 Composites*; ASM International: Material Park, OH, USA, 2001; pp. 41–45.

10. Wallenberger, F.T.; Watson, J.C.; Hong, L. Glass Fibers. In *ASM Handbook Volume 21 Composites*; ASM International: Material Park, OH, USA, 2001; pp. 27–34.

11. Hollaway, L. *Polymer Composites for Civil and structural Engineering*; Blackie Academic & Professional: Glasgow, UK, 1993.

12. Nanni, A.; Henneke, M.J.; Okamoto, T. Tensile properties of hybrid rods for concrete reinforcement. *Constr. Build. Mater.* **1994**, *8*, 27–34. [CrossRef]

13. Wu, Z.; Wang, X.; Iwashita, K.; Sasaki, T.; Hamaguchi, Y. Tensile fatigue behaviour of FRP and hybrid FRP sheets. *Compos. Part B Eng.* **2010**, *41*, 396–402. [CrossRef]

14. Kosior-Kazberuk, M. Application of basalt-FRP bars for reinforcing geotechnical concrete structures. *MATEC Web Conf.* **2019**, *265*, 05011. [CrossRef]

15. Shi, J.-W.; Zhu, H.; Wu, G.; Wu, Z.-S. Tensile behavior of FRP and hybrid FRP sheets in freeze–thaw cycling environments. *Compos. Part B Eng.* **2014**, *60*, 239–247. [CrossRef]

16. Benmokrane, B.; Elgabbas, F.; Ahmed, E.A.; Cousin, P. Characterization and comparative durability study of glass/vinylester, basalt/vinylester, and basalt/epoxy FRP bars. *J. Compos. Constr.* **2015**, *19*, 04015008. [CrossRef]

17. Alam, M.S.; Hussein, A. Experimental investigation on the effect of longitudinal reinforcement on shear strength of fibre reinforcement on shear strength of fibre reinforced polymer reinforced concrete beams. *Can. J. Civil. Eng.* **2011**, *38*, 243–251.

18. El-Sayed, A.K.; El-Salakawy, E.F.; Benmokrane, B. Shear strength of FRP-reinforced concrete beams without transverse reinforcement. *ACI Struct. J.* **2006**, *103*, 235–243.

19. El-Sayed, A.K.; El-Salakawy, E.F.; Benmokrane, B. Shear capacity of high-strength concrete beams reinforced with FRP bars. *ACI Struct. J.* **2006**, *103*, 383–389.

20. Razaqpur, A.G.; Isgor, B.O.; Greenaway, S.; Selley, A. Concrete contribution to the shear resistance of fiber reinforced polymer reinforced concrete members. *J. Compos. Constr.* **2004**, *8*, 452–460.

21. Alkhrdaji, T.; Wideman, M.; Belarbi, A.; Nanni, A. Shear strength of RC beams and slabs. In *Composites in Construction Figueiras*; Juvandes, L., Faria, R., Eds.; Balkema Publishers: Lisse, The Netherlands, 2001; pp. 409–414.

22.   Yost, J.R.; Gross, S.P.; Dinehart, D.W. Shear strength of normal strength concrete beams reinforced with deformed GFRP bars. *J. Compos. Constr.* **2001**, *5*, 263–275. [CrossRef]

23.   Khalifa, A.; Gold, W.J.; Nanni, A.; Aziz, A.M.I. Contribution of externally bonded FRP to shear capacity of RC flexural members. *ASCE J. Compos. Constr.* **1998**, *2*, 195–202. [CrossRef]

24.   Täljsten, B. Strengthening of concrete structures for shear with bonded CFRP-fabrics. In *Recent Advances in Bridge Engineering, Advanced Rehabilitation, Durable Materials, Nondestructive Evaluation and Management*; Meier, U., Betti, R., Eds.; Columbia Univ. Press: Dübendorf, Switzerland, 1997; pp. 57–64.

25.   Triantafillou, T.C. Shear strengthening of reinforced concrete beams using epoxy bonded FRP composites. *ACI Struct. J.* **1998**, *95*, 107–115.

26.   Triantafillou, T.C.; Antonopoulos, C.P. Design of concrete flexural members strengthened in shear with FRP. *ASCE J. Compos. Constr.* **2000**, *4*, 198–205. [CrossRef]

27.   Alnahhal, W.; Chiewanichakorn, M.; Aref, A.; Alampalli, S. Temporal thermal behavior and damage simulations of FRP deck. *J. Bridge Eng.* **2006**, *11*, 452–465. [CrossRef]

28.   Aiello, M.A.; Ombres, L. Structural performances of concrete beams with hybrid (fiber reinforced polymer-steel) reinforcements. *J. Compos. Constr.* **2002**, *6*, 133–140. [CrossRef]

29.   Newhook, J.P. Design of under-reinforced concrete T-sections with GFRP reinforcement. In Proceedings of the 3rd International Conference on Advanced Composite Materials in Bridges and Structures, Canadian Society for Civil Engineering, Ottawa, ON, Canada, 15–18 August 2000.

30.   Wang, X.; Wu, Z. Integrated high-performance thousand-meter scale cable-stayed bridge with hybrid FRP cables. *Compos. Part B Eng.* **2011**, *41*, 166–175. [CrossRef]

31.   Urbański, M. Compressive Strength of Modified FRP Hybrid Bars. *Materials* **2020**, *13*, 1898. [CrossRef]

32.   Lau, D. Hybrid fiber-reinforced polymer (FRP) composites for structural applications. In *Developments in Fiber-Reinforced Polymer (FRP) Composites for Civil Engineering*; Woodhead Publishing: Cambridge, UK, 2013; pp. 205–225. [CrossRef]

33.   Garbacz, A.; Szmigiera, E.; Protchenko, K.; Urbanski, M. On Mechanical Characteristics of HFRP Bars with Various Types of Hybridization. In *International Congress on Polymers in Concrete (ICPIC 2018)*; Springer: Washington, DC, USA, 2018; pp. 653–658. [CrossRef]

34.   Protchenko, K.; Młodzik, K.; Urbański, M.; Szmigiera, E.; Garbacz, A. Numerical estimation of concrete beams reinforced with FRP bars. In *MATEC Web of Conferences*; EDP Sciences: Moscow, Russia, 2016; Volume 86, p. 02011.

35.   Protchenko, K.; Szmigiera, E.D. Post-Fire Characteristics of Concrete Beams Reinforced with Hybrid FRP Bars. *Materials* **2020**, *13*, 1248. [CrossRef]

36.   Szmigiera, E.D.; Protchenko, K.; Urbański, M.; Garbacz, A. Mechanical Properties of Hybrid FRP Bars and Nano-Hybrid FRP Bars. *Arch. Civ. Eng.* **2019**, *65*, 97–110. [CrossRef]

37.   Protchenko, K.; Dobosz, J.; Urbański, M.; Garbacz, A. Wpływ substytucji włókien bazaltowych przez włókna węglowe na właściwości mechaniczne prętów B/CFRP (HFRP). [Influence of substitution of basalt fibres by carbon fibres on mechanical properties of B/CFRP (HFRP)]. *J. Civ. Eng. Environ. Archit.* **2016**, *63*, 149–156.

38.   Garbacz, A.; Urbanski, M.; Lapko, A. BFRP bars as an alternative reinforcement of concrete structures–Compatibility and adhesion issues. *Adv. Mater. Res.* **2015**, *1129*, 233–241. [CrossRef]

39.   Urbański, M.; Lapko, A.; Garbacz, A. Investigation on Concrete Beams Reinforced with Basalt Rebars as an Effective Alternative of Conventional R/C Structures. *Procedia Eng.* **2013**, *57*, 1183–1191. [CrossRef]

40.   American Concrete Institute (ACI). *Guide test methods for fiber reinforced polymers (FRPs) for reinforcing or strengthening concrete structures*; ACI: Farmington Hills, MI, USA, 2006; ISBN 9780870317811.

2

Polymer Flexible Joint as a Repair Method of Concrete Elements: Flexural Testing and Numerical Analysis

Łukasz Zdanowicz *, Szymon Seręga ⓘ, Marcin Tekieli and Arkadiusz Kwiecień ⓘ

Faculty of Civil Engineering, Cracow University of Technology, 31-155 Cracow, Poland; sserega@pk.edu.pl (S.S.); mtekieli@pk.edu.pl (M.T.); akwiecie@pk.edu.pl (A.K.)
* Correspondence: l.zdanowicz.edu@gmail.com

Abstract: Polymer Flexible Joint (PFJ) is a method for repairs of concrete elements, which enables carrying loads and large deformations effectively. This article presents the possibility of applying PFJ on beams subjected to bending and describes the influence of such joints on concrete elements. An experimental investigation was conducted to determine the behavior of concrete in a four-point bending test. The research program included flexural tests of plain concrete elements with a notch, as well as tests of elements which were repaired with PFJ after failure. Based on the experimental results, the numerical characteristics of analyzed polymer and concrete were calibrated. A nonlinear numerical model is developed, which describes the behavior of concrete elements and polymer in the experiments. The model is used to numerically analyze deformations and stresses under increasing load. The influence of flexible joint on concrete elements is described and behavior of elements repaired with PFJ is compared to original elements. Particular attention was paid to the stress redistribution in concrete. The application of flexible joint positively influences load capacity of the connected concrete elements. Furthermore, because of stress redistribution, connected elements can bear larger deformations than original ones. PFJ can therefore be considered an efficient repair method for connecting concrete elements.

Keywords: polyurethane; flexible joint; concrete repair; stress redistribution

1. Introduction

1.1. Motivation

Early-age shrinkage as well as mechanical cracks in concrete are common problems in constructions. Deformations caused by concrete shrinkage or temperature changes leads to local stress concentrations in the concrete. This results in damage of concrete components such as floors or roadways [1,2]. In addition, mechanical actions cause damage to concrete used in structural elements by overloading them. Earthquakes are the most serious excitations, resulting in concrete cracking and plastic, e.g., in infill structures (Figure 1)—visible for a specimen tested on shake table [3]. Cracked concrete is unable to transfer tensile forces thus various repair and strengthening methods of it are used [4]. Filling of cracks by external injection with bonding material such as epoxy resin or self-healing methods [5,6] use mainly stiff and high strength, but hardly deformable materials, which generate stress concentrations in concrete weakened by micro-cracks [4]. Recovering of strength and stiffness in damaged cross-sections leads also to the use of composite materials [7,8] with various adhesives [9,10].

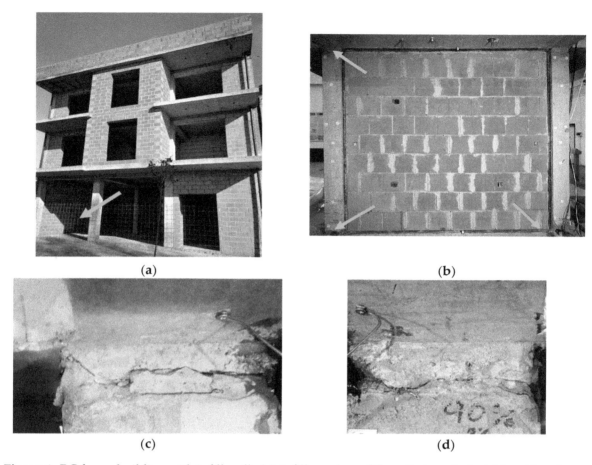

Figure 1. RC frame building with infill walls (**a**), infill specimen [3] excited on shake table with columns damaged at top and bottom (**b**), damages (hinges) of concrete in RC column tops (**c,d**).

Common joint materials, such as epoxy resins, have high stiffness and low deformability, which can be a disadvantage, since new cracks often appear after repair, close to the repaired one. On the contrary, sealants or mortars which are also used for such repairs do not have sufficient strength to carry internal forces (Figure 2) [11–13]. A good solution is to adjust the stiffness and strength of the joint material so that it is suitable for use in certain situations. This is also important considering the sustainability of buildings and Sustainable Development Goals strategy [14,15]. One option that meets these requirements is the Polymer Flexible Joint (PFJ) method.

Figure 2. Damage examples: concrete pavement with cracks caused by shrinkage (**left**) and after repair with epoxy resin (**right**).

1.2. Idea of the Polymer Flexible Joint

Polymer flexible joint is a method that allows effectively bearing the loads and absorbing large deformations at the same time [16]. Therefore, it can be suitable for repairing damaged structures (mostly fragile building materials) as well as for joining new concrete elements. PFJ is based on polyurethane, which is a material with highly effective bonding and hyper-elastic properties. Several types of PFJ are available, with tensile strength ranging from 1.7 to 20 MPa, and elongation at rupture between 10% and 120%. Currently, PFJ research is focused on the application of PM-type, PS-type, and PT-type polyurethanes, which can be used as injection or prefabricated elements, named also as PUFJ (PolyUrethane Flexible Joints) [3,17–19]. The PM-type has the lowest and the PT-type the highest values of stiffness and strength, while the properties of the PS-type are between the other two types.

This article presents experimental and numerical investigations of the most effective PT-type polymer for concrete repair. It can be applied in cracks of concrete and RC elements making elastic (or nonlinear elastic) hinge, which is able to transfer bending loads and high deformation simultaneously, also under cyclic loads. It can be used for the repair of cracked concrete surfaces (Figure 2). Moreover, damages to columns after earthquakes (forming plastic hinges) requires a quick emergency repair before an aftershock. Because of the short time between the main shock and the aftershock (hours, days), a PT-type polymer can be used as a temporary repair injection and confinement, replacing crushed concrete (Figure 1c,d) and making visco-elastic hinge (with remained steel reinforcement), withstanding dynamic loads and introducing in the structure ductile and damping properties.

1.3. Research Aim

The aim of the research presented in this article is to investigate the possibility of the application of PFJ on beams subjected to bending. Until now, PT-type polymer was not thoroughly investigated in flexure. The four main objectives of this work are: (1) to describe the influence of PFJ on concrete elements and stress redistribution, (2) to determine the behavior of concrete in four-point bending tests, including digital image correlation analysis, (3) to perform a numerical analysis of the experimental work, describe the behavior of concrete elements and polymer and analyze deformations and stresses under increasing load, and (4) to assess the influence of flexible joint on concrete elements and compare the behavior of elements repaired with PFJ to original elements, focusing on joint effectiveness.

2. Experimental Research

2.1. Materials

CONCRETE. All concrete specimens were made of one concrete mixture: normal concrete based on Portland cement with a w/c ratio of 0.45 and a maximum aggregate size of 16 mm. To determine the compressive strength of the concrete, 150 mm cubes were used, while the tensile strength was measured in the uniaxial tensile test on prisms with dimensions of $b \times h \times L = 100 \times 100 \times 200$ mm^3 with a notch $a = 25$ mm on both sides in the center of each prism. These tests were carried out 60 days after casting [20]. It can be classified as C 45/55 class.

POLYMER. A two-component PT-type polymer was used in the research program. It is an elasto-visco-plastic material with good bonding properties. The tensile test was carried out according to EN ISO 527-1 [21] at the strain ratio of 1%/min [18]. The compressive tests in accordance with ISO 7743 were conducted [22]. The Poisson ratio of the PT-type polymer is 0.495 [16]. Young modulus of PT-type polymer is $E = 700$ MPa, almost two orders lower than the applied concrete [18]. Table 1 shows the mechanical properties of concrete and polymer.

Table 1. Concrete and polymer mechanical properties [18,22,23].

Test	AV [1] [MPa]	SD [2] [MPa]	CV [3] [%]
Concrete compressive strength, f_{cm}	68.9	3.57	5.2
Concrete tensile strength, f_{ctm}	3.73	0.34	9.1
Polymer PT-type compressive strength, f_{pc}	26.8	1.29	4.8
Polymer PT-type tensile strength, f_{pt}	18.8	1.36	7.2

[1] AV—average value; [2] SD—standard deviation, [3] CV—coefficient of variation.

2.2. Testing Methodology

The experimental investigations consist of 63 four-point bending tests (4PB) with various structural materials. The test specimens were divided into 21 series, each containing 3 test specimens. The aim of this research program was to analyze the influence of the polymer PT on joint effectiveness (JE) [23,24] in concrete crack repair for one chosen series. The results of other series are not a subject of this article.

Joint effectiveness was defined as a ratio of load-bearing capacity or elongation capacity of the elements after the repair and before the repair. In this paper, experimental and numerical analysis of one series (3 test specimens before and after repair) for concrete and polymer PT cooperation is presented.

The specimens used in 4PB tests were prismatic, with a length of $L = 400$ mm and a square cross-section with $b \times h = 100 \times 100$ mm^2. Each specimen was provided with a notch of $a = 30$ mm depth at the bottom, in the middle of its length (see Figure 3). Such elements were called original elements (elements without PFJ). They were tested for bending and then repaired with a PT-type polymer after failure. The repair was carried out with a joint thickness of $t = 10$ mm, to simulate large crack with and loss of crushed concrete surface. The elements after the repair are later referred to as "repaired elements". Before the repair, the concrete surface was cleaned and a layer of primer (SIKA ZP Primer—a single component, ready to use, chemically hardening, containing solvents polyurethane primer, Sika Poland, Cracow, Poland) was applied. The bending tests of the repaired elements were carried out 16–20 h after the polymer application (a kind of emergency repair).

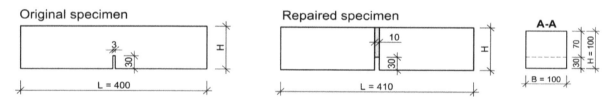

Figure 3. Specimen geometry (dimensions in mm): original specimen and specimen after repair (the roughness of the repaired concrete surface at the polymer-concrete interface is idealized in the presented scheme).

The bending tests were performed with a Zwick/Roell Z100 testing machine (Ulm, Germany). The test parameters are as follows:

- Maximum force: 100 kN,
- Initial force: 0.80 kN,
- Test displacement rate: 0.10 mm/min,
- Span length: 300 mm,
- Distance between loads: 100 mm.

Three measurement methods were used simultaneously in all 4PB tests: (1) extensometer for the crack width displacement (CMOD), (2) extensometer for the vertical deformation, (3) digital image correlation (DIC) to measure the strain field on the side surface of the specimens (Figure 4). For DIC method the CivEng Vision system (Cracow, Poland) was used [25,26].

Figure 4. Test setup for four-point bending test (**top**) with DIC field of the original specimen (**bottom left**) and the repaired specimen (**bottom right**).

2.3. Test Results

2.3.1. Load Response

Figure 5 shows the stress-strain relationship in bending tests for specimen-series $b \times h = 100 \times 100$ mm^2. The curve is linear up to the maximum load for both original and repaired elements. It is noted that repaired specimens, where the polymer was used, have a lower bending stiffness than original specimens. Therefore, higher CMOD values were observed for repaired specimens at the same load levels. The load-bearing capacity of original elements reached higher values in most cases. Before maximum nominal stress σ_{max} (in the pre-critical phase where $\sigma_{max} = 6 \times M_{max}/(b \times h'^2)$), no cracks were detected with the unarmed eye. However, crack formation was observed in the post-critical phase ($\geq \sigma_{max}$).

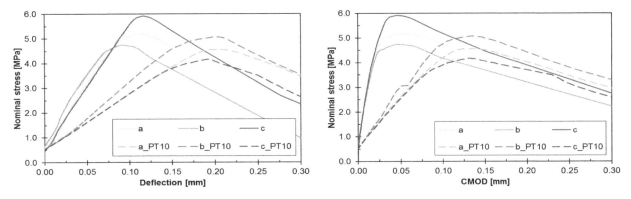

Figure 5. Load response in terms of stress-deflection (**left**) and stress-CMOD (**right**) of the notched beam in 4PB test (continuous lines for original elements, dashed lines for repaired ones).

2.3.2. Failure Mode

All specimens had a similar type of failure—a crack occurred along the notch where the stresses reached the highest values. Furthermore, the failure was sudden and brittle. Cracks run not only through the cementitious matrix but also through aggregates (Figures 6 and 7). In the repaired specimens, the failure was also caused in form of crack in the concrete and the crack developed 1–3 mm from the notch. No specimen failed due to damage through the polymer in the joint.

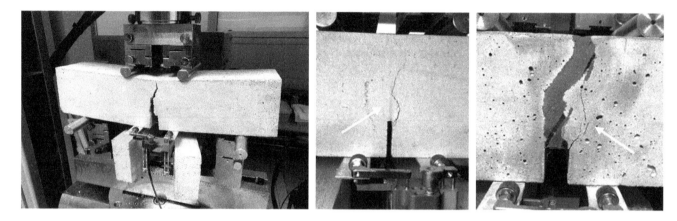

Figure 6. Failure of the original specimen (**left**) and main crack in original (**middle**) and repaired specimen (**right**).

Figure 7. Failure surface of the repaired specimens: (**a**) specimen S01, (**b**) S02, (**c**) S03.

2.3.3. Digital Image Correlation Results

Using digital image correlation method (DIC), first cracks were already observed before reaching the maximum load. A development of the horizontal strains (X-axis) of the original and repaired specimens is shown in Figure 8. In the original specimens, first cracks were visible on the left side of the notch a few steps before failure (load step 80/88). These cracks developed further and joined to form a main crack when the failure occurred (load step 88/88). The main crack followed the shortest path through the cross-section and no further cracks were visible in the damage area.

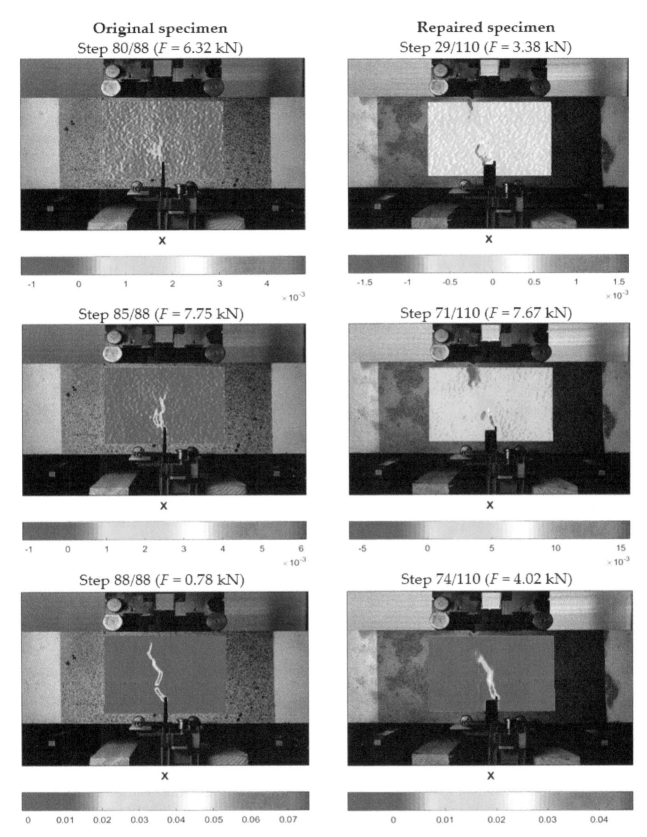

Figure 8. Crack development under applied load for original specimen (**left**) and repaired specimen with PT-type polymer joint of 10 mm (**right**); results obtained using CivEng Vision software.

Repaired specimens had a different mechanism. The first cracks appeared much earlier, which was a consequence of the already weakened cross-section (during the previous test of the original specimen). In the repaired specimen several main cracks developed further (load step 29/110) on the left side of the notch. However, halfway through the test a new crack developed on the other side of the notch (load step 71/110). From this point on, the previous, initial cracks did not develop further, and the new crack caused the failure. This sidewise shift of the damage process is visible between the last steps (load step 68/110 and 71/110). The horizontal strain at F_{max} of the repaired specimens was about 2.5-times higher than of the original specimens.

The phenomenon above confirms observation presented in [4] for tensile tests of concrete specimens repaired with PT-type polymer that the PFJ can cause mechanical closing of micro-cracks in concrete by redistribution of stress and protect the weakened (by micro-cracks) concrete zone against damage in the same place. In the analyzed case, stress redistributed by the PFJ in the failure cross-section found another place with higher level of fracture energy, where new damage occurred.

2.3.4. Joint Effectiveness

As mentioned above, the load-bearing capacity of the repaired elements was generally lower than the original ones. The average values of maximum stress for original and repaired elements are 5.28 MPa (SD = 0.59 MPa, CV = 11.1%) and 4.60 MPa (SD = 0.45 MPa, CV = 9.8%) respectively. On the other hand, the average values of CMOD are 47.7 μm (SD = 0.07 μm, CV = 1.5%) and 132.7 μm (SD = 1.1 μm, CV = 0.8%) respectively. The repaired specimens manifested load bearing capacity of 87% of the original elements, but ductile behavior increased almost three times.

A comparison of maximum stresses and maximum CMODs for the original and repaired specimens is shown in Figure 9. For each test specimen series, the maximum of average stresses and the maximum of average CMOD values are presented graphically and the effectiveness of the repair in terms of force and CMOD is calculated.

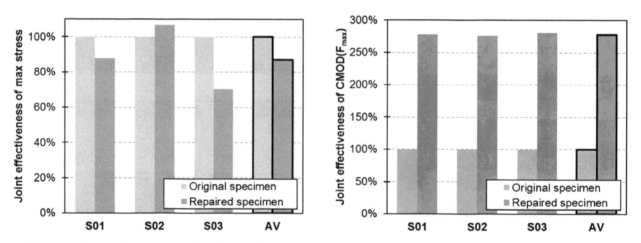

Figure 9. Joint effectiveness of polymer flexible joint with PT-type polymer in terms of maximal stresses (**left**) and CMOD at maximal load (**right**).

In terms of maximum stress, the repaired specimens achieved only 13% lower strength than the original specimens. In terms of deformability, the repaired specimens achieved significantly higher deformation at break than the original ones. This is explained by the substantially higher deformability of polymer of the hyperelastic characteristic when compared to brittle concrete. In general, the specimens repaired with a 10 mm polymer layer had the average CMOD value 2.78 times higher than the specimens before the repair. These two joint effectiveness values indicate increase in global fracture energy of the specimens after repair, what is not observed when brittle materials (cracked) are repaired (bonded) using stiff and high strength materials [27]. Typical epoxy resins are of about two orders higher stiffness (E = 30,000 ÷ 40,000 MPa) than the applied PT-type polymer (E = 700 MPa). On the other hand, ability of PT-type polymer to deform up to 10% of ultimate strain with hyperelastic characteristic results in stress redistribution, thus the PFJ can act as a load-bearing connection and absorb large deformations at the same time.

3. Numerical Analysis

3.1. Finite Element Model

The numerical analysis of the 4PB test was performed with the DIANA Finite Element Software [28] in 2D as a plane stress state. The topology of a FE-mesh used in the analysis is shown in Figure 10. The mesh consists of square isoparametric plane stress elements with square shape function (type *CQ16M*) of eight nodes [28]. The maximum size of a single element is 10 mm and the minimum size is 1 mm (elements in the notch area). To ensure the correct behavior of the concrete-polymer contact zone, additional contact elements (type *CL12I*) with zero thickness at the contact area with 95% tensile strength of the concrete were used. Boundary conditions were defined as rigid supports. The right support was modelled as a sliding type in the horizontal direction without taking any possible friction effect.

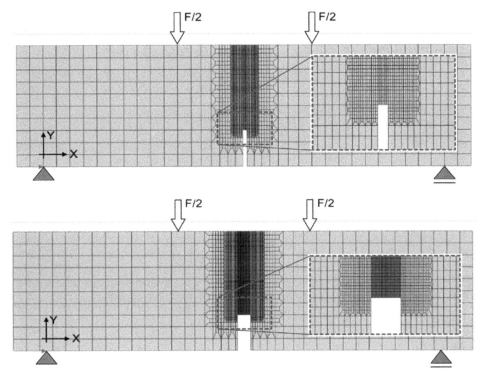

Figure 10. Finite element model for original specimen (**top**) and repaired specimen (**bottom**).

3.2. Constitutive Material Models

The constitutive model for concrete used in the analysis is based on a smeared crack model [29] and was formulated in total strains according to the concept of [30,31] and DIANA algorithm [28]. A rotating crack model is used in the analysis. A linear-elastic stress-strain relationship is assumed for concrete in compression. The concrete tensile behavior is described by exponential post-cracking behavior in the following form [28]:

$$\sigma^{cr}_{nn}(\varepsilon^{cr}_{nn}) = exp\left(-\frac{\varepsilon^{cr}_{nn}}{\varepsilon^{cr}_{nn.ult}}\right) \tag{1}$$

$$\varepsilon^{cr}_{nn.ult} = \frac{G_f}{h \cdot f_t} \tag{2}$$

where σ^{cr}_{nn} is the stress perpendicular to the crack, ε^{cr}_{nn} is the strain in the same direction and G_f is the fracture energy, f_t is the tensile strength of the concrete and h is a crack bandwidth, which is assumed here as the square root of a finite element area. Table 2 shows the material constants used for concrete and polymer in the numerical simulations.

Table 2. Mechanical properties of concrete and polymer.

Material	E [MPa]	ν [-]	f_t [MPa]	G_f [N/mm]
Concrete	36,700	0.20	3.73	0.150 [1]
Polymer PT-type	700	0.49	20.0	n/a

[1] $G_f = 73\,f_{cm}^{0.18} = 0.150$ N/mm [32].

The maximum tensile stresses in the system were equal to the concrete tensile strength; in this range, the polymer behaved as a linear-elastic material [18]. Therefore, the polymer was modelled as linear elastic ($f_{ct} << f_{pt}$).

To take this into account, the influence of the primer layer between the damaged concrete and the polymer, a zero thickness interface layer was modelled. A non-linear model of the interface was adopted with a strength equal to 95% of the concrete tensile strength. The remaining parameters were calibrated empirically on the basis of experimental results (Figure 11).

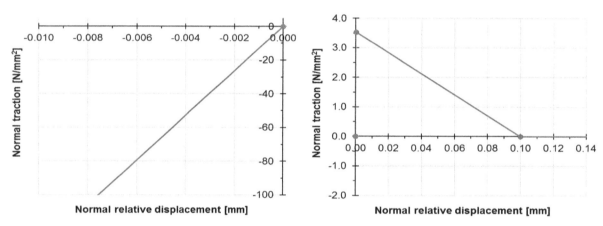

Figure 11. Mechanical parameters of interface between concrete and polymer for compression (**left**) and tension (**right**).

3.3. Solution Strategy

An incremental-iterative solution approach according to the Newton–Raphson method was used. For the first steps of analysis (up to load $F = 6000$ N) a load control procedure was carried out; the load step in this phase was equal to 200 N. After reaching load $F = 6000$ N and before the cracking load, the system switched to an arc-length control procedure with load step of 100 N. Convergence criteria assumed the force and displacement norm. Due to the relatively small deformations, the analysis only took into account material non-linearity, while non-linear geometrical analysis was excluded. Two master control nodes were located at the lower edge of the specimens next to the notch. The results of these nodes shift were called CMOD deformation.

3.4. Numerical Results

The main objective of the analysis was to compare the developed numerical model with experimental results. The values of the maximal forces from experiment differ in comparison to the values from the numerical analysis by less than 9% for the original specimens and less than 7% for the repaired specimens (Table 3). The joint effectiveness obtained from the experimental tests was thus 87% and from the numerical calculation 88%.

Table 3. Results comparison of experimental tests and numerical calculations.

Parameter	Original Specimen		Repaired Specimen	
	Experiment	FE-Analysis	Experiment	FE-Analysis
Failure load, F_{max} [kN]	8.63	7.84	7.52	7.00
CMOD at F_{max} [μm]	47.7	27.4	132.7	131.2
Strain at F_{max}, E_{knn} [-]	0.071 [1]	0.038	0.015 [1]	0.005
Damage energy [N/mm]	0.193	0.087	0.408	0.318

[1] value obtained from the results of the DIC analysis.

In relation to the CMOD at maximal load (cracking load), the difference between experimental results and calculation was 43% for the original test specimens and 1% for the repaired specimens. Such a high difference in deformation results was due to the stiffness of the test bench. It should be noted that the difference of the CMOD at cracking load between original and repaired specimen in the numerical analysis was 4.8 times.

Figure 12 shows the experimental and numerical stress-strain (CMOD) relationship. The numerical analysis of the original specimens correctly reproduced the path in the pre-critical phase. However, the difference is clearly visible in the post-critical phase, where experimental specimens behave more ductile. Possible explanation is the arc failure mechanism due to friction at both supports. Such phenomenon was not modelled in the FEM. Friction on supports increases the load-bearing capacity of bending elements due to the change in the direction of main stress. In fact, the main compressive stresses σ_1 ran along the curve line (vault effect) and near the supports their direction was not parallel to the edge of the element. It should be emphasized, however, that the numerical model correctly reproduces the pre-critical phase and failure mechanism.

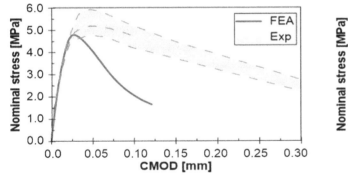 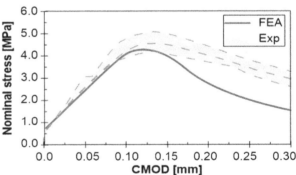

Figure 12. FE vs. experiment results—Nominal stress-CMOD diagram for original (**left**) and repaired specimen (**right**).

Similar to the original specimens, the behavior of the repaired specimens in the FE-analysis is comparable to the experimental results. It is pointed out that not only the pre-critical path was reproduced correctly, but also for the post-critical phase the numerical results are satisfactory. Probably, using of hyperelastic model for the polymeric material (instead of the linear one) [9] can predict better the post-failure path.

Comparison of initial (original specimen) and subsequent (repaired specimen) damage energy, evaluated by the area under curves in Figure 12, also confirms this observation (damage energy is defined here as an area under stress-CMOD curve up to the maximum load). The average damage energy obtained in experiment was equal to 0.193 N/mm and 0.408 N/mm for original and repaired specimens, respectively. Even if the ultimate stress was lower in the cases of repaired specimens, more ductile behavior of the specimens after repair using PFJs was advantageous.

The numerical model could also predict the crack formation (Figure 13). The specimens were damaged by cracks forming along the notch where the tensile stresses reached their highest value. It can be observed that the path of the crack pattern is similar to the strain results from DIC.

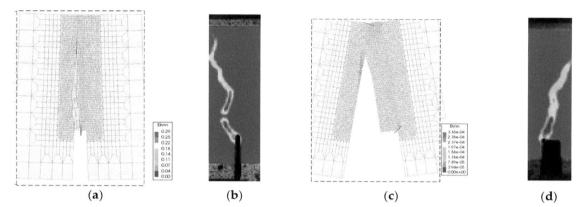

Figure 13. Failure mode and crack strains at failure in X-direction of original specimen—FE-calculation at 50× magnification (**a**), experiment acc. to DIC map (**b**) and repaired specimen—FE-calculation at 50× magnification (**c**), experiment acc. to DIC map (**d**).

4. Summary and Conclusions

The experimental results of the four-point bending tests on plain concrete beams and beams repaired with PFJ specimens were presented together with the numerical analysis based on FE-modelling. The following conclusions can be drawn from a comparative analysis of the experimental and numerical results:

(1)   The joint effectiveness of Polymer Flexible Joint with PT-type polymer in terms of load-bearing capacity is 87% (on average) in the experiments and 88% in the numerical analysis.

(2)   For specimens repaired with PT-type polymer, the polymer is decisive for the experimental and numerical results in terms of CMOD values. It was observed that the strain capacity of the repaired specimens was more than 280% higher than that of the original specimens.

(3)   The repaired specimens were able to manifest higher damage energy than the original ones; the area under stress-CMOD curve for repaired specimens was more than 2× higher than that of the original ones.

(4)   It is emphasized that the specimens repaired with PFJ have the load-bearing capacity of more than 80% of the original ones. In this way the PFJ can be used as a load-bearing connection. In addition, the higher flexibility of the connection leads to a reduction of possible imposed stresses and introduces additional ductile behavior of structural elements after repair.

(5)   Using DIC method, the phenomenon of stress redistribution around a flexible joint was documented.

(6)   The comparison between experimental and numerical analysis shows a good agreement of the results in terms of failure load and deformation (CMOD values).

(7)   The numerical model correctly reproduces the pre-critical phase and failure mechanism. It was proven that presented numerical simulations can be a useful and suitable tool for the analysis of pre-critical phase of a four-point bending test of concrete bonded with a polymer flexible compound.

(8)   However, the post-critical path was not satisfactory (lower stiffness in the results from numerical calculations than from experiments). Experimental results show much smoother curves during failure and in the post-critical phase than numerical calculations. The reason for this may be the friction of the support during the test, where a quasi-arch effect might have occurred. Further investigations are required to adequately describe the mechanism in the post-critical phase.

Author Contributions: All authors contributed extensively to this study; Ł.Z. conducted the experiments and prepared the manuscript, M.T. was responsible for the DIC analysis, S.S. and Ł.Z. performed the numerical analysis, A.K. supervised the research. All authors discussed the results, read and agreed to the amendments at all stages of the manuscript. All authors have read and agreed to the published version of the manuscript.

References

1. Houben, L. Damage on jointed plain concrete pavements: Causes and repairs. In Proceedings of the Workshop Diverse Uses of Concrete, Nairobi, Kenya, 23–27 November 2009; pp. 1–26.
2. Zilch, K.; Wingenfeld, D.; Mühlbauer, C. Experimental investigation of reinforced glued joints. In Proceedings of the Hipermat 2012, Kassel, Germany, 7–9 March 2012.
3. Rousakis, T.; Papadouli, E.; Sapalidis, A.; Vanian, V.; Ilki, A.; Halici, O.F.; Kwiecień, A.; Zając, B.; Hojdys, Ł.; Krajewski, P.; et al. Flexible Joints between RC frames and masonry infill for improved seismic performance—Shake table tests. In *Proceedings of the 17th International Brick and Block Masonry Conference (17th IB2MAC 2020), Kraków, Poland, 5–8 July 2020*; CRC Press/Balkema Taylor & Francis Group: Cracow, Poland, 2020; pp. 499–507.
4. Sánchez, M.; Faria, P.; Horszczaruk, E.; Jonkers, H.M.; Kwiecień, A.; Mosa, J.; Peled, A.; Pereira, A.S.; Snoeck, D.; Stefanidou, M.; et al. External treatments for the preventive repair of existing constructions: A review. *Constr. Build. Mater.* **2018**, *193*, 435–452. [CrossRef]
5. Šavija, B.; Feiteira, J.; Araújo, M.; Chatrabhuti, S.; Raquez, J.M.; Van Tittelboom, K.; Gruyaert, E.; De Belie, N.; Schlangen, E. Simulation-Aided Design of Tubular Polymeric Capsules for Self-Healing Concrete. *Materials* **2017**, *10*, 10. [CrossRef] [PubMed]
6. Mullem, T.V.; Anglani, G.; Dudek, M.; Vanoutrive, H.; Bumanis, G.; Litina, C.; Kwiecień, A.; Al-Tabbaa, A.; Bajare, D.; Stryszewska, T.; et al. Addressing the need for standardization of test methods for selfhealing concrete: An inter-laboratory study on concrete with macrocapsules. *Sci. Technol. Adv. Mater.* **2020**, *21*, 661–682. [CrossRef] [PubMed]
7. Soupionis, G.; Georgiou, P.; Zoumpoulakis, L. Polymer Composite Materials Fiber-Reinforced for the Reinforcement/Repair of Concrete Structures. *Polymers* **2020**, *12*, 2058. [CrossRef] [PubMed]
8. Ghatte, H.F.; Comert, M.; Demir, C.; Akbaba, M.; Ilki, A. Seismic Retrofit of Full-Scale Substandard Extended Rectangular RC Columns through CFRP Jacketing: Test Results and Design Recommendations. *J. Compos. Constr. ASCE* **2019**, *23*, 04018071. [CrossRef]
9. Kwiecień, A.; Krajewski, P.; Hojdys, Ł.; Tekieli, M.; Słoński, M. Flexible Adhesive in Composite-to-Brick Strengthening—Experimental and Numerical Study. *Polymers* **2018**, *10*, 356.
10. Cruz, R.J.; Serȩga, S.; Sena-Cruz, J.; Pereira, E.; Kwiecień, A.; Zając, B. Flexural behaviour of NSM CFRP laminate strip systems in concrete using stiff and flexible adhesives. *Compos. Part B* **2020**, *195*, 108042. [CrossRef]
11. Kwiecień, A. Highly Deformable Polymers for Repair and Strengthening of Cracked Masonry Structures. *GSTF Int. J. Eng. Technol.* **2013**, *2*, 182–196. [CrossRef]
12. Pröbster, M. *Construction Sealants. Seal Joints Successfully (in German: Baudichtstoffe. Erforgreich Fugen Abdichten)*; Vieweg+Teubner Verlag: Wiesbaden, Germany, 2011, ISBN 9783834809520.
13. Kwiecień, A.; Gruszczyński, M.; Zając, B. Tests of flexible polymer joints repairing of concrete pavements and of polymer modified concretes influenced by high deformations. *Key Eng. Mater.* **2011**, *466*, 225–239. [CrossRef]
14. United Nations. Transforming Our World: The 2030 Agenda for Sustainable Development. Available online: https://sustainabledevelopment.un.org/post2015/transformingourworld (accessed on 10 December 2020).
15. Pigliautile, I.; Marseglia, G.; Pisello, A.L. Investigation of CO_2 Variation and Mapping through Wearable Sensing Techniques for Measuring Pedestrians' Exposure in Urban Areas. *Sustainability* **2020**, *12*, 3936. [CrossRef]
16. Kwiecień, A. *Polymer Flexible Joints in Masonry and Concrete Structures (in Polish: Polimerowe Złącza Podatne w Konstrukcjach Murowych i Betonowych)*; Monography; Wydawnictwo Politechniki Krakowskiej: Cracow, Poland, 2012.
17. Kisiel, P.; Kwiecień, A. Numerical Analysis of a Polymer Flexible Joint in a Tensile Test. *Tech. Trans. Environ. Eng.* **2013**, *110*, 63–71.
18. Zdanowicz, Ł.; Kwiecień, A.; Serȩga, S. Interaction of Polymer Flexible Joint with Brittle Materials in Four-Point Bending Tests. *Procedia Eng.* **2017**, *193*, 517–524. [CrossRef]
19. Zdanowicz, Ł.; Kwiecień, A.; Tekieli, M.; Serȩga, S. Interaction of Polymer Flexible Joint with concrete elements in an uniaxial tensile test. In Proceedings of the High Tech Concrete: Where Technology and

Engineering Meet—Proceedings of the 2017 FIB Symposium, Maastricht, The Netherlands, 12–14 June 2017; pp. 1049–1057.

20. van Mier, J.G.M.; Mechtcherine, V. Minimum Demands for Deformation-Controlled Uniaxial Tensile Tests. In *Experimental Determination of the Stress-Crack Openning Curve for Concrete in Tension (RILEM 187-SOC Report)*; RILEM Publications SARL: Paris, France, 2007; pp. 5–11.

21. EN ISO 527-1. *Plastics—Determination of Tensile Properties—Part 1: General Principles*; British Standards Institution: London, UK, 2019.

22. Pelka, C. *Experimental Implementation of the Concept of a Polymer Anchorage for CFRP Rods (In German: Experimentelle Umsetzung des Konzepts Einer Polymerverankerungen für CFK-Stäbe)*; Leibniz Universität: Hannover, Germany, 2018.

23. Kwiecień, A. New repair method of cracked concrete airfield surfaces using of polymer joint. In Proceedings of the 13th International Congress of Polymers in Concrete—ICPIC 2010, Funchal, Portugal, 10–12 February 2010; pp. 657–664.

24. Malhotra, S.K. Some studies on end connection of compiste sandwich panels. In *Proceedings of the Ninth International Conference on Composite Materials (ICCM/9), Madrid, Spain, 12–16 July 1993*; Miravete, A., Ed.; Woodhead Publishing Limited: Zaragoza, Spain, 1993; pp. 384–389.

25. Tekieli, M.; De Santis, S.; de Felice, G.; Kwiecień, A.; Roscini, F.; Felice, G.D.; Kwiecień, A.; Roscini, F. Application of Digital Image Correlation to composite reinforcements testing. *Compos. Struct.* **2017**, *160*, 670–688. [CrossRef]

26. Słoński, M.; Tekieli, M. 2D Digital Image Correlation and Region-Based Convolutional Neural Network in Monitoring and Evaluation of Surface Cracks in Concrete Structural Elements. *Materials* **2020**, *13*, 3527. [CrossRef] [PubMed]

27. Jasieńko, J.; Kwiecień, A.; Skłodowski, M. New flexible intervention solutions for protection, strengthening and reconstruction of damaged heritage buildings. In Proceedings of the International Conference on Earthquake Engineering and Post Disaster Reconstruction Planning (ICEE-PDRP 2016), Bhaktapur, Nepal, 24–26 April 2016.

28. DIANA FEA. *DIANA User Manual*; DIANA FEA: Delft, The Netherlands, 2017.

29. de Borst, R. Smeared cracking, plasticity, creep, and thermal loading-A unified approach. *Comput. Methods Appl. Mech. Eng.* **1987**, *62*, 89–110. [CrossRef]

30. Vecchio, F.J. Nonlinear finite element analysis of reinforced concrete membranes. *ACI Struct. J.* **1989**, *86*, 26–35.

31. Vecchio, F.J. Reinforced Concrete Membrane Element Formulations. *J. Struct. Eng.* **1990**, *116*, 730–750. [CrossRef]

32. fib Internation Federation for Structural Concrete. *fib Model Code for Concrete Structures 2010*; International Federation for Structural Concrete (fib), Ed.; Wilheml Ernst & Sohn: Berlin, Germany, 2010.

3

Examining the Distribution of Strength across the Thickness of Reinforced Concrete Elements Subject to Sulphate Corrosion using the Ultrasonic Method

Bohdan Stawiski [1] and Tomasz Kania [2],* ⓘ

[1] Faculty of Environmental Engineering and Geodesy, Wrocław University of Environmental and Life Sciences, pl. Grunwaldzki 24, 50-363 Wrocław, Poland
[2] Faculty of Civil Engineering, Wrocław University of Science and Technology, Wybrzeże Wyspiańskiego 27, 50-370 Wrocław, Poland
* Correspondence: tomasz.kania@pwr.edu.pl

Abstract: Sulphate corrosion of concrete is a complex chemical and physical process that leads to the destruction of construction elements. Degradation of concrete results from the transportation of sulphate compounds through the pores of exposed elements and their chemical reactions with cementitious material. Sulphate corrosion can develop in all kind of structures exposed to the corrosive environment. The mechanism of the chemical reactions of sulphate ions with concrete compounds is well known and described. Furthermore, the dependence of the compressive strength of standard cubic samples on the duration of their exposure in the sulphate corrosion environment has been described. However, strength tests on standard samples presented in the scientific literature do not provide an answer to the question regarding the measurement methodology and actual distribution of compressive strength in cross-section of reinforced concrete structures exposed to sulphate ions. Since it is difficult to find any description of this type of test in the literature, the authors undertook to conduct them. The ultrasonic method using exponential heads with spot surface of contact with the material was chosen for the measurements of concrete strength in close cross-sections parallel to the corroded surface. The test was performed on samples taken from compartments of a reinforced concrete tank after five years of operation in a corrosive environment. Test measurements showed heterogeneity of strength across the entire thickness of the tested elements. It was determined that the strength of the elements in internal cross-sections of the structure was up to 80% higher than the initial strength. A drop in the mechanical properties of concrete was observed only in the close zone near the exposed surface.

Keywords: concrete elements; concrete strength; reinforced concrete tanks; concrete corrosion; sulphate corrosion; ultrasound tests

1. Introduction

There are considerable quantities of effluents generated in chemical laboratories with varied pH, which is the measure of acidity or alkalinity of an aqueous solution. Neutral solutions have a pH of approximately 7.0. Neutralisation of effluents is performed by mixing acidic and alkaline compounds (if their compositions allow it) and by adding acidic or alkaline reagents. This takes place in various types of tanks. For neutralisation of laboratory effluents, reinforced concrete tanks are also used. The ratio of acidity and alkalinity is a critical factor in the chemistry of concrete [1]. The components of concrete are cement, aggregates, and water. Cement has a very alkaline pH, in order to bind all the components, it is important for it to remain near a pH of 12 [2]. In contact with effluents, concrete corrodes. Therefore, it should be characterized by the proper strength and tightness, and should be protected from the aggressive environment by the proper lining [3].

In the literature, one may encounter the opinion that after 1989 the quality of reinforced concrete structures in Poland improved radically [4], but problems with the materials, workmanship and design still exist, as can be seen in the latest research on Polish concrete structures [5,6]. The neutralisation tank presented later was made in 2012 and became corroded, which indicates that concrete insufficiently protected from corrosion will require repair, which should be preceded by a good evaluation of the condition of the damaged structure.

Concrete corrosion not only affects laboratory tanks but develops in all kind of structures. Concrete durability is the constant subject of challenges in the fields of science, design and workmanship [5,6]. As a consequence of concrete structures' exposure to corrosive environments, various substances are being transported into the concrete, causing its expansion, cracking, and strength degradation. Among the most destructive of the numerous corrosive substances are the sulphates [7].

Recent studies on sulphate corrosion of concrete are mainly focused on the mechanism of the chemical reactions of sulphate ions with the concrete compounds [8–16] and the distribution of strength over time of cubic samples stored in sulphate solutions. The corrosive reactions of sulphates in concrete have been well studied and evaluated [17–20]. The deterioration of concrete strength under sulphate corrosion is an essential basis for the prediction of concrete performance and durability. Existing studies indicate that sulphate ions in the environment chemically react with the internal composition of concrete by entering into the concrete through diffusion, convection, capillary adsorption, and other processes to generate expansive products such as ettringite, gypsum [17–20], and sodium sulphate crystals when concrete is corroded by sulphate solution in a dry-wet cycle. The expansive products continuously fill the internal pores of concrete, making the concrete more compact with improved concrete strength before deterioration. Ions of sulphuric acid react with the cement compounds e.g., according to Equations (1), (2) and (3) [7]:

$$H_2SO_4 + CaO \cdot SiO_2 \cdot 2H_2O \rightarrow CaSO_4 + Si(OH)_4 + H_2O \tag{1}$$

$$H_2SO_4 + CaCO_3 \rightarrow CaSO_4 + H_2CO_3 \tag{2}$$

$$H_2SO_4 + Ca(OH)_2 \rightarrow CaSO_4 + 2H_2O \tag{3}$$

The formation of gypsum leads to an increase in volume of approximately 124% in comparison with $Ca(OH)_2$, the main reactant of the process [17,18]. Gypsum stone, as the product of reactions (1), (2) and (3) reacts further with tricalcium aluminates (C3A) or hydrated calcium sulfoaluminate (monosulphate) to form the final chemical product Candlot's salt (ettringite), e.g., according to Reaction (4) [18]:

$$3CaO \cdot Al_2O_3 + 3(CaSO_4 \cdot 2H_2O) + 26H_2O \rightarrow 3CaO \cdot Al_2O_3 \cdot 3CaSO_4 \cdot 32H_2O \tag{4}$$

Formation of Candlot's salt is associated with volume expansion from 230% [9] to 820% [10]. According to [21] the following prerequisites must be reached for Candlot's salt crystallization leading to the concrete expansion:

- The volume of Candlot's salt must exceed some threshold value which depends on the capillary porosity of concrete,
- Only Candlot's salt formed after the hydration of cement leads to expansion,
- Candlot's salt must be formed at the boundaries of solid phases of concrete.

Candlot's salt crystallization pressure depends on the sulphate concentration and can reach the value of 35 N/mm^2 with sulphate concentration of 350 mol/m^3 [18].

The exemplary relationships of sulphate corrosion with strength of concrete samples immersed in sulphate solution have been established and described [21–23]. Zhou et al. in [21] stated that the compressive strength of cubic samples conditioned in dry-wet cycles in sulphate solution shows the rise period and decline area. The strength of concrete samples reached its peak at the 60th day of corrosion and increased by ~6.4% on the basis of its initial strength. With the increase in degradation period, the strength of concrete decreased continuously. The compressive strength decreased by ~4.4%, 18%, and 43.1% after 90, 120, and 150 days of corrosion. This research was done under laboratory conditions on standard cubic samples with a side length of 150 mm tested on a strength machine. Shi and Wang in [22] stated that strength of concrete samples conditioned in dry-wet cycles in 15% sodium sulphate solution reached its peak at the 15th day of corrosion and increased by 29% on the basis of its initial value. Du et al. in [23] tested the C25 concrete mixed with 20% fly ash placed in a sodium sulfate solution (20%) for the full-soaking corrosion test. Samples reached peak of strength at the 100th day of corrosion and increased by 10.6% on the basis of its initial value. Due to the methodology of these tests, the distribution of strength in individual cross-sections of samples which had been previously exposed to the corrosive environment was not determined experimentally.

Laboratory tests of sulphate attack on concrete materials that are based on submerging the specimens in sulphate solution and then measuring physical properties, such as strength, are effectively collecting all of these mechanisms into a single test. The result of such research is the characterization of a particular concrete sample's performance under specific, laboratory conditions. If the field conditions are variable, the performance of the concrete can also be different. Concrete compressive strength in structures is designed to withstand the designed forces. The durability of the structure depends on the fulfilment of the limit condition of concrete strength in the section that works in the state of compression. The question arises what is the compressive strength distribution in various cross sections parallel to the surface of a reinforced concrete structure under sulphate corrosion? Since tests of such type have not been performed so far, it is difficult to find appropriate literature references regarding possible methodology or results concerning the distribution of compressive strength across the thickness of reinforced concrete elements exposed to sulphate corrosion.

This paper presents research on the concrete samples taken from a concrete neutralization tank after five years of storing chemical effluents with sulphate compounds. The main purpose of the experimental tests was to determine the compressive strength of concrete in various cross sections parallel to the corroded surface. It has been stated that compressive strength of concrete subjected to the gaseous aggressive, sulphuric environment is variable across its thickness. Values have shown the increase in strength in the internal, exposed cross sections of the walls.

2. Materials and Methods

2.1. Materials

In the described case study, a neutralisation tank in a building with numerous chemical laboratories was examined. The described tank is being used for the neutralization of liquid wastes from a research program in the field of pharmaceutical production. The chemical composition of the effluents is variable over time and depends on the actual research program in the medical laboratory. A research object selected in this way allows assessment of the corrosion condition of concrete under unplanned conditions and evaluation of the change in the condition of samples of material taken from a real object in operation. The tank was designed as a reinforced concrete box, internally divided into three chambers. The tank was constructed of reinforced, water resistant concrete class C25/30, W8. The tank was made using monolithic technology and its walls were formed in the built-in (vertical) position.

The designer anticipated protecting the concrete using epoxy chemical-resistant lining. The reinforced concrete ceiling above the chambers was made on folded sheets (as composite stay-in-place formwork). Access to each of the chambers is via a manhole covered with a stainless steel lid. The ceiling slab across its thickness in the locations of the manholes was probably not protected from corrosion because during tests there were not any traces of any layer after five years of using the tank.

The described neutralisation tank operates in the batch mode. The chemical composition of the effluents that are being neutralised is variable over time and depends on the actual research and production program in the pharmaceutical laboratory. Effluents are pumped into the chamber where an electrode measures their pH. On that basis effluents are evaluated in terms of their compliance with the set point (neutral pH value). If the pH value is out of range, chemical pumps inject an acid or caustic reagent solution as required to bring the effluent to the correct level. The agitator keep the contents of the tank mixed, so the pH probe is always measuring a representative sample of the effluent and the added reagents are quickly distributed within the tank. For the caustic reagent solutions of caustic soda ($NaOH$, Chempur, Piekary Slaskie, Poland) are used. For the acid reagent a solution of sulphuric acid (H_2SO_4, Chempur, Piekary Slaskie, Poland) is used. A schematic method diagram of the described neutralization in the tank is shown on the Figure 1.

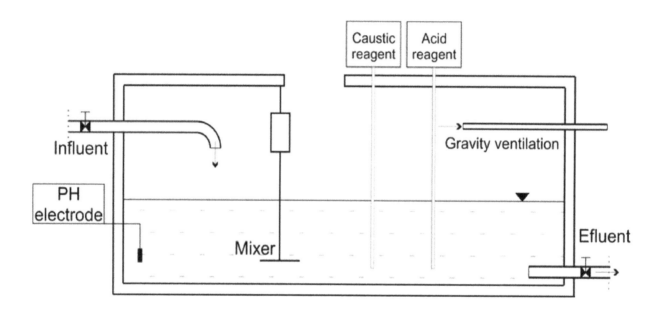

Figure 1. Method diagram of the neutralisation process in the concrete tank.

After neutralisation of effluents down to neutrality they are pumped out to the sewage system. The neutralisation process is accompanied by emission of gases which should be discharged outside, preferably using gravity ventilation or mechanical ventilation resistant to the aggressive environment. Gravity ventilation should have large cross-sections of ducts and small deviations from the vertical. Traditionally, they are openings made of brick 140 x 140 mm or round ϕ 150 mm. In the examined tank, the 'ventilation' was made of PVC pipes (Wiplast, Twardogora, Poland), diameter 50 mm, which were laid horizontally on the tank for a distance of approximately 2 m. These two parameters are sufficient to indicate its lack of effectiveness.

The tank became a natural experimental ground with respect to the effects of long-term exposure to liquid and gaseous aggressive environments on concrete, the internal surfaces of which were protected from liquid effluents using an asphalt rubber coating (probably made of Dysperbit) instead of the designed epoxy lining. No protective coating was applied for contact with the gaseous environment.

Tests were conducted on the operating tank. The lateral surfaces of the ceiling slab well visible in the manhole openings were highly corroded. The loss of concrete ranged from a dozen mm up to more than 20 mm. The folded sheet in the cutting location was also corroded. However, the corrosion rate was lower. Approximately 20 mm of sheet protrudes beyond the corroded concrete (Figure 2).

Figure 2. The folded sheet corroded more slowly than the concrete. Approximately 20 mm of sheet protrudes beyond the concrete. When built, concrete and steel were on one plane.

Such a high degree of concrete corrosion in a zone where it did not have any contact with effluents suggested that in the lower point where effluents were continuously contacting the concrete walls the situation would probably not be better, and might be even far worse. Although the tank walls had some traces of bituminous insulation, large areas lacked this coating because it had flaked off during use (Figure 3).

Figure 3. Walls below the ceiling slab were originally covered with insulation but in a considerable area of the walls this insulation had already flaked off.

At the time when the tests were performed the tank was in use and it was necessary to blind the holes after the completed tests. For this reason, in order to perform the measurements, it was decided to make one borehole with a diameter of 103 mm and other boreholes with diameters of 50 mm. The boreholes were made in the direction perpendicular to surfaces of the walls which had been formed in the built-in position (in the vertical direction). Since the boreholes were made at the same level (drilled perpendicularly to the element forming direction), in this case the variability of aggregate in the sample cross-section can be only random. The tests concerning strength distribution in concrete elements conducted by the authors and other researchers indicate the differentiation of strength with respect to the forming direction as a result of segregation of components in the gravitational field of the Earth and draining of water from concrete mix (bleeding) [24–26]. Such differentiation does not appear in the horizontal direction appropriate for the taken sample element.

2.2. Methods

The compressive strength of concrete samples in their various cross sections were determined with use of the ultrasound method on basis of longitudinal wave velocities [27]. The ultrasonic pulse velocity of a homogeneous solid can be related to its mechanical properties. Theoretical dependencies between ultrasonic wave velocity and elastic modulus and Poisson's ratio were investigated and described in the literature [28,29]. Based on the theory of elasticity applied to homogeneous and isotropic materials, for the method of testing used by the authors passing wave velocity C_L is directly proportional to the square root of the dynamic modulus of elasticity E_d, and inversely proportional to the square root of its density, ϱ, where v_d is the dynamic Poisson's ratio (5):

$$C_L = (E_d/\rho \cdot (1 - v_d)/((1 + v_d) \cdot (1 + 2v_d)))^{1/2} \text{ [km/s]} \tag{5}$$

Concrete is a heterogeneous material, so these assumptions are not strictly valid. High attenuation in concrete limits the ultrasonic pulse velocity method (UPV) to frequencies up to 100 kHz, which means that compressional waves do not interact with most concrete inhomogeneities [30]. Under this condition concrete can be regarded as a homogeneous material [31]. The tests conducted already in the seventies and eighties of the 20th century showed that there is a relationship between ultrasonic wave velocity and concrete strength. The possibility of using the correlation between these values was included both in scientific literature e.g., [28,29,32–38] as well as in norms e.g., [39–41]. In the study [38], Komlos and others compared eight basic methods of determining concrete strength based on measurements of ultrasound velocity. He concluded that the necessary requirement of such tests was to perform calibration of measurements with results of destructive tests, such are also research experiences of the authors [37,42]. The confirmation regarding the possibility of using the measurements of ultrasonic wave velocity to test compressive strength of concrete exposed to sulphate corrosion can be found in scientific literature from the beginning of this century and from the later years [43–45].

Measurements were performed using ultrasonic point probes of frequency equal to 40 kHz the testing results of which were presented in the study [37]. The structure of the probes is shown in the Figure 4. They were equipped with exponential, half wave concentrators with a length of 87 mm and base width of 42 mm. The diameter of the contact point of the concentrators was 1 mm.

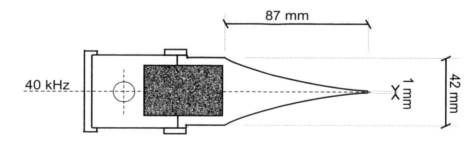

Figure 4. Ultrasonic spot head with the exponential concentrator.

The tests were performed with a UNIPAN 543 (Zaklady Aparatury Naukowej UNIPAN, Warsaw, Poland) ultrasonic pulse velocity (UPV) test instrument (Figure 5). Probe concentrators were applied from the two opposite sides of the examined concrete cylindrical samples, in planes parallel to the surface of the wall they were bored from. In that way the longitudinal wave velocities were measured.

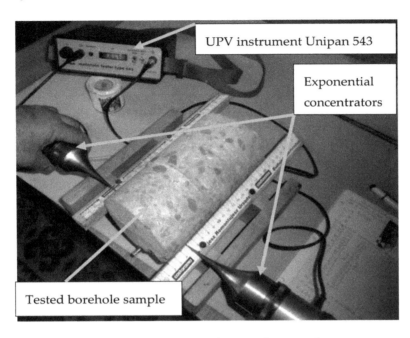

Figure 5. Borehole No. 1 during ultrasound tests.

The ultrasound rate was determined in two directions approximately perpendicular to each other, along the diameters, in planes located 10 mm from each other. Only the distance of the first measurement point from the external surface of the wall was 5 mm (Figure 6).

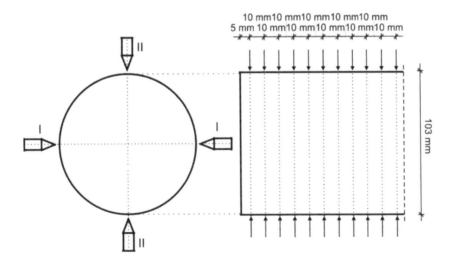

Figure 6. Layout of measuring points and sections on the tested boreholes.

The examined borehole materials were cut into samples of length equal to their diameter. Boreholes were cut thus obtaining samples with $\phi = h = 10.3$ cm and $\phi = h = 5.0$ cm. The strength of samples with $\phi = h = 10.3$ is equivalent to the strength tested on cubic samples, side 15 cm [29,41,42]. The coefficient for calculation of the strength of samples with $\phi = h = 05.0$ cm to the strength tested on cubic samples, side 15 cm is 1.08, what has been tested experimentally by Brunarski [29] and by the authors [42] in the expected strength range of the samples. To the ultrasound rate determined in the middle of the height of each sample, destructive strength was assigned as determined on the strength machine as a relation of destructive force P [N] to the surface area of cross-section A [mm^2] (6):

$$f_c = P/A \text{ [N/mm}^2] \tag{6}$$

On that basis a hypothetical scaling curve was chosen and the pairs of results were obtained: compressive strength f_c [N/mm^2]-passing wave velocity C_L [km/s] according to the methodology described in the literature [29,39].

In order to confirm the salt formation in the concrete, a colorimetric semi-quantitative method has been used with the use of Merck test strips. Then, in order to determine the distribution of sulphate salts forming along the ultrasonic measurements, a gravimetric quantitative method has been used [46–49]. Tested samples of concrete were cut, dried and crushed. In the next step, samples were extracted in one molar hydrochloric acid. As the precipitating agent barium chloride was added to the pre-heated samples extract to precipitate barium sulphate, which was weighed after washing and calcination at 800 °C to constant weight. Determination of the sulphate content in the barium sulphate precipitate was determined on the basis of mass proportions according to the Equation (7):

$$x = a\ 96.064/233.400 = a\ 0.4116 \text{ [g]} \tag{7}$$

where a is a mass of the barium sulphate [g], x is a mass of SO$_4{}^{2-}$ in the tested sample of material.

3. Results

3.1. Quantification of Sulphates in Tested Material

Concrete samples were taken from the walls and ceiling in order to determine salt content. It turned out that chlorides were available in the concrete in acceptable quantities (maximum in sample 1–0.085% of weight of the concrete), sulphates in high quantities above 1.2% by weight of the concrete. No nitrates or nitrites were found in the concrete. In the next step quantitative determination of sulfates using gravimetric method has been undertaken. Tests have been performed on the internal surface of the tank walls, and across its thickness, at distances of 20, 50 and 100 mm from the surface under sulphate attack. Quantity of the sulphates has been calculated with use of the Equation (7). The results of the analysis are presented in Table 1.

Table 1. Results of the gravimetric quantification of sulphates across the thickness of tested concrete elements.

Distance from Internal Wall Surface [mm]	Concrete Sample Mass [g]	Mass of BaSO$_4$ (a) [g]	Mass of SO$_4{}^{2-}$ (x) [g]	SO$_4{}^{2-}$ [% by Sample Weight]	SO$_4{}^{2-}$ Mean Value [%]
	10.2327	0.3796	0.1563	1.527	
0	10.1276	0.3873	0.1594	1.574	1.524
	10.3214	0.3691	0.1519	1.472	
	9.9885	0.3211	0.1321	1.323	
20	10.1445	0.3436	0.1414	1.394	1.345
	10.3417	0.3312	0.1363	1.318	
	10.1424	0.1163	0.0479	0.472	
50	9.9672	0.1480	0.0609	0.611	0.548
	10.1228	0.1382	0.0569	0.562	
	9.8276	0.0766	0.0315	0.321	
100	10.1412	0.1015	0.0418	0.412	0.315
	10.3429	0.0533	0.0219	0.212	

Performed quantitative analysis of sulphates on the thickness of the tested samples indicates their deep penetration and concentration in the entire cross-section of the walls. The highest concentration of SO$_4{}^{2-}$ has been examined in the zone near the inner wall surface (above 1.52%), at a depth of 20 mm

SO_4^{2-} percentage concentration was 1.35%, reaching 0.55% at a depth of 50 mm and 0.31% at a distance of 100 mm from the inner tank wall surface.

3.2. Calibration of Ultrasound Pulse Velocity-Compression Strength Curve Based on the Destructive Tests

Since the distribution of compressive strength across the thickness of reinforced concrete structures subject to sulphate corrosion from one side had not been tested so far, the authors performed such measurements using the ultrasonic method. In order to investigate the actual condition of concrete corrosion in the walls, three boreholes were made. The borehole location chosen was below the ceiling slab but in a manner ensuring that effluents did not overflow. The first and second borehole were made at the distance of 50 cm from the top wall edge, and the third borehole at the height of the top wall edge. Borehole No. 1 was of diameter 103 mm, and 2 & 3 diameter 50 mm. It was noted that on borehole No. 1 reinforcing meshes had moved towards the internal surface. For this reason, the thickness of lagging was reduced down to 10 mm and as a result of corrosion in the tested location 8 mm of concrete were missing, only a protective layer 2 mm thick (Figure 7) remained.

Figure 7. Thickness of non-corroded lagging in borehole No. 1 is only 2 mm.

After the measurements of ultrasonic pulse velocities of the tested samples they have been cut and tested in uniaxial loading on strength machine. On that base a hypothetical scaling curve with the following Equation (8) was chosen:

$$f_c = 53.6 \cdot C_L - 122.3 \, [\text{N/mm}^2] \tag{8}$$

where f_c is the compressive strength of concrete [N/mm²], and C_L is the ultrasound longitudinal wave velocity [km/s]. The results of destructive tests, measured pulse velocities and strength values calculated with use of Equation (8) are presented in Table 2.

Table 2. Results of uniaxial destructive tests, measured pulse velocities and strength values calculated with use of the chosen hypothetical scaling curve.

Sample No.	Core Size [cm × cm]	Ultrasound Longitudinal Wave Velocity C_L [km/s]	$f_{c,\varnothing}$-$f_{c,cube}$ Conversion Factor	Compression Strength [MPa]		
				Destructive Test		f_c from Equation (6)
				$f_{c,\varnothing}$	$f_{c,cube}$	
1	10.3 × 10.3	3.63	1.00	68.92	68.92	72.27
2	10.3 × 10.3	2.94		37.55	37.55	35.28
3	5.0 × 5.0	3.08		41.94	45.3	42.79
4	5.0 × 5.0	2.91	1.08	32.66	35.27	33.68
5	5.0 × 5.0	3.14		40.90	44.17	46.00
6	5.0 × 5.0	3.42		59.62	87.14	61.01
Mean value	-	3.19	-	46.20	48.47	48.51

In this way, scaling curves established hypothetically were used to convert the rate of ultrasound wave in the given cross-section at the borehole height into concrete compression strength in this cross-section.

3.3. Testing the Strength of Concrete across the Tank Wall Thickness

Passing times t [μs], calculated wave velocities C_L [km/s] and compressive strengths f_c [N/mm^2] in planes parallel to the surface of the boreholes are presented in Table 3. Results are presented starting with the ordinal number 1 (5 mm from the external side of the tested wall) in the direction of the internal side of the examined tank.

Table 3. Results of concrete compression strength test in borehole No. 1.

Ordinal Number	Ultrasound Netto Passing Time in Direction I-I $t_{n\,I-I}$ [μs]	Ultrasound Netto Passing Time in Direction II-II $t_{n\,II-II}$ [μs]	Mean Ultrasound Netto Passing Time t_n [μs]	Ultrasound Longitudinal Wave Velocity C_L [km/s]	Concrete Compression Strength f_c [N/mm^2]
1	33.00	34.30	33.65	3.09	43.36
2	33.40	34.20	33.80	3.08	42.62
3	34.20	35.70	34.95	2.98	37.20
4	35.30	35.40	35.35	2.94	35.39
5	33.50	32.60	33.05	3.15	46.37
6	33.00	31.20	32.10	3.24	51.36
7	30.60	33.50	32.05	3.25	51.63
8	32.20	33.80	33.00	3.15	46.62
9	30.30	32.00	31.15	3.34	56.65
10	29.60	33.80	31.70	3.28	53.55
11	28.40	29.10	28.75	3.62	71.59
12	28.50	29.00	28.75	3.62	71.59
13	28.70	29.00	28.85	3.61	70.92
14	28.70	28.60	28.65	3.63	72.27
15	28.70	28.60	28.65	3.63	72.27
16	28.70	30.30	29.50	3.53	66.66
17	32.20	32.30	32.25	3.23	50.55
18	31.40	31.00	31.20	3.33	56.37
Mean (1–18)	31.13	31.91	31.52	3.32	55.39
19	-	-	-	-	0.00

The dependencies of compressive strength as a function of depth for the borehole No. 1 are shown in Figure 8.

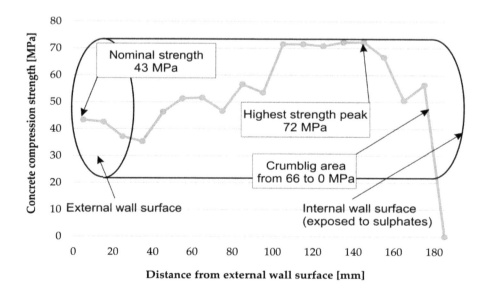

Figure 8. Change of concrete strength across the tank wall thickness in borehole No. 1.

Similar tests were performed on boreholes No. 2 & 3 which broke and reinforcement was not cut, hence their length is less than the thickness of the tank wall. The results of tests performed on borehole No. 2 are shown in Table 4.

Table 4. Results of concrete compression strength test in borehole No. 2.

Ordinal Number	Ultrasound Passing Time, Direction I-I $t_{n\,I\text{-}I}$ [μs]	Ultrasound Passing Time, Direction II-II $t_{n\,II\text{-}II}$ [μs]	Mean Ultrasound Passing Time t_n [μs]	Ultrasound Longitudinal Wave Velocity C_L [km/s]	Concrete Compression Strength f_c [N/mm²]
1	17.70	17.00	17.35	2.85	30.57
2	18.50	16.10	17.30	2.86	31.05
3	16.20	16.10	16.15	3.06	41.93
4	16.10	16.00	16.05	3.08	42.95
5	16.40	15.30	15.85	3.12	45.04
6	15.80	15.70	15.75	3.14	46.11
7	15.00	15.50	15.25	3.25	51.63
8	15.80	15.50	15.65	3.16	47.18
Mean (1–8)	16.44	15.90	16.17	3.07	42.06

Core no 2 was broken at the reinforcement mesh at a depth of 75 mm from the external surface of the wall. On the tested (not damaged) fragment of the core the growing dependency of strength as a function of depth was established. The results from Table 3 are depicted in Figure 9.

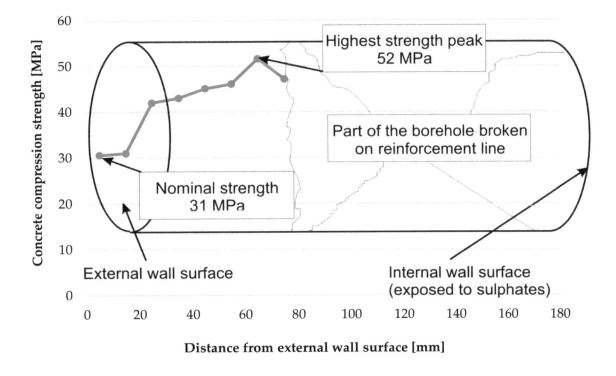

Figure 9. Change of concrete strength across the tank wall tested on borehole No. 2.

Results of tests performed on borehole No. 3 which was also broken in the middle of the tested section are shown in Table 5.

Table 5. Results of concrete compression strength test in borehole No. 3.

Ordinal Number	Ultrasound Passing Time, Direction I-I $t_{n\ I\text{-}I}$ [μs]	Ultrasound Passing Time, Direction II-II $t_{n\ II\text{-}II}$ [μs]	Mean Ultrasound Passing Time t_n [μs]	Ultrasound Longitudinal Wave Velocity C_L [km/s]	Concrete Compression Strength f_c [N/mm²]
1	16.00	17.00	16.50	3.00	38.5
2	16.30	16.80	16.55	2.99	38.0
3	16.60	17.40	17.00	2.91	33.7
4	16.00	16.60	16.30	3.04	40.4
5	16.90	17.60	17.25	2.87	31.5
6	16.90	16.90	16.90	2.93	34.7
7–9	-	-	-	-	
10	16.30	16.40	16.35	3.03	39.9
11	16.20	16.30	16.25	3.05	40.9
12	15.60	16.50	16.05	3.08	43.0
13	15.70	16.60	16.15	3.06	42.0
14	15.30	16.80	16.05	3.08	43.0
15	16.30	16.40	16.35	3.03	39.9
16	14.80	16.00	15.40	3.21	50.0
17	15.10	14.80	14.95	3.31	55.1
Mean (1–6,10–17)	16.00	16.58	16.29	3.04	40.8
18	-	-	-	-	0.00

Core No. 3 was cracked at the reinforcement at a distance of 60 to 80 mm from the external surface of the wall. For this reason the tests were not performed in this part of the core. On the tested (not damaged) fragment of the core the growing dependency of strength as a function of depth was established. The results from Table 2 are depicted in Figure 10.

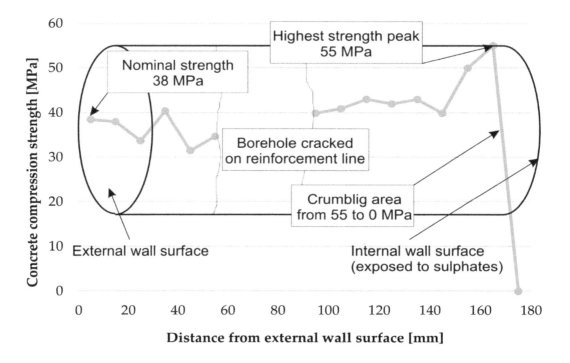

Figure 10. Change of concrete strength across the tank wall tested on borehole No. 3.

The method of measuring ultrasonic wave velocity using spot heads presented in this paper allowed determination of the distribution of strength in the cross-sections of reinforced concrete elements exposed to sulphate corrosion (with heterogeneous mechanical properties). The observations presented in this study show that the compressive strength of concrete subjected to a gaseous, aggressive, sulphuric environment is variable across its thickness. Concrete strength is variable across the wall thickness, however initially it was expected that by moving towards the tank interior, the strength would decrease, and from all boreholes the growing dependence was obtained, what has been shown in Figure 11.

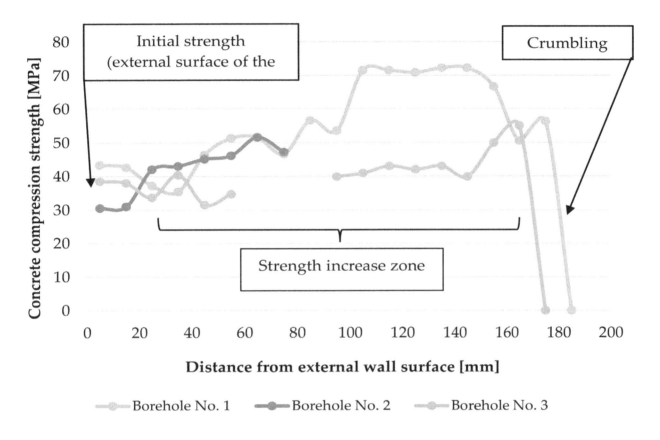

Figure 11. Change of concrete strength across the tank wall thickness in three boreholes.

Values of compressive strength of the samples taken from the tank walls show an increase in strength from 30–43 N/mm^2 in the cross sections near the external, unexposed wall layers to 55 and 72 N/mm^2 in the internal (exposed) cross sections of the walls. For the tested samples, this gives an increase in strength of 44% to 83% from its initial strength measured in the cross sections near the unexposed side of the structure. Sulphuric acid ions react with the cement compounds and the formation of gypsum leads to an increase in volume [17,18,50]. No destructive expansion of concrete takes place at this stage, and a reasonable conclusion can be drawn that the filling of pores and spaces in concrete by the calcium sulphate dihydrate causes a significant increase in the strength of the concrete, as has been shown in Figure 11. The compressive strength of the concrete samples decreases suddenly in the inner cross sections of the walls, when the salt crystallization pressure exceeds concrete tensile strength. In this stage of corrosion gypsum stone reacts with tricalcium aluminates (C3A) and hydrated calcium sulfoaluminate (monosulphate) and forms the final chemical product Candlot's salt what is associated with spalling and cracking in the surface zone of tested elements, and has been described in [51–53]. The completed tests confirmed that the drop of compressive strength took place only near internal cross-sections of the tested samples, within a distance not greater than 10 mm from their exposed surface. In the future, it is planned to carry out measurements of ultrasonic wave velocity and strength distribution in the tank walls at the height at which they are immersed in the corrosive substance and to compare them with the results presented in this article.

4. Conclusions

Based on the case study and its analysis presented in this article, the following conclusions can be made:

- Ultrasound testing methodology allowed determination of the distribution of strength as a function of depth of concrete elements under sulphate attack.
- The compressive strength of the concrete exposed to sulphate attack from one side is variable across its depth.
- The experimentally tested distribution of compressive strength at the depth of the elements showed an upward trend in the entire cross section towards the surface subject to corrosion.
- A decrease of strength appears only in the destroyed, crumbled zone of the concrete structure. The destroyed zone of tested elements did not exceed a depth of 10 mm from the surface exposed to sulphates attack.
- In the presented research, the difference in concrete strength between cross sections near the exposed and unexposed sides varied from an increase of 44% (borehole No. 3) to 83% (borehole No. 1).
- The performed tests indicate that gases may be a more corrosive environment, especially with high humidity, than liquids, therefore the coefficients of diffusion resistance or other permeability parameters, e.g., $g/m^2/24$ hours, are important parameters characterizing anticorrosive coatings. Chemical resistance to various acids, alkalis or other compounds is tested for the specific aggressive compound at a given level.
- In the tested tank, sulphur-containing gas (hydrogen sulphide) easily penetrated through the thin bituminic layer, based on the measurements at a thickness of 0.97 mm, and in contact with cement and lime formed sulphates, considerable quantities of which were found in the concrete.
- Since the lagging thickness had decreased already down to 2 mm, danger exists not only for the concrete but also for steel. The rate of concrete corrosion in the tank is probably influenced by the concentration of the gases above the liquid. If the tank had a properly built gravity ventilation, the concrete damage process would be much slower, because relative air humidity in the tank would also be much lower with effectively running ventilation.

Author Contributions: Conceptualization, B.S. and T.K.; Methodology, T.K. and B.S.; Validation, B.S., Investigation, B.S and T.K.; Resources, B.S. and T.K.; Writing-original draft preparation, B.S. and T.K.; Writing-review and editing, B.S. and T.K.; Visualization, B.S. and T.K.; Supervision, B.S.

References

1. Feldman, R.F. Porestructure, Permeability and Diffusivity as Related to Durability. In *Eighth International Congress on the Chemistry of Cement*; Bertrand Brasil: Rio de Janeiro, Brazil, 1986; Volume 1, Theme 4.
2. Gruner, M. *Corrosion and Protection of Concrete*; PWN: Warsaw, Poland, 1983. (In Polish)
3. Flaga, K. The Role of the Tightness of the Aggregate Skeleton in the Design of a Concrete Mix. *Inz. Bud.* **1984**, *7*, 14–16. (In Polish)
4. Chodor, L. Repair and Protection of Reinforced Concrete. |Chodor-Projekt|. Available online: http://chodor-projekt.net/encyclopedia/naprawa-i-ochrona-zelbetu/ (accessed on 15 July 2018).
5. Maj, M.; Ubysz, A. Cracked reinforced concrete walls of chimneys, silos and cooling towers as result of using formworks. In *MATEC Web of Conferences 146, Proceedings of Building Defects 2017, České Budějovice, Czech Republic, November 23–24, 2017*; Šenitková, I.J., Ed.; EDP Sciences: Les Ulis, France, 2018; Volume 02002, pp. 1–8. [CrossRef]
6. Trapko, T.; Musiał, M.P. Failure of pillar of sports and entertainment hall structure. In *MATEC Web of Conferences 146, Proceedings of Building Defects 2017, České Budějovice, Czech Republic, November 23–24, 2017*; Šenitková, I.J., Ed.; EDP Sciences: Les Ulis, France, 2018; Volume 02002, pp. 1–8. [CrossRef]

7. Wells, T.; Melchers, R.E.; Bond, P. Factors involved in the long term corrosion of concrete sewers. In Proceedings of the 49th Annual Conference of the Australasian Corrosion Association, Corrosion and Prevention, Coffs Harbour, Australia, 15–19 November 2009.
8. Sun, C.; Chen, J.; Zhu, J.; Zhang, M.; Ye, J. A new diffusion model of sulfate ions in concrete. *Constr. Build. Mater.* **2013**, *39*, 39–45. [CrossRef]
9. Bonakdar, A.; Mobasher, B.; Chawla, N. Diffusivity and micro-hardness of blended cement materials exposed to external sulfate attack. *Cem. Concr. Compos.* **2012**, *34*, 76–85. [CrossRef]
10. Idiart, A.E.; L'opez, C.M.; Carol, I. Chemo-mechanical analysis of concrete cracking and degradation due to external sulfate attack: A meso-scale model. *Cem. Concr. Compos.* **2011**, *33*, 411–423. [CrossRef]
11. Lorente, S.; Yssorche-Cubaynes, M.-P.; Auger, J. Sulfate transfer through concrete: Migration and diffusion results. *Cem. Concr. Compos.* **2011**, *33*, 735–741. [CrossRef]
12. Condor, J.; Asghari, K.; Unatrakarn, D. Experimental results of diffusion coefficient of sulfate ions in cement type 10 and class G. *Energy Procedia* **2011**, *4*, 5267–5274. [CrossRef]
13. Roziere, E.; Loukili, A.; Hachem, R.; Grondin, F. Durability of concrete exposed to leaching and external sulphate attacks. *Cem. Concr. Res.* **2009**, *39*, 1188–1198. [CrossRef]
14. Santhanam, M.; Cohen, M.D.; Olek, J. Modeling the effects of solution temperature and concentration during sulfate attack on cement mortars. *Cem. Concr. Res.* **2002**, *32*, 585–592. [CrossRef]
15. Pommersheim, J.M.; Clifton, J.R. Expansion of cementitious materials exposed to sulfate solutions, scientific basis for nuclear waste management. *Mater. Res. Soc.* **1994**, *333*, 363–368. [CrossRef]
16. Parande, A.K.; Ramsamy, P.L.; Ethirajan, S.; Rao, C.R.K.; Palanisamy, N. Deterioration of reinforced concrete in sewer environments. *Inst. Civ. Eng. -Munic. Eng.* **2006**, *159*, 11–20. [CrossRef]
17. Basista, M.; Weglewski, W. Chemically-assisted damage of concrete: A model of expansion under external sulfate attack. *Int. J. Damage Mech.* **2008**, *18*, 155–175. [CrossRef]
18. Basista, M.; Weglewski, W. Micromechanical modelling of sulphate corrosion in concrete: Influence of ettringite forming reaction. *Theor. Appl. Mech.* **2008**, *35*, 29–52. [CrossRef]
19. Pommersheim, J.; Clifton, J.R. *Sulphate Attack of Cementitious Materials: Volumetric Relations and Expansion*; National Institute of Standards and Technology: Gaithersburg, MD, USA, 1994; pp. 1–19.
20. Skalny, J.; Marchand, J.; Odler, I. *Sulphate Attack on Concrete*; Spon Press: London, UK, 2002.
21. Zhou, Y.; Tian, H.; Sui, L.; Xing, F.; Han, N. Strength Deterioration of Concrete in Sulfate Environment: An Experimental Study and Theoretical Modeling. *Adv. Mater. Sci. Eng.* **2015**, *2015*, 951209. [CrossRef]
22. Shi, F.; Wang, J.H. Performance degradation of cube attacked by sulfate. *Concrete* **2013**, *3*, 52–53.
23. Du, J.M.; Liang, Y.N.; Zhang, F.J. *Mechanism and Performance Degradation of Underground Structure Attacked by Sulfate*; China Railway Publishing House: Beijing, China, 2011.
24. Neville, A.M. *Properties of Concrete*; Polski Cement Sp. z o.o.: Cracow, Poland, 2000. (In Polish)
25. Petersons, N. Should standard cube test specimens be replaced by test specimens taken from structures? *Mater. Struct.* **1968**, *1*, 425–435. [CrossRef]
26. Stawiski, B. The heterogeneity of mechanical properties of concrete in formed constructions horizontally. *Arch. Civ. Mech. Eng.* **2012**, *12*, 90–94. [CrossRef]
27. Jasinski, R.; Drobiec, Ł.; Mazur, W. Validation of Selected Non-Destructive Methods for Determining the Compressive Strength of Masonry Units Made of Autoclaved Aerated Concrete. *Materials* **2019**, *12*, 389. [CrossRef]
28. Bogas, J.A.; Gomes, M.G.; Gomes, A. Compressive strength evaluation of structural lightweight concrete by non-destructive ultrasonic pulse velocity method. *Ultrasonics* **2013**, *53*, 962–972. [CrossRef]
29. Brunarski, L. Estimation of concrete strength in construction. *Build. Res. Inst. Quat.* **1998**, *2–3*, 28–45.
30. Anugonda, P.; Wiehn, J.S.; Turner, J.A. Diffusion of ultrasound in concrete. *Ultrasonics* **2001**, *39*, 429–435. [CrossRef]
31. Sansalone, M.; Streett, W.B. *Impact-Echo Nondestructive Evaluation of Concrete and Masonry*; Bullbrier Press: Ithaca, NY, USA, 1997.
32. Breysse, D. Nondestructive evaluation of concrete strength: An historical review and a new perspective by combining NDT methods. *Constr. Build. Mater.* **2012**, *33*, 139–163. [CrossRef]
33. Facaoaru, I. Contribution à i'étude de la relation entre la résistance du béton à la compression et de la vitesse de propagation longitudinale des ultrasons. *RILEM* **1961**, *22*, 125–154.

34. Leshchinsky, A. Non-destructive methods instead of specimens and cores, quality control of concrete structures. In Proceedings of the Second International RILEM/CEB Symposium, Belgium, 12–14 June 1991; pp. 377–386.

35. Bungey, J.H. The validity of ultrasonic pulse velocity testing of in-place concrete for strength. *NDT Int.* **1980**, *13*, 296–300. [CrossRef]

36. Szpetulski, J. Testing of compressive strength of concrete in construction. *Constr. Rev.* **2016**, *3*, 21–24. (In Polish)

37. Gudra, T.; Stawiski, B. Non-destructive strength characterization of concrete using surface waves. *NDT Int.* **2000**, *33*, 1–6. [CrossRef]

38. Komlos, K.; Popovics, S.; Nurnbergerova, T.; Babal, B.; Popovics, J.S. Ultrasonic Pulse Velocity Test of Concrete Properties as Specified in Various Standards. *Cem. Concr. Compos.* **1996**, *18*, 357–364. [CrossRef]

39. PN-B-06261. *Non-Destructive Testing of Structures*; Ultrasound method of testing compressive strength of concrete; Polish Committee for Standardization: Warsaw, Poland, 1974.

40. EN 12504-4. *Testing Concrete-Part 4: Determination of Ultrasonic Pulse Velocity*; European Committee for Standardization: Brussels, Belgium, 2004.

41. EN 13791. *Assessment of In-Situ Compressive Strength in Structures and Precast Concrete Components*; Polish Committee for Standardization: Brussels, Belgium, 2004.

42. Stawiski, B.; Kania, T. Determination of the influence of cylindrical samples dimensions on the evaluation of concrete and wall mortar strength using ultrasound method. *Procedia Eng.* **2013**, *57*, 1078–1085.

43. Alam, B.; Afzal, S.; Akbar, J.; Ashraf, M.; Shahzada, K.; Shabab, M.E. Mitigating Sulphate Attack in High Performance Concrete. *Int. J. Adv. Struct. Geotech. Eng.* **2013**, *2*, 11–15.

44. Genovés, V.; Vargas, F.; Gosálbez, J.; Carrión, A.; Borrachero, M.V.; Payá, J. Ultrasonic and impact spectroscopy monitoring on internal sulphate attack of cement-based materials. *Mater. Des.* **2017**, *125*, 46–54. [CrossRef]

45. Cumming, S.R. *Non-Destructive Testing to Monitor Concrete Deterioration Caused by Sulfate Attack*; University of Florida: Gainesville, FL, USA, 2004.

46. Kocjan, R. *Analytical Chemistry*; Qualitative analysis, Quantitative analysis, Instrumental analysis; PZWL: Warsawa, Poland, 2015. (In Polish)

47. Blumenthal, P.L.; Guernsey, S.C. *The Determination of Sulfur as Barium Sulfate*; Research Bulletin No. 26; AES: Ames, IA, USA, 1915.

48. Reid, J.M.; Czerewko, M.A.; Cripps, J.C. *Sulfate Specification for Structural Backfills*; Report TRL447; TRL: Berkshire, UK, 2005.

49. St. John, T.W. Quantifying acid-soluble sulfates in geological materials: A comparative study of the British Standard gravimetric method with ICP-OES/AES. In Proceedings of the 19th International Conference on Soil Mechanics and Geotechnical Engineering, Seoul, Korea, 15 September 2017.

50. Piasta, W.; Marczewska, J.; Jaworska, M. Some aspects and mechanisms of sulphate attack. *Struct. Environ.* **2014**, *6*, 19–24.

51. Brown, W.; Taylor, H. The role of ettringite in external sulfate attack. *Mater. Sci. Concr.* **1999**, *5*, 73–98.

52. Yu, C.; Sun, W.; Scrivener, K. Mechanism of expansion of mortars immersed in sodium sulfate solutions. *Cem. Concr. Res.* **2012**, *43*, 105–111.

53. Whitaker, M.; Black, L. Current knowledge of external sulfate attack. *Adv. Cem. Res.* **2015**, *27*, 1–14. [CrossRef]

4

Multi-Scale Structural Assessment of Cellulose Fibres Cement Boards Subjected to High Temperature Treatment

Tomasz Gorzelańczyk⬤, Michał Pachnicz *⬤, Adrian Różański⬤ and Krzysztof Schabowicz⬤

Faculty of Civil Engineering, Wrocław University of Science and Technology, Wybrzeże Wyspiańskiego 27, 50-370 Wrocław, Poland
* Correspondence: michal.pachnicz@pwr.edu.pl

Abstract: The methodology of multi-scale structural assessment of the different cellulose fibre cement boards subjected to high temperature treatment was proposed. Two specimens were investigated: Board A (air-dry reference specimen) and Board B (exposed to a temperature of 230 °C for 3 h). At macroscale all considered samples were subjected to the three-point bending test. Next, two methodologically different microscopic techniques were used to identify evolution (caused by temperature treatment) of geometrical and mechanical morphology of boards. For that purpose, SEM imaging with EDS analysis and nanoindentation tests were utilized. High temperature was found to have a degrading effect on the fibres contained in the boards. Most of the fibres in the board were burnt-out, or melted into the matrix, leaving cavities and grooves which were visible in all of the tested boards. Nanoindentation tests revealed significant changes of mechanical properties caused by high temperature treatment: "global" decrease of the stiffness (characterized by nanoindentation modulus) and "local" decrease of hardness. The results observed at microscale are in a very good agreement with macroscale behaviour of considered composite. It was shown that it is not sufficient to determine the degree of degradation of fibre-cement boards solely on the basis of bending strength; advanced, microscale laboratory techniques can reveal intrinsic structural changes.

Keywords: cellulose fibre cement boards; microstructure; nanoindentation; SEM-EDS analysis; temperature

1. Introduction

Fibre-cement boards were invented by the Czech engineer Ludwik Hatschek over 100 years ago. They have been used in construction as siding, ceilings, floors, roofs and tile backer boards because they are damp-proof and nonflammable light-weight, strong and durable. Nowadays the cellulose fibre cement boards belong to a special class of fibre-reinforced cementitious composites and they consist of 50–70% of cement while the other components include: mineral fibres (usually cellulose) and fillers (limestone powder, kaolin, etc.). The mechanical properties, durability and microstructure of fibre-cement boards in the various environments are widely described in the literature [1,2]. The final properties of cellulose fibre cement composites depend, aside from the fibre and the matrix components, on the manufacturing process as well as on the internal microstructure.

Chady, Schabowicz et al. [3,4] proposed various non-destructive testing methods for evaluating fibre-cement boards as to the potential occurrence of heterogeneities or defects in them. Tonoli et al. [5] analyzed the effects of natural weathering on microstructure and mineral composition of cementitious roofing tiles reinforced with fique fibre. Savastano et al. [6] tested microstructure and mechanical properties of waste fibre-cement composites. At the same time the X-ray microtopography technique has been developed to study the microstructures for non-destructive characterization of the internal structure of various materials [7,8]. Cnudde et al. [9] were using the micro-CT method to determine the impregnation depth of water repellents and consolidants inside natural building stones. Li-Ping Guo et al. [10] investigated the effects of mineral admixtures on initial defects existing in high-performance concrete microstructures using a high-resolution X-ray micro-CT. Wang et al. [11] used this technique to produce the X-ray tomography images of porous metal fibre sintered sheet with 80% porosity. 3D information about the total porosity and the pore size distribution was obtained with the combination of micro-CT and home-made 3D software [12]. Schabowicz, Ranachowski at al. [13,14] successfully used the micro computed tomography (micro-CT) and SEM in the quality control system of cellulose fibre distribution in cement composites. Ranachowski and Gorzelańczyk et al. [15–17] investigate the degradation of the microstructure and mechanical properties of fibre cement board (FCB), which was exposed to environmental hazards, resulting in thermal impact on the microstructure of the board. Visual light and scanning electron microscopy, X-ray micro tomography, flexural strength, and work of flexural test W_f measurements were used.

From the literature review presented above, it is clear that until now the mechanical parameters of the microstructure of cellulose fibres cement boards have not been studied. It is however of primary importance to identify the mechanical morphology at microscale which directly affects the mechanical behavior of material at macroscale, i.e., at the scale of engineering applications. In the case of composite materials with a cement matrix, segmentation and characterization of components' mechanical properties is impossible with the use of classical macro scale experiments [18]. Therefore, advanced laboratory techniques that allow observation of the mechanical behavior of materials at different scales are used very often nowadays. An example of such technique is a classical nanoindentation test developed by Oliver and Pharr as a method to accurately calculate hardness and elastic modulus from the "load-displacement" curve [19].

The idea of nanoindentation is to determine the mechanical properties of composite components by observing their reaction to the point load followed by continuous unload of its surface. Initially, due to hardware limitations, the nanoindentation technique was mainly used in relatively homogeneous media, or layered materials with known thickness of individual layers [20,21]. Currently, better hardware capabilities have made it possible to use this technique for microheterogeneous materials, exhibiting different mechanical morphology depending on the observation scale [22,23]. Nanoindentation tests were successfully used to identify mechanical morphologies for different types of cementitious materials in the works [24–26].

This paper presents a multiscale approach for identification of internal structural changes of cellulose fibres cement boards subjected to high temperature treatment. Two advanced laboratory techniques are used, i.e., SEM imaging with EDS analysis and nanoindentation tests. The former technique provides microphotographs of the applied fibres together with the elemental composition of tested samples. The latter (nanoindentation) reveals the mechanical morphology of the material in terms of hardness and elastic moduli. All tests were carried out on two samples: reference material at air-dried state, and the one subjected to high temperature treatment. An evolution of geometrical and mechanical morphology of investigated boards is observed and discussed.

2. Materials and Methods

In this study, two sets of specimens were prepared for examination. These were labeled A and B. The specimens were fabricated by applying the Hatschek forming method. Air-dry reference specimens (not subjected to high temperature treatment) were denoted as A. The specimens, which underwent the high temperature treatment in an electric oven for 3 h at a temperature of 230 °C, were labeled as B. The parameters of the heating procedure were chosen experimentally, after performing some preliminary tests to evoke considerable changes in the microstructure of the investigated materials. The tested specimens were cut from the different fibre cement panels of 8 mm of thickness. Prior to the main research the panels were tested using the standard procedures to assess their performance.

Comparisons of tested panels are presented in Table 1 and the set of equipment for bending strength measurements is shown in Figure 1.

Table 1. Characteristic of the tested panels.

Symbol of the Board	A	B
Type of the board	fibre-cement, exterior	fibre-cement, exterior, after 3 h of burning (230 °C)
Thickness of board [mm]	8	8
Bending strength [MPa]	23.54 *	26.86 *
Density [kg/m^3]	1600	1500
Photo of the board	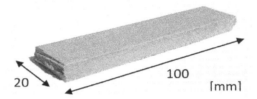	

* mean values calculated on the basis of ten independent measurements.

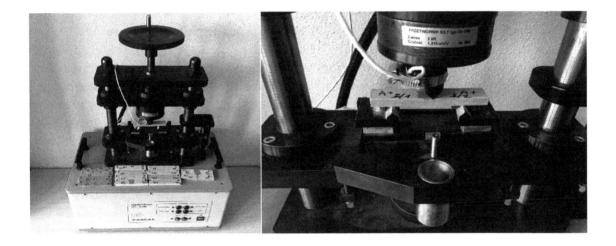

Figure 1. Test stand for bending strength measurements, and fibre-cement board during test.

Figure 2 shows an exemplary σ-ε curve of the tested fibre-cement boards under bending. The trace of flexural strength σ, bending strength MOR, limit of proportionality (LOP) and strain ε were analyzed. Bending strength MOR was calculated from the standard formula [27]:

$$MOR = \frac{3Fl_s}{2b\,e^2} \qquad (1)$$

where:

F is the loading force (N);
l_s is the length of the support span (mm);
b is the specimen width (mm); and
e is the specimen thickness (mm).

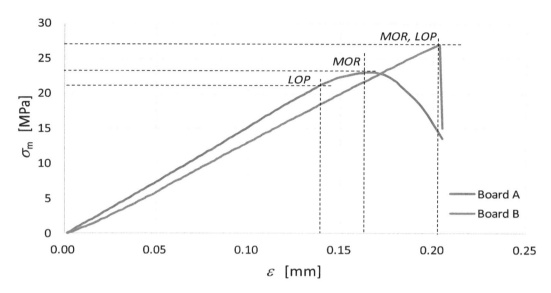

Figure 2. Diagrams of the σ-ε dependence under bending for fibre-cement boards.

In case of microscale experiments, first, the analysis was performed by SEM imaging with EDS elemental composition examination. For that purpose, the authors have prepared fractured specimens to produce the SEM micrographs applying the Quanta FEG-250 Scanning Electron Microscope (Hillsboro, OR, USA), FEI with an EDS analyser. The precondition was made at 50% of relative humidity and 22 °C to enable for different modes of decomposition of composite microstructure.

Next the boards were investigated in terms of nanoindentation approach. It is commonly known that the process of sample preparation for nanoindentation is of primary importance for getting reasonable results. In general, we can assume that the results from nanoindentation can only be as good as the sample used for testing. Hence, the aim of the preparation process was to obtain a satisfactory quality of: parallelism and roughness of surfaces, cleanness of the sample and sample tilt. We followed a common rule-of-thumb contained in the ISO standard for nanoindentation (ISO 14577) as well as requirements, concerning surface roughness criteria for cement paste nanoindentation, provided in [25]. Preparation of the samples for nanoindentation consisted of cutting fibre boards to required dimensions (approximately 1 × 1 cm), mounting specimens in the epoxy resin and thorough grinding and polishing process of the specimen surface. High-speed diamond saw Struers Labotom-5 (Copenhagen, Denmark) (Figure 3a), Struers CitoVac vacuum chamber (Copenhagen, Denmark) (Figure 3b) and Struers LaboPol-5 grinder (Copenhagen, Denmark) (Figure 3c) were used for the sample preparation. The photos of the samples after the preparation procedure are presented in Figure 4.

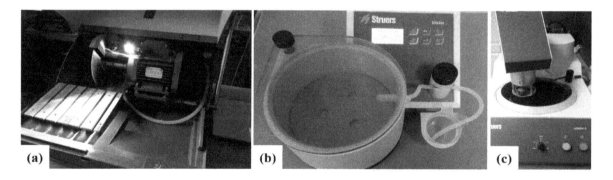

Figure 3. Equipment used for sample preparation: (**a**) Struers Labotom-5, (**b**) Struers CitoVac, (**c**) Struers LaboPol-5.

Figure 4. Samples prepared for the indentation.

Nanoindentation tests were performed using Nanoindenter CSM TTX-NHT (Neuchatel, Switzerland) (Figure 5) equipped with a diamond Berkovich tip (Poisson's ratio $\nu_i = 0.07$, elastic modulus $E_i = 1000$ GPa, $\beta = 1.034$).

Figure 5. Nanoindentation test stand.

3. Results of Multiscale Approach

3.1. SEM Analysis

In this Section the results obtained with SEM imaging are presented. In particular, Figure 6 shows the microphotographs of the applied fibres. Note that left (right) panel of Figure 6 presents the microstructure of reference (subjected to high temperature) sample.

Figure 6. *Cont.*

Figure 6. The microscopic observations of the applied fibres, in scanning electron microscope (SEM) with scale bar in sequence 1 mm, 500 μm and 50 μm: (**a**) air-dry condition (reference board), (**b**) board exposure to temperature of 230 °C-3 h.

In Figure 7 elemental composition (results of EDS analysis) of tested samples is graphically presented. The investigation was carried out separately for matrix (Figure 7a), fibre of air-dry condition (reference board, Figure 7b) "fibre" of board exposed to temperature of 230 °C for 3 h (Figure 7c).

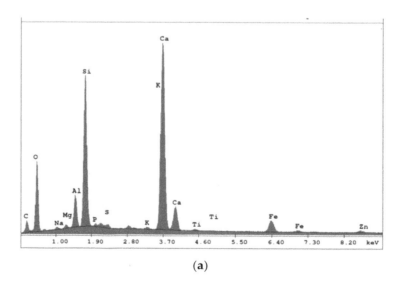

(**a**)

Figure 7. *Cont.*

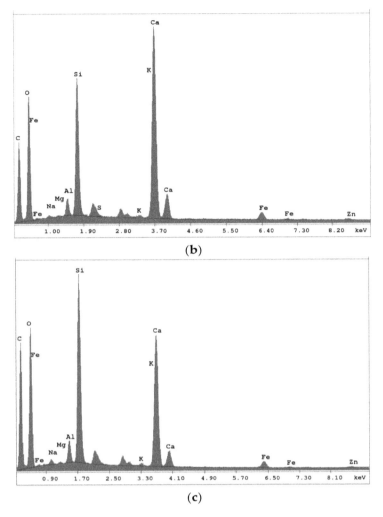

Figure 7. Composition analysis (EDS) of tested boards: (**a**) matrix, (**b**) fibre of air-dry condition (reference board) and (**c**) "fibre" of board exposure to temperature of 230 °C-3 h.

3.2. Nanoindentation

The assessment of mechanical parameters of the fibre cement boards was carried out using the method formulated for heterogeneous materials with a cement matrix. This technique, the Grid Indentation Technique (GIT), was introduced in [18]. For each specimen a regular grid of 220 tests was applied (Figure 8). The relevant grid parameters, such as the distance between indenter locations, the number of tests and the single test's maximum load were determined by trial nanoindentation tests in accordance with the recommendations presented in [18].

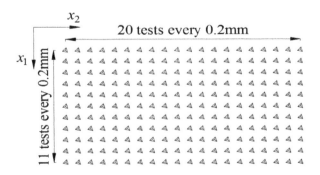

Figure 8. Applied nanoindentation grid.

In every grid point, a single standard nanoindentation test [19,28,29] was performed with the maximum load of 500 mN. The test runs as follows: continuous increase of force up to a fixed value F_{max}, then a short period in which the set maximum value of force is maintained, after that it is followed by unloading which is also carried out continuously (Figure 9a). During the test, the relation between the force F and the depth of penetration h is recorded. An example of F-h curve obtained for a single test is presented in Figure 9b. For every test two indentation parameters were calculated, namely hardness (H_{IT}) defined as follows:

$$H_{IT} = \frac{F_{max}}{A} \tag{2}$$

and the indentation modulus (M_{IT}):

$$M_{IT} = \frac{S}{2\beta} \frac{\sqrt{\pi}}{\sqrt{A}} \tag{3}$$

where, F_{max} is the maximum force of indentation, A is the projection of the contact area on the surface of the sample. This value is usually defined as a function of the maximum indentation depth h_{max} [18] and S is the initial slope of the unloading curve according to [18].

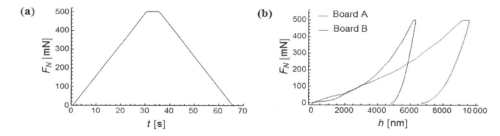

Figure 9. (**a**) Load function used for single nanoindentation, (**b**) and example of F-h curve.

As a consequence of performed tests, distribution of the mechanical parameters on the surface of the samples was evaluated. Contour maps of hardness and indentation modulus distribution are presented in Figures 10 and 11 for Board A and Board B, respectively.

All obtained values are presented in the form of histograms in Figure 12a,b. The results corresponding to the case of high temperature heating (230 °C-3 h, Board B) are displayed in grey. Red and blue colors are corresponding to the results obtained for reference samples (Board A). In particular, the red color represents the distribution of the indentation modulus, whereas the blue one represents the frequency of hardness values. The vertical dashed lines refer to the mean values averaged over 220 individual results. Furthermore, exact values of statistical measures, i.e., mean values μ and standard deviations σ, are summarized in Table 2.

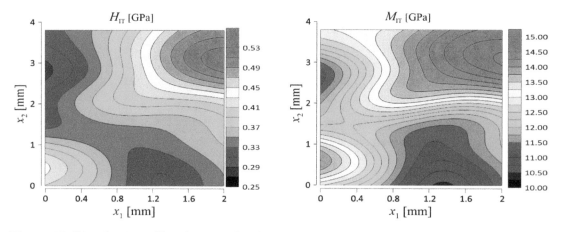

Figure 10. Distribution of hardness and indentation modulus for Board A (reference material).

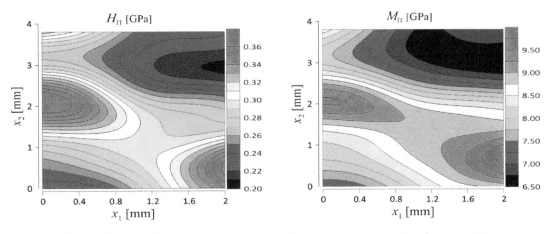

Figure 11. Distribution of hardness and indentation modulus for Board B.

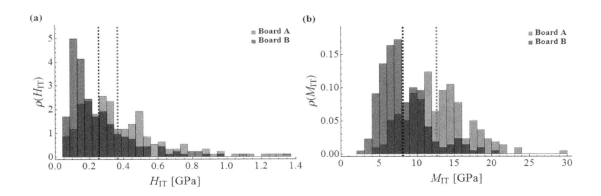

Figure 12. (**a**) Histograms of hardness; (**b**) histograms of indentation modulus.

Table 2. Summary of nanoindentation results.

Parameter		Specimen before High Temperature Treatment (Board A)	Specimen after High Temperature Treatment (Board B)
Mean value of hardness	$\mu^{H_{IT}}$ [GPa]	0.382	0.285
Standard deviation of hardness	$\sigma^{H_{IT}}$ [GPa]	0.226	0.201
Coefficient of variation * of hardness	$c.o.v^{H_{IT}}$	0.592	0.705
Mean value of indentation modulus	$\mu^{M_{IT}}$ [GPa]	12.686	8.147
Standard deviation of indentation modulus	$\sigma^{M_{IT}}$ [GPa]	4.093	3.138
Coefficient of variation * of indentation modulus	$c.o.v^{M_{IT}}$	0.318	0.385

* Coefficient of variation is defined as the ratio of the standard deviation to the mean value.

In Figures 13 and 14 the variation of mean values averaged along two independent directions, respectively for x_1 and x_2 (see coordinate system shown in Figure 8), is presented. The black color corresponds to the board subjected to heating and blue (hardness) and red (indentation modulus) colors are representing the results obtained for the reference sample.

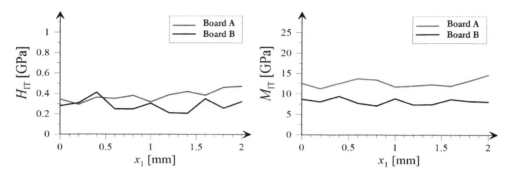

Figure 13. Average values of hardness (H_{IT}) and indentation modulus (M_{IT}) in the x_2 direction.

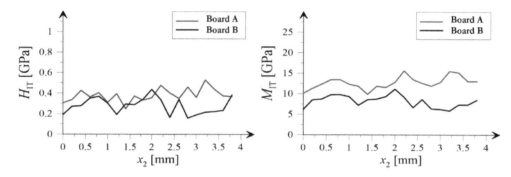

Figure 14. Average values of hardness (H_{IT}) and indentation modulus (M_{IT}) in the x_1 direction.

4. Discussion of Results

The results presented in Table 1 and Figure 2 show that, under high temperature, the bending strength *MOR* increases. The increase (in average sense) from 23.54 to 26.86 MPa is observed. For fibre-cement Board B, it was also observed that under the influence of a temperature of 230 °C, the structure of the Board Becomes more brittle; after peak value of stress, a sudden drop of strength is observed (Figure 2). An analysis of the classical macroscale results showed that for fibre-cement Board B the extent of the nonlinear increase in bending stress is reduced until the bending strength *MOR* comes to be level with the proportionality limit *LOP*. In the case of the reference board, the *MOR* and *LOP* values were clearly separated. It should be noted that the reference samples were under an air-dry condition and their bulk moisture content amounted to 6–8%.

Based on the macroscale results (bending strength) one can draw the conclusion that after exposure to the temperature (230 °C for 3 h) there is no degradation of the board, and even the strengthening effect is obtained. Nevertheless, advanced microscale laboratory techniques have revealed damaging and irreversible structural changes in both geometrical and mechanical morphology of microstructure.

An analysis of the images obtained from the scanning electron microscope and the EDS analyzer shows that the fibre cement Board A has a compact microstructure (Figure 6a). Microscopic examinations revealed a fine-pore structure, with pores of up to 50 μm in size. Cavities and grooves, up to 500 μm wide, were visible in the fracture areas where the fibres had been pulled out. Cellulose fibres and PVA fibres, are clearly visible in the images. Various forms of hydrated calcium silicates of the C-S-H type occur. Both an "amorphous" phase and a phase built of strongly-adhering particles predominate. An analysis of the fibre composition showed that fibre elements and some cement elements were present. An analysis of the chemical composition of the matrix showed elements that are typical of cement (Figure 7). The surface of the fibres was covered with a thin layer of cement paste and hydration products. The fact that there are very few areas with a space between the fibres and the cement paste, indicates that the fibre-cement bond is strong.

A microscopic analysis of the fibre cement Board B, which was exposed to a temperature of 230 °C for 3 h, shows a clear change in the colour of the samples (Figure 6b). Most of the fibres in the board were found to be burnt-out, or melted into the matrix, leaving cavities and grooves which were visible in all of the tested boards. The structure of the few remaining fibres was strongly degraded. An examination of the cement particles on the fracture surface revealed burning-out of their structure. The structure of the matrix was found to be more granular, showing many delaminations (Figure 6b). Numerous caverns and grooves left by the pulled-out fibres, as well as the pulled-out cement particles, were observed.

Noticeable changes of mechanical morphology of the analyzed boards due to the influence of high temperature is observed within the nanoindentation approach. The average value of the indentation modulus, being the measure of the elastic response of the material, after the exposure to the temperature of 230 °C for 3 hours, significantly decreased (see Figure 12b). The histogram for the Board B (grey colour) concentrates around lower values of modulus compared with the histogram for Board A. In addition, according to Table 2 the average value of the indentation modulus, after heating at the temperature of 230 °C, decreased from 12.686 GPa to 8.147 GPa which equals to an almost 50% drop in the elastic stiffness of examined material. Furthermore, as shown in the right panels of Figures 13 and 14, the mean value of indentation modulus is significantly smaller for Board B compared to the values obtained for Board A in the entire range of x_1 and x_2 values. These observations are in a very good agreement with SEM analysis. A decrease of elastic response of material's microstructure, its stiffness in fact, is mostly due to the changes in the microstructure geometry observed in SEM. The presence of cavities and grooves in Board B can be a direct cause of decrease in M_{IT} values. It is worth noticing that the changes of geometrical (SEM analysis) and mechanical (nanoindentation) morphology of Board B is a direct cause of the macroscale behavior; by observing Figure 2, one can simply notice that the stiffness of Board B is reduced compared to the one evaluated for Board A (the slope of σ-ε curve decreased after exposure to the temperature); and it was observed in case of all boards under bending.

Slightly different conclusions can be drawn for the hardness of the material. In general, the average hardness value, as a result of exposure to the temperature 230 °C, decreased, from 0.382 GPa for the Board A, to 0.285 GPa for the Board B (see Figure 12a and Table 2). On the other hand, observing the left panels of Figures 13 and 14 one can notice that "locally", i.e., for given ranges of x_1 or x_2 values, the decrease of H_{IT} is not observed. As shown in [30,31] it is the hardness H_{IT} which is the parameter determining the strength of the microstructure. This is due to fact that the strength is proportional to the hardness of individual microstructure components; the higher the hardness, the higher the strength of microstructure. Therefore the phenomenon observed e.g., in the left panels of Figure 13 (in the range of x_1 from 0 to 0.5 mm) and Figure 14 (in the range of x_2 from 0.5 to 2.0 mm), where no decrease in H_{IT} is observed, can justify the macroscale bending strength results. As shown in Table 1, the bending strength of Board B is slightly higher than the one evaluated for Board A; this is in the average sense since the values in Table 2 represent mean values. However, during individual bending tests we also observed the cases when the strength was not increased or even slightly decreased. This is the effect of the fact that "locally" hardness does not change under the influence of temperature, and hence, this can cause such macroscopic behavior of the boards.

It should also be mentioned that the standard deviations for both the indentation modulus and the hardness of the specimen subjected to temperature 230 °C slightly decreased. However, as mentioned above, the mean values of both properties decreased in a more evident manner, and hence the mechanical morphology of Board B seems to have more heterogeneous nature. This is clearly revealed by plots shown in Figures 13 and 14 as well as by c.o.v. (coefficient of variation) values summarized in Table 2. For Board B, the coefficient of variation of both hardness and indentation modulus increased.

5. Final Conclusions

Two methodologically different microscopic techniques were used to identify the evolution of geometrical and mechanical morphology of boards. The methodology of multi-scale structural assessment of the different cellulose fibre cement boards subjected to high temperature treatment was presented. For that purpose, two specimens were investigated: Board A (air-dry reference specimen) and Board B (exposed to a temperature of 230 °C for 3 h).

SEM examinations were carried out to get a better insight into the changes taking place in the structure of the tested boards. Significant changes take place in the structure of the boards, especially after the high temperature treatment in an electric oven for 3 h at a temperature of 230 °C. Most of the fibres in the board were burnt-out, or melted into the matrix, leaving cavities and grooves which were visible in all of the tested boards. The structure of the few remaining fibres was strongly degraded, as confirmed by the nanoindentation tests. Nanoindentation tests revealed significant changes of mechanical properties caused by high temperature treatment: "global" decrease of the stiffness (characterized by nanoindentation modulus) and "local" decrease of hardness. The results observed at microscale are in a very good agreement with macroscale behaviour of considered composite.

In the authors' opinion, the above findings are important for building practice because it was clearly shown that it is not sufficient to determine the degree of degradation of fibre-cement boards solely on the basis of bending strength. Based only on the macroscale results (bending strength) one can draw the conclusion that after exposure to the temperature (230 °C for 3 h) there is no degradation of the board. Moreover, in the average sense, some strengthening effect can also be observed. Microscale laboratory techniques adapted in this work, however, reveal damaging and irreversible microstructural changes of boards caused by the high temperature treatment.

It should be also noticed that the presented results are preliminary and starting a research cycle. Based on them, changes in mechanical parameters, especially of fibres in the fibre cement board, after exposed to a temperature of 230 °C for 3 h have been demonstrated. Currently, studies are being carried out to show the impact of high-temperature on the fibre cement board, but in a much shorter time. The authors hope to publish promising results soon.

Author Contributions: Conceptualization, K.S. and A.R.; Investigation, T.G. and M.P.; Methodology, T.G. and A.R.; Software, T.G. and M.P.; Supervision, K.S.; Visualization, T.G. and M.P.; Writing—original draft, K.S. and A.R.; Writing—review & editing, T.G. and M.P.

References

1. Akhavan, A.; Catchmark, J.; Rajabipour, F. Ductility enhancement of autoclaved cellulose fiber reinforced cement boards manufactured using a laboratory method simulating the Hatschek process. *Constr. Build. Mater.* **2017**, *135*, 251–259. [CrossRef]
2. Ardanuy, M.; Claramunt, J.; Toledo Filho, R.D. Cellulosic fiber reinforced cement-based composites: A review of recent research. *Constr. Build. Mater.* **2015**, *79*, 115–128. [CrossRef]
3. Chady, T.; Schabowicz, K.; Szymków, M. Automated multisource electromagnetic inspection of fibre-cement boards. *Autom. Constr.* **2018**, *94*, 383–394. [CrossRef]
4. Schabowicz, K.; Gorzelańczyk, T. A nondestructive methodology for the testing of fibre cement boards by means of a non-contact ultrasound scanner. *Constr. Build. Mater.* **2016**, *102*, 200–207. [CrossRef]
5. Tonoli, G.H.D.; Santos, S.F.; Savastano, H.; Delvasto, S.; Mejía de Gutiérrez, R.; Del M. Lopez de Murphy, M. Effects of natural weathering on microstructure and mineral composition of cementitious roofing tiles reinforced with fique fibre. *Cem. Concr. Compos.* **2011**, *33*, 225–232. [CrossRef]
6. Savastano, H.; Warden, P.G.; Coutts, R.S.P. Microstructure and mechanical properties of waste fibre-cement composites. *Cem. Concr. Compos.* **2005**, *27*, 583–592. [CrossRef]
7. Schabowicz, K.; Ranachowski, Z.; Jóźwiak-Niedźwiedzka, D.; Radzik, Ł.; Kudela, S.; Dvorak, T. Application of X-ray microtomography to quality assessment of fibre cement boards. *Constr. Build. Mater.* **2016**, *110*, 182–188. [CrossRef]
8. Nowak, T.; Karolak, A.; Sobótka, M.; Wyjadłowski, M. Assessment of the Condition of Wharf Timber Sheet Wall Material by Means of Selected Non-Destructive Methods. *Materials* **2019**, *12*, 1532. [CrossRef] [PubMed]
9. Cnudde, V.; Cnudde, J.P.; Dupuis, C.; Jacobs, P.J.S. X-ray micro-CT used for the localization of water repellents and consolidants inside natural building stones. *Mater. Charact.* **2004**, *53*, 259–271. [CrossRef]
10. Guo, L.P.; Carpinteri, A.; Sun, W.; Qin, W.C. Measurement and analysis of defects in high-performance concrete with three-dimensional micro-computer tomography. *J. Southeast Univ.* **2009**, *25*, 83–88.

11. Wang, Q.; Huang, X.; Zhou, W.; Li, J. Three-dimensional reconstruction and morphologic characteristics of porous metal fiber sintered sheet. *Mater. Charact.* **2013**, *86*, 49–58. [CrossRef]

12. Liu, J.; Li, C.; Liu, J.; Cui, G.; Yang, Z. Study on 3D spatial distribution of steel fibers in fiber reinforced cementitious composites through micro-CT technique. *Constr. Build. Mater.* **2013**, *48*, 656–661. [CrossRef]

13. Ranachowski, Z.; Schabowicz, K. The contribution of fiber reinforcement system to the overall toughness of cellulose fiber concrete panels. *Constr. Build. Mater.* **2017**, *156*, 1028–1034. [CrossRef]

14. Schabowicz, K.; Jóźwiak-Niedźwiedzka, D.; Ranachowski, Z.; Kudela, S.; Dvorak, T. Microstructural characterization of cellulose fibres in reinforced cement boards. *Arch. Civ. Mech. Eng.* **2018**, *18*, 1068–1078. [CrossRef]

15. Ranachowski, Z.; Ranachowski, P.; Dębowski, T.; Gorzelańczyk, T.; Schabowicz, K. Investigation of Structural Degradation of Fiber Cement Boards Due to Thermal Impact. *Materials* **2019**, *12*, 944. [CrossRef] [PubMed]

16. Schabowicz, K.; Gorzelańczyk, T.; Szymków, M. Identification of the degree of fibre-cement boards degradation under the influence of high temperature. *Autom. Constr.* **2019**, *101*, 190–198. [CrossRef]

17. Schabowicz, K.; Gorzelańczyk, T.; Szymków, M. Identification of the Degree of Degradation of Fibre-Cement Boards Exposed to Fire by Means of the Acoustic Emission Method and Artificial Neural Networks. *Materials* **2019**, *12*, 656. [CrossRef]

18. Constantinides, G.; Ravi Chandran, K.S.; Ulm, F.J.; Van Vliet, K.J. Grid indentation analysis of composite microstructure and mechanics: Principles and validation. *Mater. Sci. Eng. A* **2006**, *430*, 189–202. [CrossRef]

19. Oliver, W.C.; Pharr, G.M. An improved technique for determining hardness and elastic modulus using load and displacement sensing indentation experiments. *J. Mater. Res.* **1992**, *7*, 1564–1583. [CrossRef]

20. Cheng, Y.T.; Cheng, C.M. Scaling, dimensional analysis, and indentation measurements. *Mater. Sci. Eng. R Rep.* **2004**, *44*, 91–149. [CrossRef]

21. Oliver, W.C.; Pharr, G.M. Measurement of hardness and elastic modulus by instrumented indentation: Advances in understanding and refinements to methodology. *J. Mater. Res.* **2004**, *19*, 3–20. [CrossRef]

22. Randall, N.X.; Vandamme, M.; Ulm, F.J. Nanoindentation analysis as a two-dimensional tool for mapping the mechanical properties of complex surfaces. *J. Mater. Res.* **2009**, *24*, 679–690. [CrossRef]

23. Giannakopoulos, A.E.; Suresh, S. Determination of elastoplastic properties by instrumented sharp indentation. *Scr. Mater.* **1999**, *40*, 1191–1198. [CrossRef]

24. Constantinides, G.; Ulm, F.J.; Van Vliet, K. On the use of nanoindentation for cementitious materials. *Mat. Struct.* **2003**, *36*, 191–196. [CrossRef]

25. Miller, M.; Bobko, C.; Vandamme, M.; Ulm, F.J. Surface roughness criteria for cement paste nanoindentation. *Cem. Concr. Res.* **2008**, *38*, 467–476. [CrossRef]

26. Ulm, F.J.; Vandamme, M.; Bobko, C.; Ortega, J.A.; Tai, K.; Ortiz, C. Statistical Indentation Techniques for Hydrated Nanocomposites: Concrete, Bone, and Shale. *J. Am. Ceram. Soc.* **2007**, *90*, 2677–2692. [CrossRef]

27. *Cellulose Fibre Cement Flat Sheets-Product Specification and Test Methods*; EN 12467; British Standards Institution: London, UK, 2018.

28. Rajczakowska, M.; Stefaniuk, D.; Łydżba, D. Microstructure Characterization by Means of X-ray Micro-CT and Nanoindentation Measurements. *Studi. Geotech. Mech.* **2015**, *37*, 75–84. [CrossRef]

29. Rajczakowska, M.; Łydżba, D. Durability of crystalline phase in concrete microstructure modified by the mineral powders: Evaluation by nanoindentation tests. *Studi. Geotech. Mech.* **2016**, *38*, 65–74. [CrossRef]

30. Ganneau, F.P.; Constantinides, G.; Ulm, F.J. Dual-indentation technique for the assessment of strength properties of cohesive-frictional materials. *Int. J. Solids Struct.* **2006**, *43*, 1727–1745. [CrossRef]

31. Cariou, S.; Ulm, F.J.; Dormieux, L. Hardness–packing density scaling relations for cohesive-frictional porous materials. *J. Mech. Phys. Solids* **2008**, *56*, 924–952. [CrossRef]

5

Pore Structure Damages in Cement-Based Materials by Mercury Intrusion: A Non-Destructive Assessment by X-Ray Computed Tomography

Xiaohu Wang, Yu Peng, Jiyang Wang and Qiang Zeng *⬤

College of Civil Engineering and Architecture, Zhejiang University, Hangzhou 310058, China
* Correspondence: cengq14@zju.edu.cn

Abstract: Mercury intrusion porosimetry (MIP) is questioned for possibly damaging the micro structure of cement-based materials (CBMs), but this theme still has a lack of quantitative evidence. By using X-ray computed tomography (XCT), this study reported an experimental investigation on probing the pore structure damages in paste and mortar samples after a standard MIP test. XCT scans were performed on the samples before and after mercury intrusion. Because of its very high mass attenuation coefficient, mercury can greatly enhance the contrast of XCT images, paving a path to probe the same pores with and without mercury fillings. The paste and mortar showed the different MIP pore size distributions but similar intrusion processes. A grey value inverse for the pores and material skeletons before and after MIP was found. With the features of excellent data reliability and robustness verified by a threshold analysis, the XCT results characterized the surface structure of voids, and diagnosed the pore structure damages in terms of pore volume and size of the paste and mortar samples. The findings of this study deepen the understandings in pore structure damages in CBMs by mercury intrusion, and provide methodological insights in the microstructure characterization of CBMs by XCT.

Keywords: non-destructive method; damage; mercury intrusion porosimetry; X-ray computed tomography

1. Introduction

Pore structure characteristics of cement-based materials (CBMs) importantly indicate their mechanical property and durability performance. Determining the pore structure of CBMs, however, still faces big challenges because (1) pore structure testing methods, more or less, have intrinsic shortages, and (2) the microstructure of cement hydrates is rather sensitive to environments [1]. Mercury intrusion porosimetry (MIP) is probably one of the most widely used techniques for characterizing the pore structure of CBMs due to its advantages of simple physic principle, broad pore rang (depending on the maximum pressure applied), and low costs in time and manpower (fast and easy operation and sample preparation). With those features, the pore structure characteristics by MIP may become a benchmark when assessing the pore structure of CBMs with different methods [2].

Because liquid mercury is hydrophobic to most solids, it cannot invade into pores spontaneously without sufficient external pressures. The surface forces of the mercury fronts in pores, inversely depending on the pore curvatures, will resist the forces applied. By recording the stepwise increased mercury volume with pressure, the volume–pressure data can be obtained. Those original data generally are not directly used without the specific relation between pressure and pore size. With the assumptions of cylindrical, size-graded and connected pores, the pore size-pressure relation gives: $D = -4\gamma\cos\theta/P$, with D: the pore diameter; P: the applied pressure; γ: the surface tension of mercury;

and θ: the contact angle between mercury and pore wall. This relation is known as the Washburn equation [3]. Through this simple equation, MIP provides various size-related pore parameters, such as accumulative and differential pore size distributions (PSDs), mean size, threshold size and fractal dimension [4–7].

Whilst MIP has been popularly used, its accuracy in pore structure characterization is always debatable [8–14]. Generally, the debates of MIP pore data focus on: (1) the microstructure damages by sample pretreatment (drying), (2) the oversimplifications of pore topology (geometry and connectivity), (3) the constants of MIP parameters, (4) the conformance effect of samples, and (5) the pore damages by high pressures during mercury intrusion. The first term on the microstructure damages by pretreatment is inevitable but may be mitigated by using the relatively mild drying methods (e.g., solvent exchange [1,4,15]). The second term on the oversimplifications of pore topology that are intrinsically related to the physical bases of MIP would yield the so-called "ink-bottle" effect [8]. To compensate for this shortage, a multi-cycled intrusion-extrusion test scheme was often applied [14,16,17]. The third term on the constant MIP parameters is argued with the significant influences of the contact angle between mercury fronts and pore walls [11,18]. The parameters of MIP may be greatly different when the tested pores are narrowed to nano sizes [18]. However, due to the lack of data at nano sizes, the improvements in size-associated MIP parameters for pore structure characterization are limited. The fourth term on the conformance effect of samples is rarely mentioned because the samples of CBMs generally have no big differences. Our recent tests, however, showed that the surface conformance control may greatly narrow the threshold pore size [13]. The last term on the pore structure damages of CBMs during mercury intrusion, albeit noticed by previous researcher [19–21], can hardly be quantitatively characterized.

Feldman [20] used a repeat-intrusion testing scheme to detect the possible pore structure alterations of blend cement pastes. Note that the repeat-intrusion testing scheme used by Feldman [20] is different from the multi-cycled intrusion-extrusion test that is often operated in a stepwise loading-unloading way without expelling the mercury entrapped in the pores. In his tests, the second mercury intrusion was operated after the entrapped mercury during the first intrusion was completely removed, so the differences in PSDs between the first and second MIP tests can reflect the sizes of the pores damaged. It was observed by Feldman that damages to the pore structure occurred at 70 MPa in the hydrated blends. Olson et al. [21] employed an environmental scanning electron microscopy (ESEM) to in situ observe the damages to pore structure of a hardened Portland cement paste by mercury intrusion. Analyses indicated that the connectivity of the pores between 1–10 μm was raised after the intrusion pressure reached the threshold value. However, the results of Feldman [20] may fail to capture the real damaged pore sizes due to the oversimplifications of pore topology by MIP as mentioned above. Despite of the direct and obvious evidence of pore structure damages by MIP documented by Olson et al. [21], the ESEM tests on the open samples without specific treatments would be unsafe to the operators because mercury is highly evaporable at room temperature and poison to humans. Therefore, seeking a non-destructive method to assess the microstructure damages of CBMs before and after MIP is urgently wanted. X-ray computed tomography (XCT) may be a preferable candidate because XCT not only is a non-destructive method, but also provides the component and spatial information of the object tested.

By delivering X-ray beams at different angles, numerous 2D radiographic projections of a scanned object can be gathered and treated in a digital geometry processing to construct the 3D digital structures of the object [22]. The continual methodological developments of XCT test make it extensively used for the phase characterization in CBMs for predicting their mechanical and transport performances [23–26]. Because mercury has much stronger X-ray absorptivity than any constituent in CBMs and other nonmetallic materials, the combination of XCT and MIP may provide an effective way to enhance the ability of XCT to detect the pores beyond the normal voxel resolution [27]. This further provides a routine to detect the pore damages of CBMs after MIP because mercury drops can be entrapped in the damaged and/or undamaged pores [13]. The method also generates significances for pore structure characterization because mercury entrapment may also, to some extent, reflect the connectivity of pores [28].

In the present study, XCT tests were operated on ordinary Portland cement (OPC) paste and mortar before and after mercury intrusion to evaluate their pore structure changes with deepened analyses and discussions. The findings of this study provide a new and effective routine to non-destructively characterize the pore structure damages of CBMs by MIP.

2. Materials and Experiments

2.1. Materials and Sample Preparation

A PI 42.5 OPC cement (corresponding to ASTM Type I) was used as the only binding phase to prepare the porous paste and mortar samples. The chemical component and physical properties of the cement are shown in Table 1. When preparing the samples, no agents were used to control the fluidity of the fresh paste and mortar slurries. However, to obtain the similar fluidity, the water-to-cement (w/c) ratios of 0.4 and 0.5 were adopted for the paste and mortar, respectively. Commercial standard quartz sands with the fineness modulus of 2.6 and the SiO_2 content above 95% (Xiamen ISO Standard Sand Co., Ltd., Xiamen, China) were used as the fine aggregates to prepare the mortar. The cement/sand ratio was controlled as 1/3. Following standard casting, moulding and demoulding procedures, macro paste and mortar specimens were prepared, and then cured in a chamber with temperature at $20 \pm 2\,^\circ C$ and relative humidity above 95%. After 28 days, the well cured specimens were crushed into small pieces (around 1 mL in volume or 2 g in mass) for further experiments.

Table 1. Chemical component and physical properties of cement.

Oxides	Content (%)	Minerals	Content (%)	Physical Properties	Value
SiO_2	21.68	C_3S	57.34	Density (g/mL)	3.10
Al_2O_3	4.80	C_2S	18.09	Specific area (m^2/kg)	345
Fe_2O_3	3.70	C_3A	6.47	Mean size (μm)	11
CaO	64.90	C_4AF	11.25		
MgO	2.76	Others	6.04		
SO3	0.29				
Na_2O(eq)	0.56				
CaO(f)	0.93				

The crushed OPC paste and mortar pieces were then immersed into pure ethanol to cease the hydration of cement. After two days of immersion in ethanol, the samples were then removed into an oven at 105 °C to expel the water physically absorbed in the pores. After 24 h, the samples were then stored in a sealed desiccator to eliminate the possible influences of water and carbon oxide in the air on the microstructure, and were readily prepared for XCT and MIP tests. Note that the drying temperature used here may be too severe to preserve the microstructure of the samples because the rapid water loss in the pores may yield high capillary stresses to damage the material matrix [1] and to alter the status of calcium-silicate-hydrate (C-S-H) gels [29]. However, because the microstructure alterations by drying are stable and will not recover during the following MIP and XCT tests, those alterations can be treated as the intrinsic microstructure features, and thus would be not considered here.

2.2. MIP Tests

The pre-dried samples were then placed into a sample chamber for MIP test in a device of Autopore IV 9510 (Micromeritics, Norcross, GA, USA). After a pre-equilibrium step to fill the gaps between the sample and chamber wall at 0.5 psi (3.45 KPa), pressures on mercury were automatically and stepwise raised to 60,000 psi (413.69 MPa) and then unloaded to certain values. With the equilibrium time of 10 s, an complete MIP test lasted about 130 min.

Immediately after the MIP tests finished, the samples were carefully and rapidly removed into small plastic tubes. A quick-hardening epoxy resin was rapidly poured into the tubes to completely cover the samples, and those tubes were quickly and tightly lidded (Figure 1). Those steps were to cease the leakages of mercury from the mercury-filled samples for diminishing the possible dangers to the technicians when handling those CBM samples.

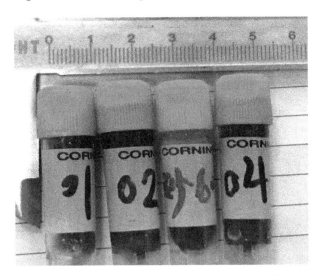

Figure 1. Paste and mortar samples encased in epoxy resin after mercury intrusion.

2.3. XCT Tests

Before and after the MIP tests, XCT scans were performed on the samples by an X-CT scanner of Nikon XTH 255/320 LC (Nikon, Tokyo, Japan). For the XCT tests before and after MIP processes, the voltages to deliver the X-ray beams were set as 100 KeV and 150 KeV, respectively (Table 2; see below for detailed explanations). The penetrated X-ray beams were detected by a high-sensitive detector (DRZplus Scintillator with the pixels of 2000(h) × 2000(v)) at the back of the objects synchronously. During testing, the samples rotated in the rate of 12 °/min. The collected data were then loaded into the software of VGStudio Max (version 3.1, Volume Graphics, Inc., Charlotte, NC, USA) for further analyses. Because the pre-MIP sample size was smaller than the post-MIP one (CBM sample plus tube), the pixel resolution of the former case was slightly higher (5.02 μm for the pre-MIP sample versus 5.60 μm for the post-MIP sample).

Table 2. XCT parameters used before and after MIP tests.

Condition	Voltage (KeV)	Pixel Resolution (μm)
Before MIP (Pre-MIP)	150	5.02
After MIP (Post-MIP)	100	5.60

The reason for using different delivering voltages for the pre-and post-MIP tests was because mercury has far higher X-ray mass attenuation coefficient (MAC) than the main solid phases in the CBM samples. Figure 2 shows the MAC curves of mercury, SiO_2, C-S-H and cement in the generally used photon energy interval (90–160 KeV). At 100 KeV (the photon energy used for the pre-MIP XCT tests), the MACs of SiO_2, C-S-H and cement are rather close (0.17–0.21 m^2/g). The slight MAC differences among those phases will even become less at a higher photon energy (see the inserted panel in Figure 2), so to obtain the high quality images of the pre-MIP samples, the photon energy of 100 KeV was selected. When mercury was intruded into the CBM samples, the MACs are greatly altered. As displayed in Figure 2, the MAC of mercury is 25–30 times higher than that of SiO_2, C-S-H and cement at 100 KeV. Such strong X-ray absorption by mercury would induce significant beam hardening artifacts [30]. Since the MAC gaps between mercury and the other phases will be narrowed with increasing photon energy (Figure 2), it is thus expected that a higher photon energy may bring less beam hardening artifacts. Our practices indeed indicated the images at the photon energy of 150 KeV would achieve high quality images for further analysis.

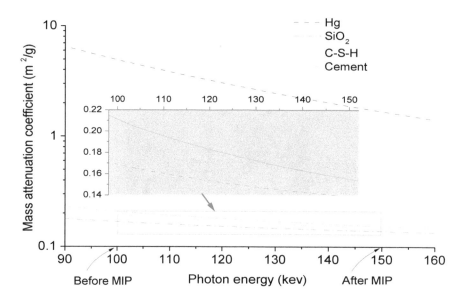

Figure 2. Spectra of mass attenuation coefficient of Hg, SiO_2, C-S-H and cement between 90 KeV and 160 KeV (Data from Ref. [31]).

Because the samples used have irregular shapes and rough surfaces, it is unrealistic to analyse the entire volume of the samples. Instead, some volumes of interest (VOI) inside the samples (or region of interest in 2D analysis) were selected for image analysis, which can prevent edge effects and increase data efficiency. Due to the heterogeneity in microstructure and the chaos in pore structure of CBMs [6], a random selection of VOI that would be preferred for obtaining representative and reproductive results was not adopted here. In order to easily and precisely identify the same VOI of the samples before and after mercury intrusion, the microstructure of the VOI must have distinguished characteristics. In this study, we intentionally selected the VOIs containing big voids. Figure 3 shows an example of the best fit registration of two cubic VOIs in the pre-MIP paste sample. Clearly, with these big voids (dark circles in the VOIs shown in Figure 3), the same VOIs can be easily identified for the same sample after mercury intrusion.

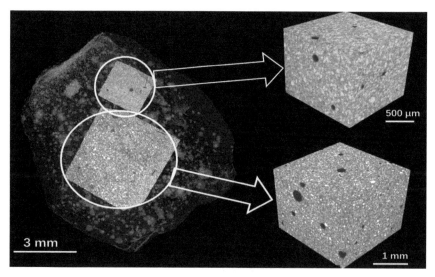

Figure 3. An example of VOI selections from a pre-MIP paste sample.

3. Results and Discussion

3.1. MIP Outcomes

For pore structure characterization, the classic Washburn equation was used to interpret the MIP data with the mercury surface tension of 485 mN/m and the contact angle between mercury and substrate of 130°. With those data, some characteristic pore parameters of the paste and mortar samples, i.e., total porosity, volume-median pore size, specific surface area and threshold pore size, can be evaluated (Table 3). Obviously, compared with mortar, paste showed the higher total porosity and specific surface area, the similar threshold pore size, but the lower volume-median pore size. The results are reasonably due to the fact that the impermeable sands in the mortar occupied more than 60% of the total volume. The looser compactness and more porous cement hydrates of the mortar induced by the higher w/c ratio, as well as the porous interfacial transition zones (ITZs) between cement matrix and aggregates, caused the higher volume-median pore size (independent of the absolute pore volumes), but remained unable to compensate for the reductions in porosity and specific surface area (Table 3). The similar threshold pore sizes between the paste and mortar suggested that the connected throats formed from the interparticle continuum had the similar widths.

Table 3. Characteristic pore parameters of paste and mortar form MIP.

Sample	Total Porosity (%)	Volume-Median Pore Size (nm)	Specific Surface Area (m²/g)	Threshold Pore Size (nm)
Paste	20.0	66.1	12.3	76.5
Mortar	15.1	85.1	9.3	76.9

Figure 4 shows the (top) accumulative and (bottom) differential PSDs of the paste and mortar. While the PSDs of the paste and mortar had the different shapes, they both displayed the similar five-stage characteristics: surface conformance, non-channel stage, capillaries by flaws and ITZs, capillaries by interparticle space, and gel pores.

1. Mercury first covered the open cracks, gaps, cavities and irregularities on the sample surfaces in relatively low pressures (termed as the surface conformance effect) [13,32]. Our previous study [13] suggested that the surface conformance effect might not be avoided because these cavities, cracks and flaws can be inevitably induced during sample pretreatments such as cutting and drying [1,15]. However, the volume increases at the very beginning stage of MIP by the surface conformance effect can be mitigated by controlling the exposed areas of the MIP samples [13]. In this study, the surface conformance effect was insignificant for both the paste and mortar samples (<0.005 mL/g).

2. Later, almost no mercury intrusion was recorded between 2 μm and 100 μm (Figure 4). This meant that no open channels (not the pores inside the materials) in such size interval can be recognized by MIP, which was termed as the non-channel stage.

3. As the size decreased further, the mercury increases of both the paste and mortar became obvious (Figure 4). Generally, for normally cured cement paste, these increases can be rarely observed [8]. In this study, the very severe drying scheme (105 °C) was used, so the microstructure flaws or damages by drying [1] would account for the abnormal mercury rises in this stage. For the mortar sample, the porous ITZs, together with the capillary flaws by drying, were responsible for the higher PSD data (see the shadowed areas shown in Figure 4).

4. After that, the intrusion volumes rose rapidly and significantly with obvious peaks around 70 nm (Figure 4). The peak size was identical to the threshold pore size form the percolated pore continuum [29,33,34]. Because of the 'ink-bottle' effect [8], the volumes at or below the threshold size could partially represent the capillaries of the interparticle space that remained unfilled by cement hydration. Compared with the mortar sample, the paste sample showed the faster raising rate and higher peak intensity because of the higher capillary pores.

5.   Under the higher pressures, the mercury rising rates became slower and the differential PSDs were depressed (Figure 4) because only limited space (mainly gel pores) was available to accommodate the mercury after the capillaries were filled. Since the MIP parameters in nano scales remained debatable [11,18], those data would not shed much light on gel pore characterization.

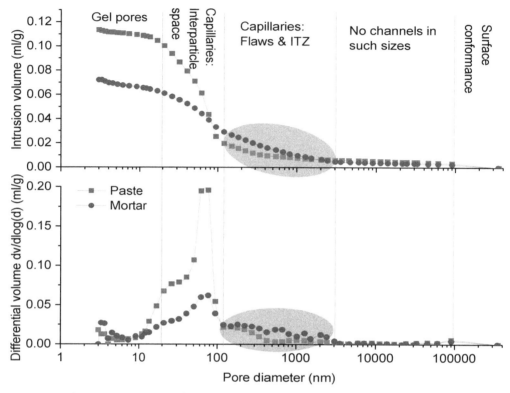

Figure 4. Accumulative (**top**) and differential (**bottom**) pore size distributions of paste and mortar samples (The specific contribution of ITZ in the mortar was singled out in the shadowed areas).

3.2. Threshold Analysis

Despite the fact that XCT is a powerful tool to non-destructively characterize the microstructure of various materials, the results, as cautioned elsewhere [22,27], are highly depending on the process of threshold segmentation. In this section, the effect of threshold segmentation on the reconstructed results of the pore phase was discussed, so the reliability and robustness of the damage diagnoses by XCT could be guaranteed.

Figure 5 shows the voxel-grey value distributions of a VOI in the paste sample before and after mercury intrusion. Generally, due to the lower X-ray absorption, the empty pores in a CBM sample were captured by the low grey value area, and the cement skeletons (including the hydration products and unhydrated cement clinkers) that can absorb more X-rays thus were represented by the high grey value area (Figure 5a). When mercury was intruded into the empty pores, the characteristic areas of grey value were exchanged. Specifically, the high grey value area represented the mercury-filled pores, while the low grey value area denoted the cement matrix (Figure 5b). This feature of grey-value inverse was recently used to determine the mercury drops entrapped in the pores of HCP samples with/without surface conformance control for pore structure characterization [13].

The voxel-grey value distributions displayed in Figure 5 clearly showed two peaks, so the threshold segmentation should be operated at the minimum between the two peaks. Here, to discuss the threshold sensitivity, three threshold points were selected, i.e., the middle, low (−5%) and high (+5%) threshold values shown in Figure 5a. In Figure 6, the 2D and 3D images of the pores segmented from the paste skeleton are comparatively plotted with the designed three threshold values. Apparently, no obvious differences can be seen from those images. To specifically compare the pore information in different scales, the volume-size plots of all the extracted pores in a VOI of the cement with three threshold values were illustrated in Figure 7. The results showed that both the number and size of those objects had no obvious differences (Figure 7). A similar threshold analysis was also performed on the post-MIP samples to obtain the appropriate threshold grey values with the reliability data.

Overall, the data of Figures 6 and 7 implied that the XCT analyses used in this study provided reliable and robust pore structure information for identifying the pore structure damages of CBMs by MIP. In the following contexts, all discussions were based on the XCT results with the middle threshold segmentation.

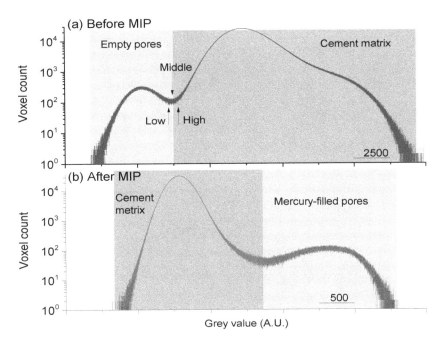

Figure 5. Voxel-grey value distributions of a VOI in the paste sample (**a**) before and (**b**) after mercury intrusion. Three threshold values (low, middle and high) were selected to test the influence of threshold process on pore segmentation.

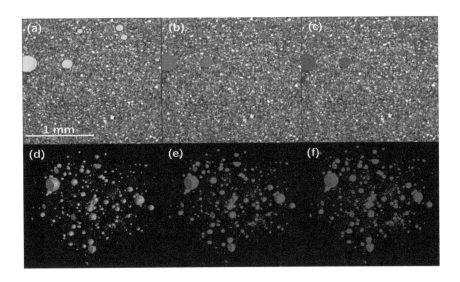

Figure 6. 2D (**a**–**c**) and 3D (**d**–**f**) images of pores segmented from paste skeleton at (**a,c**) low, (**b,d**) middle, and (**c,f**) high threshold values from Figure 5.

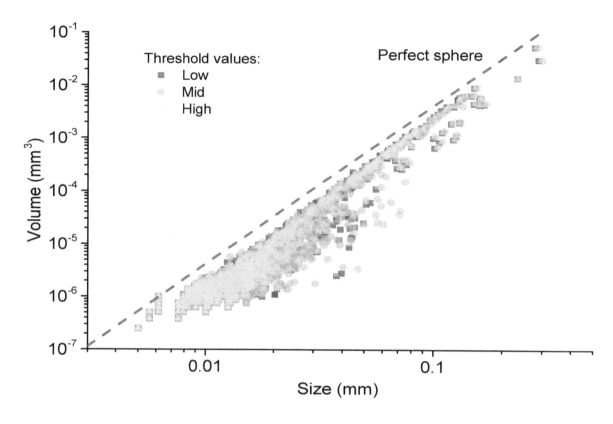

Figure 7. Volume-size distributions of the pores segmented from paste skeleton at low, middle, and high threshold values from Figure 5.

3.3. Characteristics of XCT Results

Figures 8 and 9 show the 2D and 3D representative images of a VOI and a localized big pore of the cement and mortar sample, respectively. In a much clearer way, the grey-value inverse of the pores before and after mercury intrusion can be displayed. For instance, the same pores were illustrated in the darkest color before MIP and the brightest color after MIP (Figure 9b,e).

Some features in microstructure characterization of CBMs by the combination of MIP and XCT can be pointed out from Figures 8 and 9. Firstly, all the visible air voids in the paste and mortar samples were fully filled with mercury. For ordinary CBMs, the air voids with the size range of 10–500 μm and the content less then 2.5% were reported [8]. Those air voids, however, can not be detected by MIP because it only measures the open channels rather than the pore chambers [13]. This finding again made evident that MIP fails to detect the 'ink-bottle' like pores [8]. Secondly, the air voids in the paste and mortar samples showed different surface structures. Specifically, the surfaces of the voids in the paste were rather rough with and without mercury intrusion (Figure 8c,f), while those in the mortar were much smoother (Figure 9c,f). Although the akin rough pore surfaces of CBMs were documented [13,35], the mechanisms for the surface structure differences between paste and mortar remained unclear and deserved further rigorous studies. Thirdly, the mercury intrusion process enhanced the contrast of the mortar images to figure out the aggregates (quartz). As shown in Figure 9d, the aggregates were illustrated as the darkest phase (due to the lowest X-ray absorption, Figure 2) embedded in the much brighter cement matrix and the brightest mercury-filled voids. Because of the closely valued MACs of quartz, C-S-H and cement clinkers (Figure 2), separating quartz from the other two phases would be difficult. The post-MIP XCT test used in this study may provide an effective way to obtain the packing pattern of the aggregates in mortar.

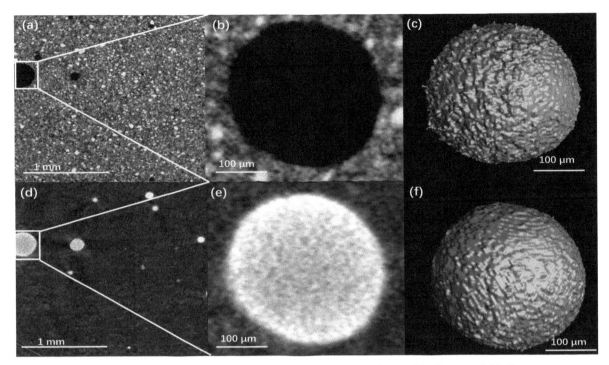

Figure 8. Representative images of a VOI in the cement sample (**a–c**) before and (**d–f**) after mercury intrusion: (**a,d**) the 2D sectional view with (**b,e**) the magnified pore and (**c,f**) its 3D structure.

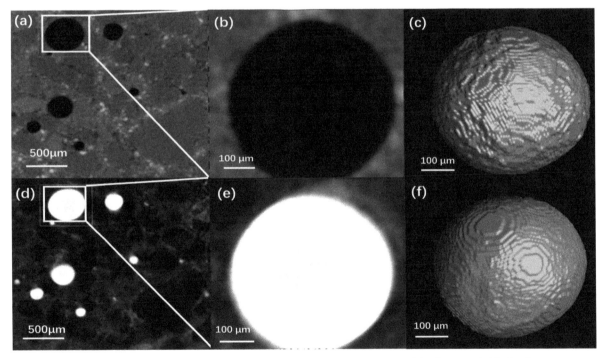

Figure 9. Representative images of a VOI in the mortar sample (**a–c**) before and (**d–f**) after mercury intrusion: (**a,d**) the 2D sectional view with (**b,e**) the magnified pore and (**c,f**) its 3D structure.

Last but not least, the mercury intrusion paths were identified from the XCT images of the post-MIP mortar. For instance, Figure 10 displays a 2D XCT image of a local area of the mortar sample after mercury intrusion, where the bright areas along the ITZs around the aggregates indicated the thoroughly penetrated path to the air void. However, these mercury penetration paths cannot be identified from the XCT images of the post-MIP paste sample (Figure 8d) because the penetration sizes in the paste (0.2 μm [8]) would be beyond the resolution of our XCT tests.

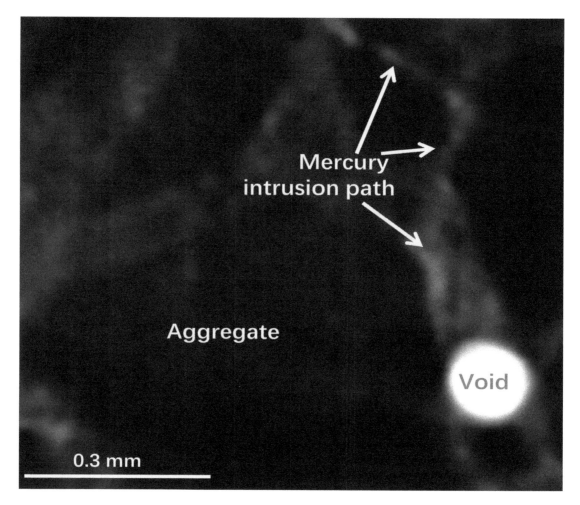

Figure 10. A 2D sectional image of a local area in the mortar showing a mercury intrusion path along the ITZs between aggregates and cement matrix to an air void.

3.4. Damage Diagnosis

We then used the XCT data of the pre-and post-MIP samples to diagnose whether or not the microstructure of those samples was damaged after MIP and in what sizes the damages occurred.

Figures 11 and 12 comparatively plot the PSDs and the statistics in pore volume of VOIs, respectively, in the paste and mortar samples before and after mercury intrusion. Note that the minimum diameters of the reconstructed pores (around 10 µm) shown in Figures 11 and 12 were higher than the minimum detectable pixel sizes (Table 2) because the voxel resolution for a 3D reconstructed object (depending on the geometry of the object) would be always lower than the 2D pixel resolution [22]. After the MIP tests finished, clearly, the total pore volume was increased from 0.29 mm^3 to 0.31 mm^3 by 6.7% for the paste sample (Figure 11a), and from 0.13 mm^3 to 0.14 mm^3 by 7.6% for the mortar sample (Figure 12a). In a statistic manner, the mean pore volume was largely augmented from $5 \times 10^{-6} \text{ mm}^3$ to $8 \times 10^{-5} \text{ mm}^3$ by around 16 times for the paste sample (Figure 11b), and slightly from $5 \times 10^{-5} \text{ mm}^3$ to $6 \times 10^{-5} \text{ mm}^3$ by 20% for the mortar sample (Figure 12b). However, the heavy increases in mean pore volume shown in Figure 11b may be misleading because the numbers of the pores recognized (especially the thin-sized pores) were largely decreased. The increases in pore volume and decreases in pore number for CBMs were in line with the results reported by Olson et al. [21] through ESEM observations. From Figures 11 and 12, one could further read the sizes of the pores damaged directly. Two obvious pore volume increases shown in Figure 11a indicated that the damages mainly occurred to the pores of 100–200 µm and 300–500 µm for the paste sample. For the mortar sample, the damages concentrated at the size interval of 100–400 µm (Figure 12b).

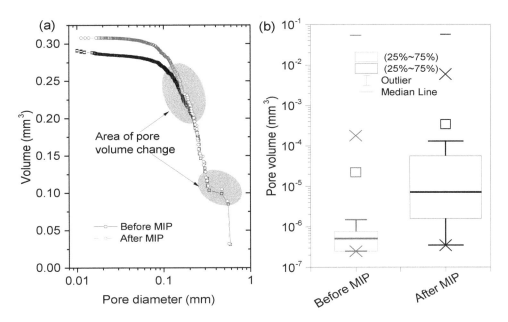

Figure 11. (**a**) comparative plots of pore size distribution of a VOI in the paste sample before and after mercury intrusion, and (**b**) the statistic results of the pore phase.

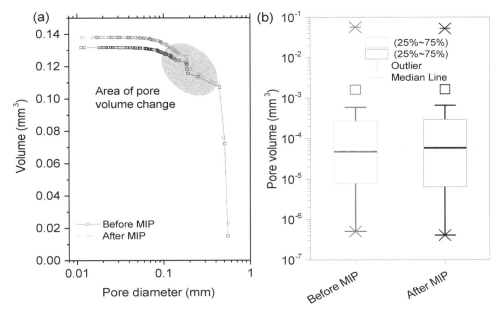

Figure 12. (**a**) comparative plots of pore size distribution of a VOI in the mortar sample before and after mercury intrusion, and (**b**) the statistic results of the pore phase.

3.5. *Further Discussion*

When mercury is enforced to invade into a porous CBM sample, the capillaries among the compacted particles and the porous cement hydrates deform to sustain the applied pressures. If the pressures are (even locally) higher than the strength of the phases in contact with the stressed mercury, damages take place. Those mechanically reasonable damages to CBMs by mercury intrusion can be schematically illustrated in Figure 13. Before MIP, the voids in a CBM sample (generally in 10–500 μm [8]) may be isolated by the material matrix consisting of the closely compacted cement particles and their hydration products (Figure 13a). The capillaries (channels) connecting those voids are generally too thin to be diagnosed by normal XCT. However, after mercury intrusion under sufficient pressures, the voids as well as these throats will be filled with mercury. Since mercury can strongly absorb the X-ray penetrated, the signals of the mercury-entrapped channels (albeit below the resolution of the XCT) may become detectable (Figure 13b). This is the regime applied in this study to diagnose the pore damages of CBM samples after mercury intrusion.

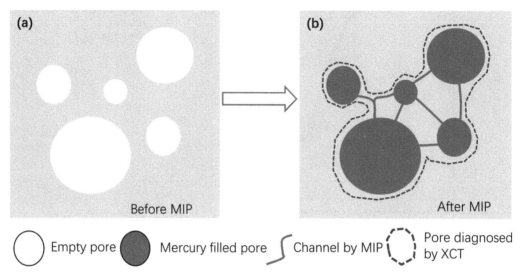

Figure 13. Mechanisms of pore damages induced by mercury intrusion under high pressures with the possibility of an overestimation on the pore size of a CBM sample after MIP (right) than that before MIP (left) by XCT.

In our tests, the damaged pore sizes measured by XCT were much higher than the data obtained by Olson et al. [21], who found that the connectivity of the pores in the 1–10 μm size range was greatly increased, and the average size was enlarged from 1.60 μm to 2.36 μm after MIP. A much lower size of the damaged pores was reported by Feldman [20] (around 18 nm corresponding to the applied pressure of 70 MPa). The size differences between our data and those reported in the literature [20,21] were mainly due to the different methods used and different objects concerned. In Feldman's tests [20], the pore damages were assessed by the PSD differences of cement blends before and after mercury intrusion, so the obtained sizes were always underestimated due to the intrinsic biases in the pore sizes of MIP, e.g., the 'ink-bottle' effect [8]. In the tests by Olson et al. [21], 2D images from limited local areas were obtained from ESEM. In our tests because of the limited resolution of the XCT used, the pores below 10 μm, mainly the throats to connect the voids [8], can not be detected. Instead, the damaged voids under high pressures were diagnosed.

While the present study reported obvious damages to the pore structure of the paste and mortar samples, several themes remained to be discussed further. Firstly, one must understand that the obvious enhancement in image contrast by mercury may induce some biases in pore structure characterization due to the beam hardening artifacts [27,36]. For example, if the voids were neighbours, the signals of X-ray beams may overlap so the individual pores as well as the connecting channels may be diagnosed as a big pore. This would significantly decrease the detected pore numbers but increase the pore volumes (Figure 13b). This regime may also partially account for the results shown in Figures 11 and 12. Secondly, the pore structure alterations by the severe drying process (105 °C) may induce additional variances when assessing the pore damages by mercury intrusion. It has been recognized that severe drying can greatly impact the packing patterns of C-S-H gels and the connectivity of pores [15,37]. Those may increase the difficulties in diagnosing the pore damages to CBMs after mercury intrusion. Future tests on the CBM samples with the milder drying schemes are preferred to mitigate this effect.

4. Conclusions

- XCT is a powerful technique to non-destructively characterize the microstructure of CBMs. The significant differences in X-ray MACs between mercury and the phases in CBMs can greatly enhance the contrast gradients in XCT images and facilitate the reconstruction of 3D microstructure.

- MIP tests indicated that, compared with the mortar sample, the paste sample had the higher porosity and specific surface area, similar threshold pore size, but lower median pore size. The MIP PSDs of the paste and mortar samples showed the similarly five-stage intrusion curves but the different specific spectra. The drying at 105 °C brought additional flaws just before the threshold stage to the paste and mortar samples.

- The grey values for the pores and material skeletons in the CBM samples were inversely distributed due to the shifts in X-ray absorptivity when the pores were filled with mercury.

- A threshold analysis indicated that the obtained XCT results showed good reliability and robustness in pore phase segmentation.

- The surfaces of the voids in the paste were rough, while those in the paste were smooth. Mercury intrusion paths along the ITZs around aggregates in the mortar sample were visible in the post-MIP XCT images.

- Mercury intrusion in the paste and mortar samples caused the increases in pore volume and the decreases in pore number as determined by XCT. The results were consistent with those reported in the literature.

Overall, the damages to the pore structure of CBMs after mercury intrusion can be non-destructively diagnosed by XCT with quantitative parameters. Going beyond this, the combination of MIP and XCT may provide a powerful tool to probe the pore structure alterations in CBMs under different environments.

Author Contributions: X.W. and Y.P. conducted the experiments and analysed the data, Q.Z., Y.P. and J.W. designed this work, and Q.Z. wrote this paper.

References

1. Zhang, Z.; Scherer, G.W. Evaluation of drying methods by nitrogen adsorption. *Cem. Concr. Res.* **2019**, *120*, 13–26. [CrossRef]
2. Zuo, Y.; Ye, G. Pore structure characterization of sodium hydroxide activated slag using mercury intrusion porosimetry, nitrogen adsorption, and image analysis. *Materials* **2018**, *11*, 1035. [CrossRef]
3. Washburn, E.W. Note on a method of determining the distribution of pore sizes in a porous material. *Proc. Natl. Acad. Sci. USA* **1921**, *7*, 115–116. [CrossRef] [PubMed]
4. Zeng, Q.; Li, K.; Fen-Chong, T.; Dangla, P. Pore structure characterization of cement pastes blended with high-volume fly-ash. *Cem. Concr. Res.* **2012**, *42*, 194–204. [CrossRef]
5. Zeng, Q.; Li, K.; Fen-Chong, T.; Dangla, P. Surface fractal analysis of pore structure of high-volume fly-ash cement pastes. *Appl. Surf. Sci.* **2010**, *257*, 762–768. [CrossRef]
6. Zeng, Q.; Luo, M.; Pang, X.; Li, L.; Li, K. Surface fractal dimension: An indicator to characterize the microstructure of cement-based porous materials. *Appl. Surf. Sci.* **2013**, *282*, 302–307. [CrossRef]
7. Leóny León, C.A. New perspectives in mercury porosimetry. *Adv. Colloid Interface Sci.* **1998**, *76*, 341–372. [CrossRef]
8. Diamond, S. Mercury porosimetry: An inappropriate method for the measurement of pore size distributions in cement-based materials. *Cem. Concr. Res.* **2000**, *30*, 1517–1525. [CrossRef]
9. Moro, F.; Boehni, H. Ink-Bottle effect in mercury intrusion porosimetry of cement-based materials. *J. Colloid Interface Sci.* **2002**, *246*, 135–149. [CrossRef]
10. Ma, H. Mercury intrusion porosimetry in concrete technology: Tips in measurement, pore structure parameter acquisition and application. *J. Porous Mater.* **2014**, *21*, 207–215. [CrossRef]
11. Muller, A.C.A.; Scrivener, K.L. A reassessment of mercury intrusion porosimetry by comparison with 1H nmR relaxometry. *Cem. Concr. Res.* **2017**, *100*, 350–360. [CrossRef]
12. Dong, H.; Zhang, H.; Zuo, Y.; Gao, P.; Ye, G. Relationship between the Size of the Samples and the Interpretation of the Mercury Intrusion Results of an Artificial Sandstone. *Materials* **2018**, *11*, 201. [CrossRef] [PubMed]

13. Zeng, Q.; Wang, X.; Yang, P.; Wang, J.; Zhou, C. Tracing mercury entrapment in porous cement paste after mercury intrusion test by X-ray computed tomography and implications for pore structure characterization. *Mater. Charact.* **2019**, *151*, 203-215. [CrossRef]

14. Zhang, Y.; Yang, B.; Yang, Z.; Ye, G. Ink-bottle effect and pore size distribution of cementitious materials identified by pressurization–depressurization cycling mercury intrusion porosimetry. *Materials* **2019**, *12*, 1454. [CrossRef] [PubMed]

15. Galle, C. Effect of drying on cement-based materials pore structure as identified by mercury intrusion porosimetry: A comparative study between oven-, vacuum-, and freeze-drying. *Cem. Concr. Res.* **2001**, *31*, 1467–1477. [CrossRef]

16. Zhou, J.; Ye, G.; Breugel, K.V. Characterization of pore structure in cement-based materials using pressurization depressurization cycling mercury intrusion porosimetry (PDC-MIP). *Cem. Concr. Res.* **2010**, *40*, 1120–1128. [CrossRef]

17. Gao, Z.; Hu, Q.; Hamamoto, S. Using multicycle mercury intrusion porosimetry to investigate hysteresis of different porous media. *J. Porous Med.* **2018**, *21*, 607–622. [CrossRef]

18. Wang, S.; Javadpour, F.; Feng, Q. Confinement correction to mercury intrusion capillary pressure of shale nanopores. *Sci. Rep.* **2016**, *6*, 20160. [CrossRef]

19. Shi, D.; Winslow, D.N. Contact angle and damage during mercury intrusion into cement paste. *Cem. Concr. Res.* **1985**, *15* , 645–654. [CrossRef]

20. Feldman, R.F. Pore structure damage in blended cements caused by mercury intrusion. *J. Am. Ceram. Soc.* **1984**, *67*, 30–33. [CrossRef]

21. Olson, R.A.; Neubauer, C.M.; Jennings, H.M. Damage to the pore structure of hardened Portland cement paste by mercury intrusion. *J. Am. Ceram. Soc.* **1997**, *80*, 2454–2458. [CrossRef]

22. Cnudde, V.; Boone, M.N. High-resolution X-ray computed tomography in geosciences: A review of the current technology and applications. *Earth-Sci. Rev.* **2013**, *123*, 1–17. [CrossRef]

23. Promentilla, M.; Cortez, S.; Papel, R.; Tablada, B.; Sugiyama, T. Evaluation of microstructure and transport properties of deteriorated cementitious materials from their X-ray computed tomography (CT) images. *Materials* **2016**, *9*, 388. [CrossRef] [PubMed]

24. Erdem, S.; Gürbüz, E.; Uysal, M. Micro-mechanical analysis and X-ray computed tomography quantification of damage in concrete with industrial by-products and construction waste. *J. Clean. Prod.* **2018**, *189*, 933–940. [CrossRef]

25. Yang, S.; Cui, H.; Poon, C.S. Assessment of in-situ alkali-silica reaction (ASR) development of glass aggregate concrete prepared with dry-mix and conventional wet-mix methods by X-ray computed micro-tomography. *Cem. Concr. Compos.* **2018**, *90*, 266–276. [CrossRef]

26. Buljak, V.; Oesch, T.; Bruno, G. Simulating fiber-reinforced concrete mechanical performance using CT-based fiber orientation data. *Materials* **2019**, *12*, 717. [CrossRef] [PubMed]

27. Fusi, N.; Martinez-Martinez, J. Mercury porosimetry as a tool for improving quality of micro-CT images in low porosity carbonate rocks. *Eng. Geol.* **2013**, *166*, 272–282. [CrossRef]

28. Zeng, Q.; Li, K.; Fen-Chong, T.; Dangla, P. Analysis of pore structure, contact angle and pore entrapment of blended cement pastes from mercury porosimetry data. *Cem. Concr. Compos.* **2012**, *34*, 1053–1061. [CrossRef]

29. Zhou, C.; Ren, F.; Zeng, Q,; Xiao, L.; Wang, W. Pore-size resolved water vapor adsorption kinetics of white cement mortars as viewed from proton nmR relaxation. *Cem. Concr. Res.* **2018**, *105*, 31–43. [CrossRef]

30. Katsura, M.; Sato, J.; Akahane, M.; Kunimatsu, A.; Abe, O. Current and novel techniques for metal artifact reduction at CT: Practical guide for radiologists. *Radiographics* **2018**, *38*, 450–461. [CrossRef]

31. National Institute of Standards and Technology. Avaliable on line: https://physics.nist.gov/PhysRefData/FFast/html/form.html (accessed on 11 May 2019).

32. Peng, S.; Zhang, T.; Loucks, R.G.; Shultz, J. Application of mercury injection capillary pressure to mudrocks: Conformance and compression corrections. *Mar. Pet. Geol.* **2018**, *88*, 30–40. [CrossRef]

33. Katz, A.J.; Thompson, A.H. Quantitative prediction of permeability in porous rock. *Phys. Rev. B* **1986**, *34*, 8179. [CrossRef] [PubMed]

34. Zhou, C.; Ren, F.; Wang, Z.; Chen, W.; Wang, W. Why permeability to water is anomalously lower than that to many other fluids for cement-based material? *Cem. Concr. Res.* **2017**, *100*, 373–384. [CrossRef]

35. Wang, Z.; Zeng, Q.; Wang, L; Li, X.; Xu, S.; Yao, Y. Characterizing frost damages of concrete with flatbed scanner. *Constr. Build. Mater.* **2016**, *102*, 872–883. [CrossRef]

36. Hiller, J.; Hornberger, P. Measurement accuracy in X-ray computed tomography metrology: Toward a systematic analysis of interference effects in tomographic imaging. *Precis. Eng.* **2016**, *45*, 18–32. [CrossRef]

37. Gajewicz, A.M.; Gartner, E.; Kang, K.; McDonald, P.J.; Yermakou, V. A 1H nmR relaxometry investigation of gel-pore drying shrinkage in cement pastes. *Cem. Concr. Res.* **2016**, *86*, 12–19. [CrossRef]

6

Wave Frequency Effects on Damage Imaging in Adhesive Joints using Lamb Waves and RMS

Erwin Wojtczak *[ID] and Magdalena Rucka[ID]

Department of Mechanics of Materials and Structures, Faculty of Civil and Environmental Engineering, Gdansk University of Technology, Narutowicza 11/12, 80-233 Gdansk, Poland; magdalena.rucka@pg.edu.pl or mrucka@pg.edu.pl
* Correspondence: erwin.wojtczak@pg.edu.pl

Abstract: Structural adhesive joints have numerous applications in many fields of industry. The gradual deterioration of adhesive material over time causes a possibility of unexpected failure and the need for non-destructive testing of existing joints. The Lamb wave propagation method is one of the most promising techniques for the damage identification of such connections. The aim of this study was experimental and numerical research on the effects of the wave frequency on damage identification in a single-lap adhesive joint of steel plates. The ultrasonic waves were excited at one point of an analyzed specimen and then measured in a certain area of the joint. The recorded wave velocity signals were processed by the way of a root mean square (RMS) calculation, giving the actual position and geometry of defects. In addition to the visual assessment of damage maps, a statistical analysis was conducted. The influence of an excitation frequency value on the obtained visualizations was considered experimentally and numerically in the wide range for a single defect. Supplementary finite element method (FEM) calculations were performed for three additional damage variants. The results revealed some limitations of the proposed method. The main conclusion was that the effectiveness of measurements strongly depends on the chosen wave frequency value.

Keywords: Lamb waves; scanning laser vibrometry; adhesive joints; non-destructive testing; damage detection; excitation frequency

1. Introduction

Adhesive bonding is one of the effective methods for joining elements in metallic structures, besides welding, riveting, and bolting [1]. It has dozens of applications in the aerospace, machine, automotive, military, and electronics industries [2]. Structural adhesive joints have numerous advantages in comparison to other joining techniques. Firstly, adhesives do not interfere with the structure of adherends (joined elements), which is what happens in bolted joints (openings weakening joined parts) or welded joints (internal stresses after welding). Moreover, gluing enables the creation of heterogenic connections, especially when welding or hole drilling is forbidden. There are also some disadvantages; from these, among the most significant is high vulnerability to accuracy in the processes of preparation and manufacturing. Particularly, the most important issues for the strength of the joint are the accurate surface treatment [3] and the protection against any contamination [4]. Any inaccuracy may lead to the formation of kissing defects or voids [5,6]. Their presence can cause a significant decrease in the strength of the joint and, as a result, its failure. The problematic issue is that kissing defects are not detectable in the visual assessment, because of their existence in the internal structure of the joint. This creates the necessity of application of non-destructive testing (NDT). There are a number of promising methods that have also been successfully applied for damage identification in adhesive joints, using ultrasounds [7–9], thermography [10], radiography [11], laser-induced breakdown spectroscopy [12], or electric time-domain reflectometry [13]. These methods are the basis for structural health monitoring (SHM) systems that provide a real-time evaluation of analyzed structures of different types, such as bridges [14–16], tunnels [17], or marine structures [18]. Nowadays, SHM strategies are becoming more and more popular for composite materials also [19–21].

The guided wave propagation phenomenon is commonly used for damage identification in structures of different types (e.g., [22–27]). Lamb waves are a specific type of guided waves that propagate in plate-like elements. It is worth noticing that they are multimodal; i.e., in general, an infinite number of different modes (symmetric and antisymmetric) can propagate in each medium. Another significant feature is the dispersive nature, which means that wave characteristics such as the wavenumber and propagation velocities of each mode are frequency-dependent. These properties make the question of wave propagation a complex problem. For certain frequency ranges, some modes do not propagate, whereas for a different range, the same modes can travel with certain velocities; thus, they influence the wave propagation. For this reason, the appropriate choice of the excitation frequency is an essential issue for the effectiveness of obtained results. High sensitivity to any disruption of geometry and changes in material properties create many applications of guided waves in non-destructive diagnostics of existing structures. Previous studies prove their usefulness for the identification of damages of different types, such as cracks in metallic beams and plates [28,29] or delamination and flaws in composites [30–32]. With regard to the adhesive joints, guided waves were efficiently used for the identification of disbond areas in the single-lap joints of plates made from different materials. Ren and Lissenden [7] detected damaged areas in the adhesive film in a CFRP (carbon fiber reinforced polymer) plate stiffened with a stringer using the adjustable angle beam transducers. Nicassio et al. [33] analyzed debonding in an adhesive joint of aluminum plates using piezo sensors. Sunarsa et al. [34] used air-coupled ultrasonic transducers to detect debonding and weakened bonding areas of different shapes in adhesively bonded aluminum plates. The time of flight of the measured signals was estimated with the support of the wavelet transform. Parodi et al. [35] analyzed numerically and experimentally wave propagation in a wall of a composite pressure vessel with flaws in the interface between the aluminum layer and CFRP coating, considering the excitation frequency in a range of 20 to 100 kHz. Ultrasonic waves are also successfully used for the evaluation of adhesion levels between adhesive and adherends. Gauthier et al. [36] analyzed the influence of different adherend surface treatment methods on the guided wave propagation in a single aluminum plate covered with epoxy-based adhesive. Castaings [37] considered a contamination of the overlap surface by an oil pollutant in the single-lap adhesive joints of aluminum plates.

The guided wave propagation method usually consists of the excitation of waves in one point of an analyzed structure and a collection of signals in some other points. If the number of measurement points (fitted with ultrasonic or piezoelectric transducers) is relatively small, the actual state of the considered structure is determined by the analysis of registered time histories. For a greater number of measurements, the non-contact methods are beneficial, allowing to sense the guided wave field in a considered area. The scanning laser Doppler vibrometry (SLDV) is one of the methods that provide a more accurate analysis [38–41]. As the effect, the plane representation of propagating waves (the so-called SLDV map) can be obtained. The existence of any defect in the scanned area results in the disturbance of the wave front shape, but its actual position and shape are indeterminable. Therefore, further signal processing is required to obtain a useful defect image. For example, Sohn et al. [42] detected delamination and disbond in composite plates based on the SLDV maps processed with the use of different techniques such as Laplacian image filtering. Another quite simple but effective method of damage imaging is based on the vibration energy distribution, and requires root mean square (RMS) calculations or its alternative weighted variant (WRMS). Recently, it has been successfully applied for the damage identification of different structures [43–49]. Saravanan et al. [43] detected missing bolts, attached masses, and openings in aluminum specimens assuming the excitation frequency as 50 kHz. Radzieński et al. [44] analyzed the detection of additional mass in aluminum and composite plates for frequencies of 35 and 10 kHz, respectively. In another work [45], they examined aluminum plates strengthened with riveted L-shape stiffeners, considering different excitation frequencies (5, 35, and 100 kHz). Aluminum plates with notches of different directions were studied by Lee and Park [46], who proved that the orientation of defects to the incident wave front was significant. In another research study, Lee et al. [47] investigated the notches and corrosion defects of different areas using the weighted root mean square and edge detection algorithms. Rucka et al. [48] studied the influence of a weighting factor on the efficiency of WRMS maps. Aryan et al. [49] visualized defects in the form of corrosion, surface cracks, and dents in aluminum plates and the delamination in a composite beam using scanning laser Doppler vibrometer and RMS calculations. The excitation frequencies were chosen from a range of 100 to 300 kHz. To sum up, the above-mentioned works present the application of root mean square calculations of registered guided wave signals without extensive consideration of the influence of the excitation frequency. This parameter was usually arbitrary assumed; notwithstanding, it can significantly affect the legibility of obtained RMS maps.

The aim of the study is damage imaging in a Lamb-wave based inspection of adhesively bonded joints. Particular attention was paid to the influence of the excitation frequency on the efficiency of obtained results. The guided wave signals were collected by the scanning laser Doppler vibrometer and further processed using root mean square calculations. The experimental research was conducted on a real-scale physical model of a single-lap joint of metal plates bonded with the epoxy-based adhesive. The verification of measurements was provided by numerical analyses carried out on finite element method (FEM) models. A novel element of the study is the proposition of choosing the adequate excitation frequency by the qualitative measure of the effectiveness of RMS damage imaging. The hypothesis is that an efficient frequency range for experimental measurements can be determined in the way of initial FEM calculations for artificial defects. The relative difference between RMS values in the damaged and intact areas of the joint can be assumed as the measure of the efficiency.

2. Materials and Methods

2.1. Specimen Description

The investigations were conducted on the single-lap adhesive joints of steel plates. The geometry of the specimens is presented in Figure 1. The dimensions of each plate were 270 mm × 120 mm × 3 mm. The overlap surface was 120 mm × 60 mm. The internal defect in the form of partial debonding was designed in the adhesive film in four variants (#1 to #4, as shown in Figure 2), from which the first one was chosen for experimental measurements. The defect (#1) was obtained by sticking a PTFE (polytetrafluoroethylene) tape of 0.2-mm thickness in the middle of the overlap before manufacturing the connection. To avoid creating unintended debonding areas, the overlap surface of each adherend was treated with fine sandpaper (grit size 120) and degreased with Loctite-7063 cleaner just before joining. The epoxy-based adhesive Loctite Hysol 9461 (Henkel, Düsseldorf, Germany) was used to join the plates. The measured bondline thickness was equal to approximately 0.2 mm. To control the expected geometry of the defect, the adherends were disconnected after the experiments. The separated plates are presented in Figure 3. The failure occurred in the interfaces between the glue layer and the steel plates (mainly the lower one); it has a purely adhesive character. There are visible leaks of adhesive into the area of the intended defect (on the upper plate). Moreover, the defect edges are irregular (visible mainly on the lower plate).

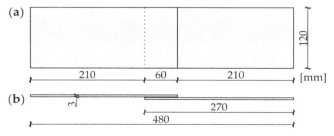

Figure 1. Geometry of investigated specimen: (**a**) plane view; (**b**) side view.

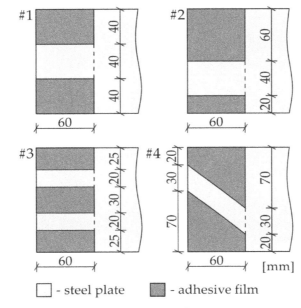

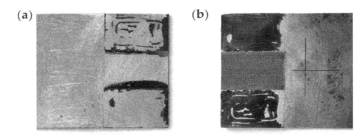

Figure 2. Variants of defects (#1 to #4).

Figure 3. Photograph of experimental specimen after separation: (**a**) upper plate; (**b**) lower plate.

2.2. Experimental Setup

The experimental examination of prepared specimen #1 consisted of the excitation and the acquisition of the Lamb wave propagation signals in the specified area of the joint by the scanning laser Doppler vibrometry method. The experimental setup is presented in Figure 4a. The generation of the input wave signal was provided by the arbitrary function generator AFG 3022 (Tektronix, Inc., Beaverton, OR, USA) with the support of the high-voltage amplifier PPA 2000 (EC Electronics, Krakow, Poland). The plate piezoelectric actuator NAC2024 (Noliac, Kvistgaard, Denmark) with dimensions of 3 mm × 3 mm × 3 mm was used for the excitation of the guided wave field in one of the adherends. The actuator was attached to the top surface of the specimen by the petro wax 080A109 (PCB Piezotronics, Inc., Depew, NY, USA). The input signal was a wave packet obtained from the five periods of the sinusoidal function by the Hanning window modulation. The excitation frequency was individual for each measurement and varied from 20 to 350 kHz. The signals of the guided wave field were recorded by the scanning head of the laser vibrometer PSV-3D-400-M (Polytec GmbH, Berlin, Germany) equipped with a VD-07 velocity decoder. The sampling frequency was assumed to be 2.56 MHz. The improvement of light backscatter was provided by covering the scanned surface with a retro-reflective sheeting. The out-of-plane components of velocity values were acquired in the time domain in 3721 points distributed over the area featuring the overlap surface and the part of the plate after it (at the top side of the specimen); see Figure 4b,c. The scanning was performed point by point in the quadratic mesh of 61 rows and 61 columns, resulting in the resolution of about 1.93 mm. The representations of acquired signals for specific time instances show the propagation of the full guided wave field (SLDV maps).

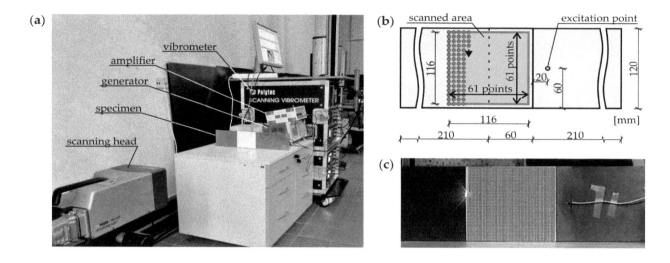

Figure 4. Experimental measurements: (**a**) experimental setup for the generation and acquisition of Lamb waves; (**b**) view of a specimen with the position of a scanned area and excitation point; (**c**) investigated specimen with indicated scanning points.

2.3. FEM Modeling

The numerical modeling of elastic wave propagation in composite structures, such as adhesive joints, is a complex problem, mainly because of material inhomogeneity and an uncertainty of contact at the interfaces between different materials. An effective contribution to this issue was made by Chronopoulos [50] and Apalowo and Chronopoulos [51]. In the present paper, numerical analysis of the guided wave propagation in the considered adhesive joints (#1 to #4) was conducted using the finite element method in Abaqus/Explicit software. Some assumptions were made to simplify the modeling process and shorten the calculations. Three-dimensional FEM models were prepared for a transient dynamic analysis. Each structure was discretized by eight-node solid elements with reduced integration (C3D8R) from the explicit element library. The appropriate mapping of the wave behavior requires at least 20 nodes for the shortest wavelength of interest [52]. According to this limitation, the mesh was initially assumed to be regular and consisted of cube-shaped elements with a global size of 1 mm, which was reduced to 0.2 mm for the thickness of the adhesive layer (see Figure 5). The mesh convergence test was conducted taking into account a few refined meshes. The out-of-plane velocity values in some randomly chosen points at specific time instances were assumed as the measure of the convergence. The relative differences between results were negligible; thus, the exact calculations were conducted with the use of the above-mentioned mesh with the global element size of 1 mm. The boundary conditions were free at all the edges. The materials were adopted to meet the assumptions of a homogenous, isotropic material model. The material parameters were: for steel $E_s = 195.2$ GPa, $v_s = 0.30$, $\rho_s = 7741.7$ kg/m^3, and for adhesive $E_a = 5$ GPa, $v_a = 0.35$, and $\rho_a = 1330$ kg/m^3. The material damping was neglected because of its marginal influence on the RMS damage imaging. Both adherends and the adhesive film were assumed to be independent structures combined rigidly at the part of their surfaces by means of a tie connection (compatibility of translational degrees of freedom at all the contacting nodes). The excitation of guided waves was applied at the lower adherend in the form of the concentrated force surface load with the amplitude varying in time in accordance with the wave packet signal. The excitation frequency range was extended in comparison to experimental measurements (20 to 500 kHz). The dynamic analysis was conducted with the use of the central difference method with a fixed time step of 10^{-7} s. This value meets the recommendation of at least 20 points per each cycle of the wave with the higher frequency [52]. The results of the analysis were out-of-plane velocity signals collected at 3721 points spread over the area of the joint corresponding with experimental measurements.

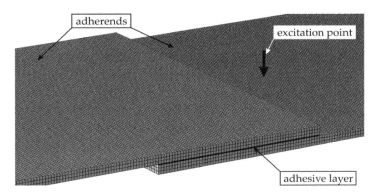

Figure 5. View of a discretized model for numerical calculations with an indicated excitation point.

2.4. RMS Damage Imaging

The signals of propagating waves acquired during scanning with a laser vibrometer need further processing techniques that allow detecting damaged areas and defining their actual shape. The essential point of damage imaging is to show the differences between the undamaged and damaged part of an analyzed structure. One of the simplest method consists of the calculation of the root mean square (RMS) for each recorded signal. The RMS value for the continuous time signal $s(t)$ can be calculated with the following formula:

$$RMS = \sqrt{\frac{1}{t_2 - t_1} \int_{t_1}^{t_2} s(t)^2 dt} \qquad (1)$$

where t_1 is the beginning and t_2 is the end of the time window, which is defined as the difference between these two values. For a discrete signal $s_k = s(t_k)$ recorded with the time interval Δt, the RMS value can be calculated as follows:

$$RMS = \sqrt{\frac{1}{n} \sum_{k=1}^{n} s_k^2} \qquad (2)$$

where n is the number of samples, and the time window is defined as $T = \Delta t(n - 1) = t_2 - t_1$. The map prepared from the calculated RMS values allows identifying and determining the geometry of any possible defects existing in the scanned area. Overall, for damaged areas of an analyzed element (e.g., delamination, crack, opening), different RMS values are attained because of different characteristics of Lamb wave propagation (changes due to material stiffness or geometry disturbance).

3. Results and Discussion

3.1. Dispersion Curves

The initial step in the damage detection of adhesive joints was the comparison of Lamb wave characteristics in a three-layer medium (steel–adhesive–steel, simulating a properly prepared adhesive joint) and in a single-layer medium (single steel plate or disbonded area of the adhesive joint). For this purpose, dispersion curves were prepared experimentally and numerically for a steel plate with dimensions of 240 mm × 300 mm × 3 mm (sample D1) and for two plates with an adhesive film with a thickness of 0.2 mm bonding them together (sample D2). In each specimen, a wave packet in the form of a single-cycle Hanning windowed sinusoidal function was excited. The carrier frequency was changing in the range from 50 to 300 kHz with a step of 50 kHz. Additionally, for each frequency, symmetric and antisymmetric Lamb modes were excited independently. The velocity signals (out-of-plane components) were acquired in 101 points distributed along the straight line with a total length of 100 mm. The dispersion curves in the form of the maps representing wavenumber–frequency relations were obtained in the way of 2D-FFT (two-dimensional fast Fourier transform) calculations for each of 12 measurements (cf. [6,36,53]). The final result was the superposition of all the compound maps (Figure 6).

A comparison of experimental and numerical curves led to the conclusion that both approaches gave consistent results. This also proved the appropriateness of the assumption of adhesive material parameters. To track the theoretical dispersion curves of Lamb waves in the investigated media, our own code was developed in the Matlab® software (9.3.0.713597, The MathWorks, Inc., Natick, MA, USA), implementing the transfer matrix method [54,55]. Figure 7a shows wavenumber–frequency relations for both media (samples D1 and D2). The shape of the curves is approximately the same as that shown in the maps in Figure 6. However, the possibility of the effective excitation of certain modes was not the same in the results of the measurements (in both experimental and numerical curves for the single plate and the joint). In sample D1 (Figures 6a and 7, black curves), only fundamental S_0 and A_0 modes can propagate in the considered frequency range, notwithstanding that the S_0 curve is not as strongly exposed as A_0 in the maps, which may be the result of the acquisition of only out-of-plane components on the upper surface that are related mainly to antisymmetric modes. In sample D2 (Figures 6b and 7, red curves), in addition to the fundamental pair, A_1 and S_1 modes are present starting from the frequencies of about 130 and 260 kHz, respectively. Moreover, the shape of the S_0 curve changes meaningfully compared with the single-layer plate. The shape of the A_0 curve does not change significantly in comparison to sample D1. The differences between the two types of media are also clearly visible on group velocity–frequency relations (Figure 7b), which will be useful in further considerations.

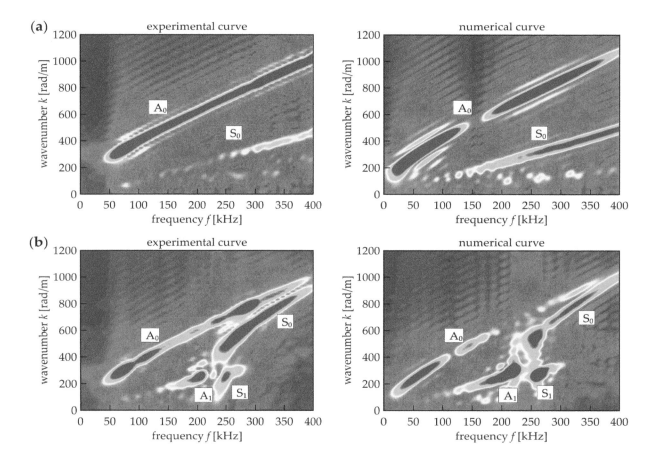

Figure 6. Experimental and numerical dispersion curves: (**a**) steel plate (sample D1)—single layer medium ($d_s = 3$ mm, $E_s = 195.2$ GPa, vs. $= 0.3$, $\rho_s = 7741.7$ kg/m^3); (**b**) adhesive joint (sample D2)—three-layer medium consisted of two steel plates (parameters same as in (**a**)) and adhesive film ($d_a = 0.2$ mm, $E_a = 5$ GPa, $v_a = 0.35$, $\rho_a = 1330$ kg/m^3).

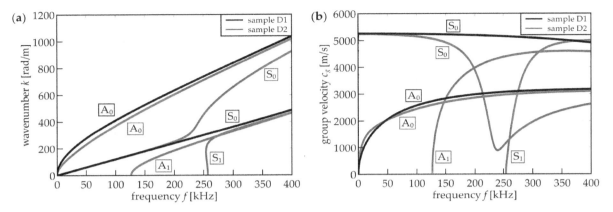

Figure 7. Theoretical dispersion curves for a steel plate (D1, black curves)–single layer medium (d_s = 3 mm, E_s = 195.2 GPa, vs. = 0.3, ρ_s = 7741.7 kg/m^3) and adhesive joint (D2, red curves)–three-layer medium consisting of two steel plates (parameters same as above) and adhesive film (d_a = 0.2 mm, E_a = 5 GPa, v_a = 0.35, ρ_a = 1330 kg/m^3): (**a**) wavenumber–frequency relations; (**b**) group velocity–frequency relations.

3.2. Influence of Excitation Frequency on the RMS Damage Imaging

The analysis of the influence of the excitation frequency on the effectiveness of RMS damage imaging was performed for specimen #1. Experimental and numerical approaches were applied for an analysis of guided Lamb wave fields and RMS maps.

3.2.1. Guided Wave Fields

Guided wave fields representing out-of-plane velocity values were prepared for a specific time instance t = 30 µs. Certain frequencies (50, 100, 150, 210, 300, and 350 kHz) were chosen for a comparative analysis of experimental and numerical results. The maps of propagating waves are presented in Figure 8. The comparison of presented snapshots revealed the variability of group velocity in relation to the excitation frequency. For the lowest frequency (50 kHz, Figure 8a), the wavefront is moderately visible (disturbance only at the initial part of the overlap). In the case of higher frequencies, the wavefront moved to the left side of the overlap, which suggests the greater speed of the excited wave packet. In fact, the individual selection of the time instance for each measurement can reduce these differences. Minor differences were observed between higher frequencies, because the group velocity was similar. Knowing that the excitation has an antisymmetric character, the A_0 mode is expected to be dominant. These observations agree with the dispersion curves (Figure 7b). The A_0 curve for the three-layer plate indicates significant growth in the group velocity value in the initial frequency range, and almost no variations in the further range. This explains why there are meaningful differences between snapshots for lower frequencies (50, 100, and 150 kHz), but wave fields are comparable for higher ones (210, 300, and 350 kHz), neglecting considerable changes in periods of wave packets.

Comparing the experimental and numerical results, there are some slight differences. Firstly, the numerical maps are symmetric, whereas the symmetry of the experimental wave fields is vaguely disturbed, probably by an imperfect preparation of the specimen and an inaccurate assumption of the scanning area for measurements. Moreover, the wavefronts are disturbed sharply at the edges of the defect in the numerical snapshots, but this effect is not that demonstrable in the experimental results because of the irregularities in the shape of defect edges (cf. Figure 3). Additionally, the group velocity is slightly higher for the experimental maps. The reason might lie in the differences between the mechanical properties of both materials (steel, adhesive) or the geometry of plates and the adhesive film (especially thickness). The observation of each snapshot allows identifying the defect. Significant disturbances of the wavefront indicate the intended lack of the adhesive in the middle of the overlap. Nonetheless, the determination of the actual geometry of the damaged area is not possible, and additional signal processing is required.

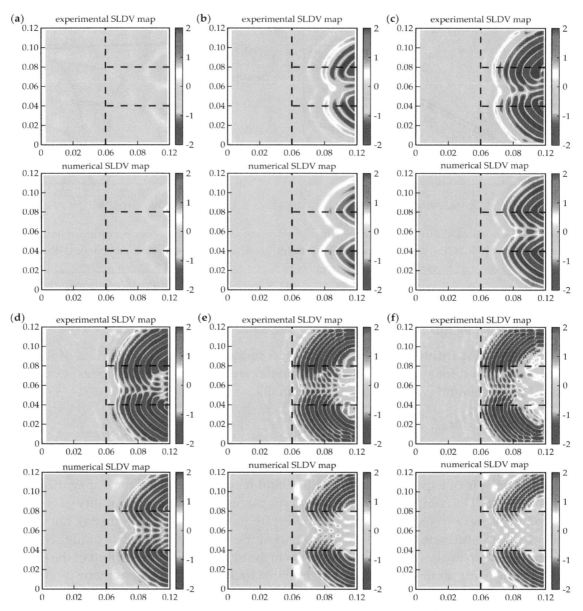

Figure 8. Experimental and numerical guided wave fields (values in m/s·10^{-3}) for a specific time, 30 μs, and different excitation frequencies: (**a**) $f = 50$ kHz; (**b**) $f = 100$ kHz; (**c**) $f = 150$ kHz; (**d**) $f = 210$ kHz; (**e**) $f = 300$ kHz; and (**f**) $f = 350$ kHz.

3.2.2. RMS Imaging

Figure 9 shows the RMS maps normalized to unity for experimental and numerical signals collected for specimen #1. The chosen frequencies were the same as those for the SLDV maps. Each RMS value was calculated with respect to Equation (2). The time window covered the whole time of signal acquisition, i.e., $T = 3.2$ ms. The individual characterization of a single-layer medium (steel plate, such as the damage area) and a three-layer medium (properly prepared adhesive joint) should result in the clear difference of the calculated RMS values. However, it is undeniable that excitation frequency is an essential factor affecting the effectiveness of RMS damage imaging. For the lower frequencies (50 kHz, Figure 9a; 100 kHz, Figure 9b; 150 kHz, Figure 9c), the damaged area is characterized by highest RMS values rather than an appropriately prepared joint (similar to the single plate after the joint). The difference between these two areas is clear. Some differences between experimental and numerical maps result from irregularities in defect geometry (cf. Figure 3). It is worth noticing that at the lower excitation frequency, the lower resolution can be obtained in the map and, as a result, the larger defects can be omitted. This may be very important in the case of small defects; however, for the considered damage area, it is not essential. The frequency 210 kHz (Figure 9d) give an ineffectual result: there is almost no difference between the defect and intact joint, especially in the numerical map. In the experimental RMS, some boundary effects (intensification of the wave energy on the irregular edge) led to higher RMS values. This example shows that the invalid choice of the excitation frequency can make the measurement results useless. For the frequencies higher than 210 kHz (300 kHz, Figure 9e; 350 kHz, Figure 9f) the RMS values are lower in the damaged area than in the intact joint. The correlation between RMS values for these two areas is inverted. What is important is that the visual assessment of RMS maps shows that the distinction between the damaged and intact area of the joint is much more pronounced in the lower frequency range, especially for experimental maps, where the whole area of the adhesive layer does not have the same value. This may be the effect of the limitations of the used experimental setup. What is more, the increase in the excitation frequency is related to the increase in the wave attenuation.

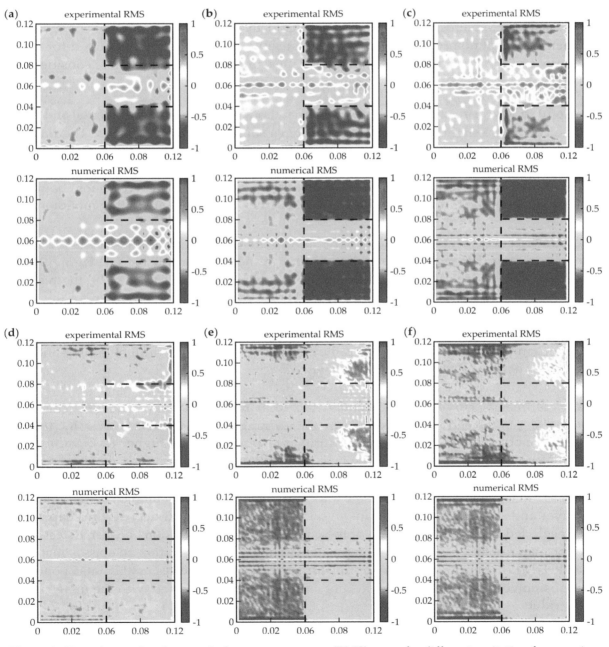

Figure 9. Experimental and numerical root mean square (RMS) maps for different excitation frequencies: (**a**) $f = 50$ kHz; (**b**) $f = 100$ kHz; (**c**) $f = 150$ kHz; (**d**) $f = 210$ kHz; (**e**) $f = 300$ kHz; and (**f**) $f = 350$ kHz.

The effectiveness of RMS damage identification is an important issue, so there is the need for a qualitative measure of contradistinction between the damaged and intact areas of the joint. The proposition is the relative difference between the level of the RMS in these two areas, which can be expressed by the relation:

$$R_d = \frac{l_d - l_i}{l_i} \qquad\qquad (3)$$

where l_i and l_d denote the mean RMS value in the properly prepared area of the overlap and in the damaged area, respectively. The definition of the R_d value induces that the damaged area is defined, so it cannot be used if the joint has any unknown defects. Nevertheless, the aim of R_d calculations is only the demonstration of changes in the effectiveness of RMS imaging in relation to the excitation frequency. The area of the overlap was divided before the calculations into two parts (damaged and intact) with a rejection of points localized on the edges of the defect and on the longitudinal axis of the joint, because for these points, the RMS values are distinctly high (intensification of energy evoked by the symmetry and boundaries, cf. Figure 9). The mean RMS values were calculated for the points of both areas, and the R_d value was calculated for measurements over the whole considered range of frequency.

Figure 10 shows the relation between R_d and excitation frequency for experimental and numerical results. The curves are slightly different, but both have some characteristic points. The first one is a local maximum for 50 kHz. For this value, the A_0 modes for single and three-layer plates are crossing on dispersion curves (cf. Figure 7b). Then, there are some fluctuations that differ between the two curves. Another peak repeating for both experimental and numerical curves is for about 120 kHz, when the A_1 mode for the three-layer plate appears. Further, the curves are falling monotonically.

The numerical curve has the root equal to approximately 215 kHz, and this is the frequency value for which the damaged area and intact joint are not distinguishable (cf. Figure 9). For the experimental curve, the root is translated to approximately 250 kHz. This may be the result of the energy intensification on the edges of the defect (cf. Figure 9d). The global minimum for the numerical curve is attained for approximately 260 kHz (the appearance of the S_1 mode for the three-layer plate). This is the frequency for which the joint and the defect can be distinguished with maximal efficiency in the frequency range above 210 kHz. Further, the curve is slightly rising until obtaining another root (about 500 kHz) at the end of the frequency range.

The experimental curve does not obtain the local minimum above 210 kHz; instead, it is constantly falling to the end of the assumed frequency range up to 350 kHz. Generally, the positive values of R_d are obtained when the damaged area is characterized by higher RMS values than the intact joint. Negative values indicate on the inverted relation. If R_d equals zero, there is no possibility of identifying the defects. What is important is that the absolute values of R_d are higher below the first root (about 215 kHz), which suggests that the lower frequencies allow obtaining a better differentiation of three-layer and single layer media. However, the above-mentioned decrease in map resolution cannot be neglected. To compromise both of these factors (the differentiation between defect and intact joint and the map resolution), the excitation frequency should be chosen from an approximate range of 120 to 180 kHz. The determination of an optimal frequency value requires mathematical optimization and a proposition of an objective function containing components linked with the image resolution and the R_d value.

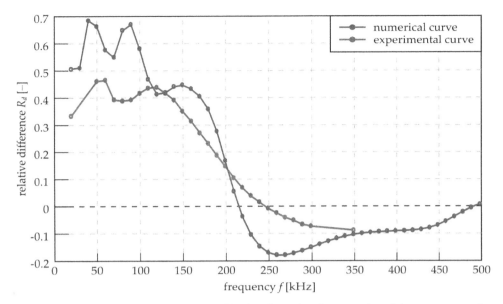

Figure 10. Relative difference between mean RMS values for damaged and intact areas of overlap for experimental (frequency range from 20 to 350 kHz) and numerical results (frequency range from 20 to 500 kHz).

3.2.3. Statistical Analysis of RMS Values

In addition to the foregoing considerations, the statistical analysis was conducted. All the RMS values calculated for the whole overlap surface, rejecting points at the edges of the defect and on the longitudinal axis (as for R_d calculations), were treated as the single series of values. Histograms were calculated for each dataset. If there are no defects in the analyzed area, the purely unimodal distribution would be obtained, because all the RMS values should accumulate over a single value, which is symbolized above by l_i. The presence of a defect in the adhesive layer should result in the bimodal distribution caused by the existence of two dominant values for the intact joint l_i and for defect l_d.

Figure 11 presents RMS histograms that have been prepared for certain frequencies (the same as for the RMS maps). The results are normalized, both for the RMS value and the quantity axes. For the lower frequencies (50, 100, and 150 kHz), bimodal distributions were obtained for the experimental and numerical data. The first mode is characterized by the lower RMS values and related to the intact joint area (cf. Figure 9a–c). The second mode indicates the defect existence (higher RMS values), and it obtains less quantity than the first mode, because the defect surface is twice as small as the intact joint surface. For 210 kHz (Figure 9d), the histograms are unimodal, which results from the equality of mean RMS values calculated for the defect and intact joint ($R_d = 0$, not efficient damage imaging). The experimental histogram is not as narrow as the numerical one, which is related to the translation of root of curves from Figure 10. For higher frequencies (300 kHz, 350 kHz), the distributions are not unimodal; the damage can be identified, and its mode is related to the lower RMS values. The dissociation of modes is not as clear as for lower frequencies—this effect is related to the lower absolute values of R_d for higher frequencies. In numerical histograms, the defect mode is characterized by the lower intensity than the intact joint mode (compatibly with the relation of surfaces of damaged and intact areas). The experimental histograms do not cover the same rule; the intact joint mode has a lower intensity because higher RMS values are not obtained in the whole area of the properly prepared joint (cf. Figure 9e,f). To sum up, the histogram analysis can reveal the existence of damaged areas, but it has some limitations. Firstly, the geometry of the defects cannot be determined. Moreover, the method is not efficient for small defects, because the damage modes would be of small quantity, which makes it impossible to identify them on the histograms.

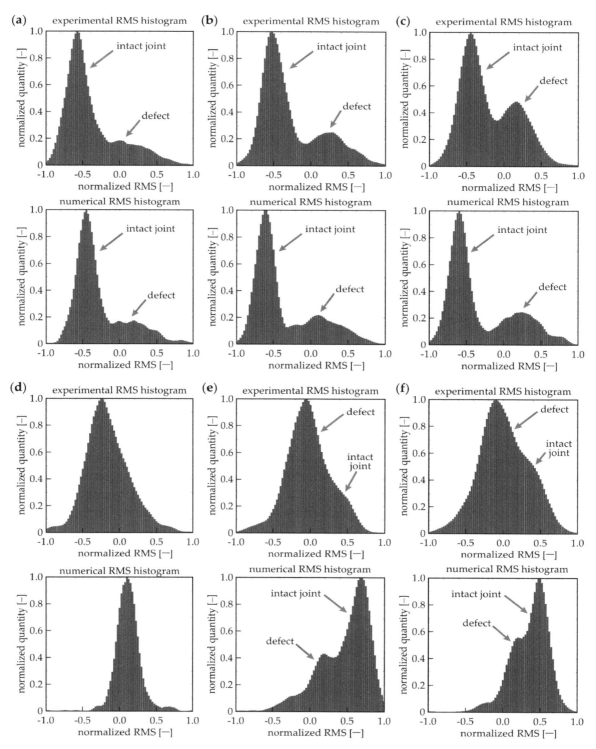

Figure 11. Experimental and numerical RMS histograms for different excitation frequencies: (**a**) f = 50 kHz; (**b**) f = 100 kHz; (**c**) f = 150 kHz; (**d**) f = 210 kHz; (**e**) f = 300 kHz; and (**f**) f = 350 kHz.

3.3. Influence of Different Defect Geometry

The above considerations were conducted only for a single joint, #1. Next, the observed effects were verified on specimens #2 to #4 by the way of numerical calculations for three certain frequencies (100 kHz, 210 kHz, and 300 kHz). The normalized RMS maps are presented in Figure 12. It is visible that for the frequency of 100 kHz, higher RMS values were obtained for the damaged areas than for the intact joint (for all the analyzed specimens). The frequency of 210 kHz appeared to be inefficient for RMS damage imaging, i.e., the difference between the damaged and properly prepared area was not significant. The frequency of 300 kHz resulted in lower RMS values in the damaged area. Moving to the histograms (Figure 13), the frequency 210 kHz gave the unimodal distribution (no difference between the defect and intact joint). The histograms for 100 and 300 kHz gave bimodal distributions, but for lower frequencies, the defect mode was related to higher RMS values, whereas for higher frequencies, it was related to lower RMS values. The quantity for the defect mode was always smaller than that for an intact joint. The considerations for joints #2 to #4 provided the same results as for specimen #1. Summarizing, the efficiency of a measurement with a specific excitation frequency does not change with the geometry of a damaged area. It is a satisfying conclusion, because generally, the geometry of the damaged area is unknown. This means that an additional preliminary study consisting of numerical calculations can reveal the appropriate excitation frequency value and reduce the number of measurements. However, the possibility of the damage identification due to a frequency value depends strongly on the characteristics of the considered media, so they need to be determined.

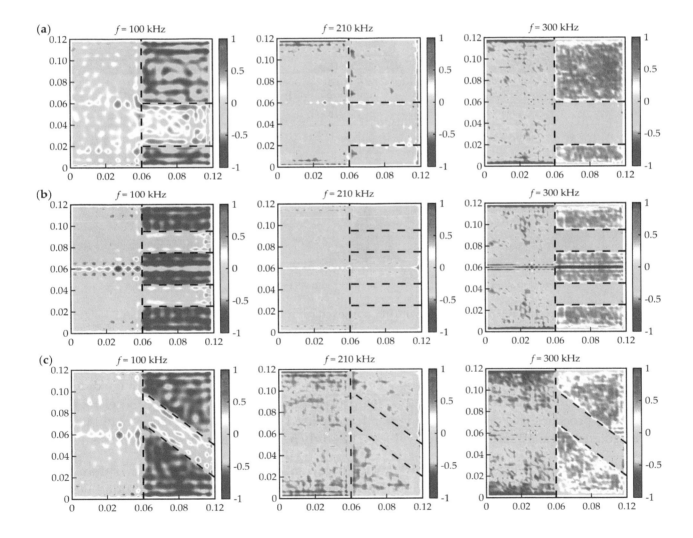

Figure 12. Numerical RMS maps for different excitation frequencies (100, 210, and 300 kHz) and different defects: (**a**) #2; (**b**) #3; and (**c**) #4.

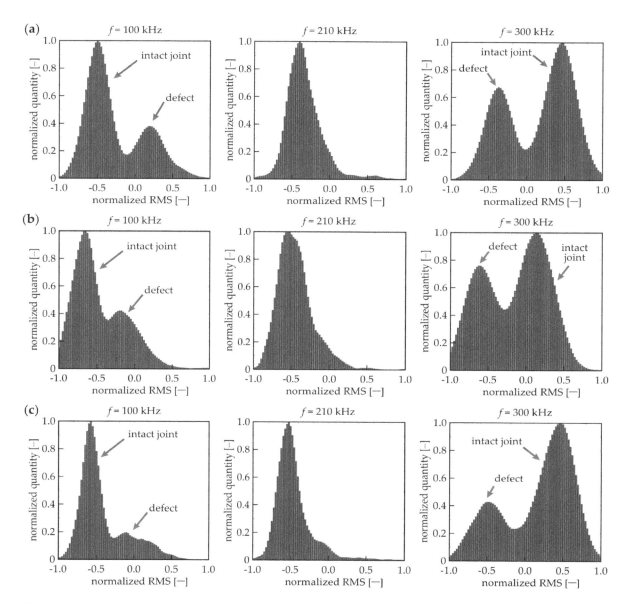

Figure 13. Numerical RMS histograms for different excitation frequencies (100, 210, and 300 kHz) and different defects: (**a**) #2; (**b**) #3; and (**c**) #4.

4. Conclusions

The paper discussed the effects of the wave frequency on the efficiency of damage detection in adhesive joints of steel plates using Lamb wave propagation and RMS imaging. Experimental and numerical approaches were applied. The research comprised the visual appreciation of obtained RMS maps and statistical analysis of calculated values. The study resulted in the conclusions presented below.

- The guided wave fields enabled identifying the occurrence of the defect regardless of the excitation frequency. However, the actual location and shape are indeterminable; thus, guided wave field measurements can only be an initial step for further analyses.
- The RMS maps allowed determining the geometry of the damaged areas. The effectiveness of damage visualization was strongly dependent on the excitation frequency.
- The variability of the relative difference between the mean RMS values for the intact joint and the damage was fully compatible with the clarity of the RMS maps. Some analogies between the relative RMS difference and dispersion curves were observed.

- The statistical analysis was successfully used to determine the effectiveness of the results obtained for different excitation frequencies based on the RMS histograms. The important advantage of this approach is the independence of the defect geometry.
- The statistical analysis in a certain frequency range on the single numerical model with a random defect can be sufficient for the determination of the adequate frequency for the further experimental testing of samples with an unknown state.

The guiding conclusion was that the Lamb wave-based inspection of adhesive joints with the use of scanning laser Doppler vibrometry and signal processing, such as root mean square calculations, provides a successful method for damage imaging. To obtain valuable results, some initial analyses need to be conducted before the exact measurements. The main factor is the choice of an appropriate excitation frequency, which can be conducted using numerical calculations supported by statistical analysis.

Author Contributions: Conceptualization and Methodology, E.W. and M.R.; Experimental Investigations, E.W. and M.R.; FEM Calculations, E.W.; Formal Analysis, E.W.; Visualization, E.W.; Writing—Original Draft Preparation, E.W.; Writing—Review and Editing, M.R.; Supervision, Project Administration and Funding Acquisition, M.R.

Acknowledgments: Abaqus calculations were carried out at the Academic Computer Centre in Gdańsk.

References

1. Adams, R.D.; Wake, W.C. *Structural Adhesive Joints in Engineering*; Elsevier Applied Science Publishers: London, UK, 1986; ISBN 978-94-010-8977-7.

2. Dillard, D.A. *Advances in Structural Adhesive Bonding*, 1st ed.; Woodhead Publishing: Cambridge, UK, 2010; ISBN 9781845694357.

3. Martínez-Landeros, V.H.; Vargas-Islas, S.Y.; Cruz-González, C.E.; Barrera, S.; Mourtazov, K.; Ramírez-Bon, R. Studies on the influence of surface treatment type, in the effectiveness of structural adhesive bonding, for carbon fiber reinforced composites. *J. Manuf. Process.* **2019**, *39*, 160–166. [CrossRef]

4. Jeenjitkaew, C.; Guild, F.J. The analysis of kissing bonds in adhesive joints. *Int. J. Adhes. Adhes.* **2017**, *75*, 101–107. [CrossRef]

5. Sengab, A.; Talreja, R. A numerical study of failure of an adhesive joint influenced by a void in the adhesive. *Compos. Struct.* **2016**, *156*, 165–170. [CrossRef]

6. Ong, W.H.; Rajic, N.; Chiu, W.K.; Rosalie, C. Lamb wave–based detection of a controlled disbond in a lap joint. *Struct. Health Monit.* **2018**, *17*, 668–683. [CrossRef]

7. Ren, B.; Lissenden, C.J. Ultrasonic guided wave inspection of adhesive bonds between composite laminates. *Int. J. Adhes. Adhes.* **2013**, *45*, 59–68. [CrossRef]

8. Puthillath, P.K.; Yan, F.; Kannajosyula, H. Inspection of Adhesively Bonded Joints Using Ultrasonic Guided Waves. In Proceedings of the 17th World Conference on Nondestructive Testing, Shanghai, China, 25–28 October 2008; pp. 25–28.

9. Korzeniowski, M.; Piwowarczyk, T.; Maev, R.G. Application of ultrasonic method for quality evaluation of adhesive layers. *Arch. Civ. Mech. Eng.* **2014**, *14*, 661–670. [CrossRef]

10. Tighe, R.C.; Dulieu-Barton, J.M.; Quinn, S. Identification of kissing defects in adhesive bonds using infrared thermography. *Int. J. Adhes. Adhes.* **2016**, *64*, 168–178. [CrossRef]

11. Opdam, N.J.M.; Roeters, F.J.M.; Verdonschot, E.H. Adaptation and radiographic evaluation of four adhesive systems. *J. Dent.* **1997**, *25*, 391–397. [CrossRef]

12. Sato, T.; Tashiro, K.; Kawaguchi, Y.; Ohmura, H.; Akiyama, H. Pre-bond surface inspection using laser-induced breakdown spectroscopy for the adhesive bonding of multiple materials. *Int. J. Adhes. Adhes.* **2019**, 1–9. [CrossRef]

13. Steinbild, P.J.; Höhne, R.; Füßel, R.; Modler, N. A sensor detecting kissing bonds in adhesively bonded joints using electric time domain reflectometry. *NDT E Int.* **2019**, *102*, 114–119. [CrossRef]

14. Malik, H.; Zatar, W. Software Agents to Support Structural Health Monitoring (SHM)-Informed Intelligent Transportation System (ITS) for Bridge Condition Assessment. *Procedia Comput. Sci.* **2018**, *130*, 675–682. [CrossRef]

15. dos Reis, J.; Oliveira Costa, C.; Sá da Costa, J. Local validation of structural health monitoring strain measurements. *Meas. J. Int. Meas. Confed.* **2019** *136*, 143–153. [CrossRef]

16. Comisu, C.C.; Taranu, N.; Boaca, G.; Scutaru, M.C. Structural health monitoring system of bridges. *Procedia Eng.* **2017**, *199*, 2054–2059. [CrossRef]

17. Yang, J.P.; Chen, W.Z.; Li, M.; Tan, X.J.; Yu, J.X. Structural health monitoring and analysis of an underwater TBM tunnel. *Tunn. Undergr. Space Technol.* **2018**, *82*, 235–247. [CrossRef]

18. Miśkiewicz, M.; Pyrzowski, Ł.; Wilde, K.; Mitrosz, O. Technical Monitoring System for a New Part of Gdańsk Deepwater Container Terminal. *Polish Marit. Res.* **2017**, *24*, 149–155. [CrossRef]

19. Gomes, G.F.; Mendéz, Y.A.D.; da Silva Lopes Alexandrino, P.; da Cunha, S.S.; Ancelotti, A.C. The use of intelligent computational tools for damage detection and identification with an emphasis on composites—A review. *Compos. Struct.* **2018**, *196*, 44–54. [CrossRef]

20. Martins, A.T.; Aboura, Z.; Harizi, W.; Laksimi, A.; Khellil, K. Structural health monitoring for GFRP composite by the piezoresistive response in the tufted reinforcements. *Compos. Struct.* **2019**, *209*, 103–111. [CrossRef]

21. Chroscielewski, J.; Miskiewicz, M.; Pyrzowski, L.; Rucka, M.; Sobczyk, B.; Wilde, K. Dynamic Tests and Technical Monitoring of a Novel Sandwich Footbridge. In *Dynamics of Civil Structures, Volume 2*; Pakzad, S., Ed.; Conference Proceedings of the Society for Experimental Mechanics Series; Springer: Berlin, Germany, 2019; pp. 55–60.

22. Ostachowicz, W.; Kudela, P.; Krawczuk, M.; Zak, A. *Guided Waves in Structures for SHM: The Time-Domain Spectral Element Method*; Wiley: Hoboken, NJ, USA, 2012; ISBN 9781119965855.

23. Rose, J.L. *Ultrasonic Guided Waves in Solid Media*; Cambridge University Press: New York, NY, USA, 2014; ISBN 9781107273610.

24. Yu, X.; Zuo, P.; Xiao, J.; Fan, Z. Detection of damage in welded joints using high order feature guided ultrasonic waves. *Mech. Syst. Signal Process.* **2019**, *126*, 176–192. [CrossRef]

25. Zhang, W.; Hao, H.; Wu, J.; Li, J.; Ma, H.; Li, C. Detection of minor damage in structures with guided wave signals and nonlinear oscillator. *Meas. J. Int. Meas. Confed.* **2018**, *122*, 532–544. [CrossRef]

26. Pan, W.; Sun, X.; Wu, L.; Yang, K.; Tang, N. Damage Detection of Asphalt Concrete Using Piezo-Ultrasonic Wave Technology. *Materials (Basel)* **2019**, *12*, 443. [CrossRef]

27. Schabowicz, K. Ultrasonic tomography - The latest nondestructive technique for testing concrete members - Description, test methodology, application example. *Arch. Civ. Mech. Eng.* **2014**, *14*, 295–303. [CrossRef]

28. He, S.; Ng, C.T. Guided wave-based identification of multiple cracks in beams using a Bayesian approach. *Mech. Syst. Signal Process.* **2017**, *84*, 324–345. [CrossRef]

29. Pahlavan, L.; Blacquière, G. Fatigue crack sizing in steel bridge decks using ultrasonic guided waves. *NDT E Int.* **2016**, *77*, 49–62. [CrossRef]

30. Munian, R.K.; Mahapatra, D.R.; Gopalakrishnan, S. Lamb wave interaction with composite delamination. *Compos. Struct.* **2018**, *206*, 484–498. [CrossRef]

31. Shoja, S.; Berbyuk, V.; Boström, A. Delamination detection in composite laminates using low frequency guided waves: Numerical simulations. *Compos. Struct.* **2018**, *203*, 826–834. [CrossRef]

32. Xiao, H.; Shen, Y.; Xiao, L.; Qu, W.; Lu, Y. Damage detection in composite structures with high-damping materials using time reversal method. *Nondestruct. Test. Eval.* **2018**, *33*, 329–345. [CrossRef]

33. Nicassio, F.; Carrino, S.; Scarselli, G. Elastic waves interference for the analysis of disbonds in single lap joints. *Mech. Syst. Signal Process.* **2019**, *128*, 340–351. [CrossRef]

34. Sunarsa, T.Y.; Aryan, P.; Jeon, I.; Park, B.; Liu, P.; Sohn, H. A reference-free and non-contact method for detecting and imaging damage in adhesive-bonded structures using air-coupled ultrasonic transducers. *Materials (Basel)* **2017**, *10*, 1402. [CrossRef]

35. Parodi, M.; Fiaschi, C.; Memmolo, V.; Ricci, F.; Maio, L. Interaction of Guided Waves with Delamination in a Bilayered Aluminum-Composite Pressure Vessel. *J. Mater. Eng. Perform.* **2019**, 1–11. [CrossRef]

36. Gauthier, C.; Ech-Cherif El-Kettani, M.; Galy, J.; Predoi, M.; Leduc, D.; Izbicki, J.L. Lamb waves characterization of adhesion levels in aluminum/epoxy bi-layers with different cohesive and adhesive properties. *Int. J. Adhes. Adhes.* **2017**, *74*, 15–20. [CrossRef]

37. Castaings, M. SH ultrasonic guided waves for the evaluation of interfacial adhesion. *Ultrasonics* **2014**, *54*, 1760–1775. [CrossRef] [PubMed]

38. Kudela, P.; Wandowski, T.; Malinowski, P.; Ostachowicz, W. Application of scanning laser Doppler vibrometry for delamination detection in composite structures. *Opt. Lasers Eng.* **2016**, *99*, 46–57. [CrossRef]

39. Rothberg, S.J.; Allen, M.S.; Castellini, P.; Di Maio, D.; Dirckx, J.J.J.; Ewins, D.J.; Halkon, B.J.; Muyshondt, P.; Paone, N.; Ryan, T.; et al. An international review of laser Doppler vibrometry: Making light work of vibration measurement. *Opt. Lasers Eng.* **2017**, *99*, 11–22. [CrossRef]

40. Derusova, D.; Vavilov, V.; Sfarra, S.; Sarasini, F.; Krasnoveikin, V.; Chulkov, A.; Pawar, S. Ultrasonic spectroscopic analysis of impact damage in composites by using laser vibrometry. *Compos. Struct.* **2019**, *211*, 221–228. [CrossRef]

41. Pieczonka, Ł.; Ambroziński, Ł.; Staszewski, W.J.; Barnoncel, D.; Pérès, P. Damage detection in composite panels based on mode-converted Lamb waves sensed using 3D laser scanning vibrometer. *Opt. Lasers Eng.* **2017**, *99*, 80–87. [CrossRef]

42. Sohn, H.; Dutta, D.; Yang, J.Y.; Desimio, M.; Olson, S.; Swenson, E. Automated detection of delamination and disbond from wavefield images obtained using a scanning laser vibrometer. *Smart Mater. Struct.* **2011**, *20*, 045017. [CrossRef]

43. Saravanan, T.J.; Gopalakrishnan, N.; Rao, N.P. Damage detection in structural element through propagating waves using radially weighted and factored RMS. *Measurement* **2015**, *73*, 520–538. [CrossRef]

44. Radzieński, M.; Doliński, L.; Krawczuk, M.; Zak, A.; Ostachowicz, W. Application of RMS for damage detection by guided elastic waves. *J. Phys. Conf. Ser.* **2011**, *305*, 1–10. [CrossRef]

45. Radzieński, M.; Doliński, Ł.; Krawczuk, M.; Palacz, M. Damage localisation in a stiffened plate structure using a propagating wave. *Mech. Syst. Signal Process.* **2013**, *39*, 388–395. [CrossRef]

46. Lee, C.; Park, S. Flaw Imaging Technique for Plate-Like Structures Using Scanning Laser Source Actuation. *Shock Vib.* **2014**, *2014*, 725030. [CrossRef]

47. Lee, C.; Zhang, A.; Yu, B.; Park, S. Comparison study between RMS and edge detection image processing algorithms for a pulsed laser UWPI (Ultrasonic wave propagation imaging)-based NDT technique. *Sensors (Switzerland)* **2017**, *17*, 1224. [CrossRef] [PubMed]

48. Rucka, M.; Wojtczak, E.; Lachowicz, J. Damage imaging in Lamb wave-based inspection of adhesive joints. *Appl. Sci.* **2018**, *8*, 522. [CrossRef]

49. Aryan, P.; Kotousov, A.; Ng, C.T.; Cazzolato, B.S. A baseline-free and non-contact method for detection and imaging of structural damage using 3D laser vibrometry. *Struct. Control Health Monit.* **2017**, *24*, 1–13. [CrossRef]

50. Chronopoulos, D. Calculation of guided wave interaction with nonlinearities and generation of harmonics in composite structures through a wave finite element method. *Compos. Struct.* **2018**, *186*, 375–384. [CrossRef]

51. Apalowo, R.K.; Chronopoulos, D. A wave-based numerical scheme for damage detection and identification in two-dimensional composite structures. *Compos. Struct.* **2019**, *214*, 164–182. [CrossRef]

52. Moser, F.; Jacobs, L.J.; Qu, J. Modeling elastic wave propagation in waveguides with the finite element method. *NDT E Int.* **1999**, *32*, 225–234. [CrossRef]

53. Gauthier, C.; Galy, J.; Ech-Cherif El-Kettani, M.; Leduc, D.; Izbicki, J.L. Evaluation of epoxy crosslinking using ultrasonic Lamb waves. *Int. J. Adhes. Adhes.* **2018**, *80*, 1–6. [CrossRef]

54. Lowe, M.J.S. Matrix Techniques for Modeling Ultrasonic-Waves in Multilayered Media. *IEEE Trans. Ultrason. Ferroelectr. Freq. Control* **1995**, *42*, 525–542. [CrossRef]

55. Maghsoodi, A.; Ohadi, A.; Sadighi, M. Calculation of Wave Dispersion Curves in Multilayered Composite-Metal Plates. *Shock Vib.* **2014**, *2014*, 1–6. [CrossRef]

7

Identification of the Destruction Process in Quasi Brittle Concrete with Dispersed Fibers based on Acoustic Emission and Sound Spectrum

Dominik Logoń[ID]

Faculty of Civil Engineering, Wrocław University of Science and Technology, 50-377 Wrocław, Poland; dominik.logon@pwr.edu.pl

Abstract: The paper presents the identification of the destruction process in a quasi-brittle composite based on acoustic emission and the sound spectrum. The tests were conducted on a quasi-brittle composite. The sample was made from ordinary concrete with dispersed polypropylene fibers. The possibility of identifying the destruction process based on the acoustic emission and sound spectrum was confirmed and the ability to identify the destruction process was demonstrated. It was noted that in order to recognize the failure mechanisms accurately, it is necessary to first identify them separately. Three- and two-dimensional spectra were used to identify the destruction process. The three-dimensional spectrum provides additional information, enabling a better recognition of changes in the structure of the samples on the basis of the analysis of sound intensity, amplitudes, and frequencies. The paper shows the possibility of constructing quasi-brittle composites to limit the risk of catastrophic destruction processes and the possibility of identifying those processes with the use of acoustic emission at different stages of destruction.

Keywords: acoustic emission AE; acoustic spectrum; quasi brittle cement composites; destruction process

1. Introduction

The application of acoustic emission (AE) measurements in determining the cracks, maximum load, and failure of reinforcement in cement composites has been widely presented in the literature.

The continuous AE evaluation in composites was earlier reported [1–3] and this technique has been applied to determine crack propagation in the fracture process in cement composites with and without reinforcement [4,5]. The acoustic emission (AE) events sum was also recorded for easier recognition of the first crack and crack propagation process [6–8].

It was also noticed that at the preliminary stage of degradation, the damage of the concrete elements was possible to detect with the application of the AE method [9,10]. The effectiveness of acoustic emission (AE) measurements in determining the critical stress of cement composites was tested [11], which enables the accurate definition of the elastic range corresponding to Hook's law. Previously conducted tests have shown that AE is a good method for crack formation monitoring in mechanically loaded specimens [12–18] and has been successfully used to monitor structures [19,20]. Most of the papers have used AE to identify the destruction process of materials in structures [21–28] including crack orientation [29–31]. The AE method is still used and improved for the purpose of the identification of failure processes [32–34].

Previous works, however, have not focused on the correlation between AE and the individual failure processes of each of the different composite components based on the sound spectrum. These papers [10,11] showed that for the accurate recognition of composite failure processes, the AE (and the

AE events sum) recording should be expanded to include the analysis of each sound separately and the analysis of the range of sounds corresponding to a given mechanical effect with the use of acoustic spectrum. The acoustic spectrum should be correlated with the load-deflection curve and with other acoustic effects, which enables the identification of the failure process (of the structure or the applied reinforcement) [10,11,18]. The quasi-brittle ESD cement composites (ESD—elastic range, strengthening, deflection control) are characterized by a higher load and absorbed energy in the elastic range when compared to the sample without reinforcement (E/E$_0$) (Figure 2). Additionally, those composites are distinguished by a highly deflected structure damaged with macrocracks, multicracking effects, and the ability to carry additional stress in the strengthening area. Moreover, in the deflection control area, the samples' ability to carry stress is higher than in the elastic range area. This paper focuses on determining the relation between the acoustic and mechanical ESD effects, in other words, reinforcement breaking, pull-out, macrocracks, and microcracking with the use of space spectrum. In [11], it was noted that in order to assess the destruction process, the analysis of a single signal and the AE events sum with the acoustic spectrum was required (each kind of the mechanical effect results in a different acoustic spectrum). In order to conduct a more in-depth analysis of the composite destruction process, what should be taken into account when interpreting the acoustic spectrum is not only the range of signals corresponding to a given mechanical effect (in a wide range of frequencies corresponding to the sound intensity), but also a single signal in a very small range of frequencies.

It was confirmed that there is a possibility of correlation between AE and the failure process in ESD composites. That correlation enables a determination of the stage of damage in cement composites increasing the safety in the use of the composite and the decision of whether or not the damaged composite can be repaired.

2. Materials and Methods

The materials for the concrete (matrix—sample without reinforcement) consisted of: Portland cement CEM I 42.5R—368.7 kg/m^3, silica fume 73.75 kg/m^3, fly ash 73.75 kg/m^3, sand and coarse aggregates 0/16 mm–1640 kg/m^3, superplasticizer (SP), tap water 188.6 kg/m^3, w/c = 0.51.

The ESD concrete was reinforced with polypropylene fibers (curved/wave), minimum tensile strength 490 MPa, E = 3.5 GPa, equivalent diameter d = 0.8–1.2 mm, l = 54 mm. The reinforcement was randomly dispersed V$_f$ = 1.5%.

Concrete was mixed in the concrete mixer and then used to mold samples. Beams (600 mm × 150 mm × 150 mm) were cast in slabs and then cured in water at 20 ± 2 °C. After 180 days of ageing, beams were prepared for the bending test (Figure 1). The samples were not notched.

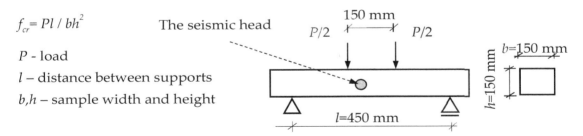

$$f_{cr} = Pl \, / \, bh^2$$

P - load

l – distance between supports

b,h – sample width and height

Figure 1. Four-point bending test.

Acoustic emission effects were recorded in order to monitor the progress of the fracture process in correlation with the load-deflection curve. The crosshead displacement was continuous and the rate was 1 mm/min. A seismic head HY919 (Spy Electronics Ltd.) was used to record the acoustic emission effects in the range from 0.2–20 kHz. The head was placed on the side in the central part of the loaded beams (Figure 1). The acoustic emission effects were presented as a 2D and 3D acoustic spectrum (amplitude of the frequency depending on sound intensity). The mechanical effects of the ESD composites were correlated with the recorded acoustic spectrum effects. The 2D sound spectrum was achieved with the use of the Audacity program (free digital audio editor) and the 3D spectrum using SpectraPLUS-SC (Pioneer Hill Software LLC, USA).

Figure 2 presents the mechanical effects of the ESD (Eng. elastic range, strengthening, deflection control) cement composites with the corresponding acoustic effects and compiled acoustic spectra with various amplitudes corresponding to different mechanical effects (reinforcement breaking, pull-out, macrocracks, and microcracking).

The ESD reinforcement effect is presented by characteristic points f_x (F_x-load, ε_x-deflection, W_x-work) and areas A_X under the load-deflection curve.

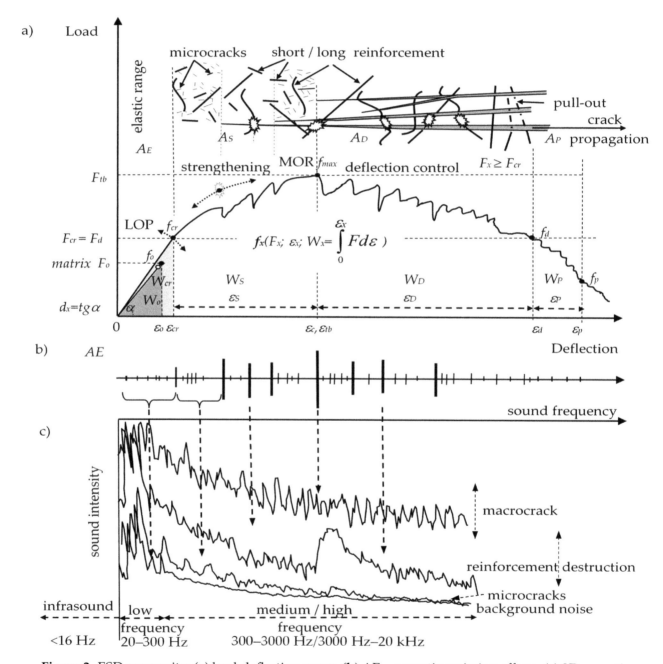

Figure 2. ESD composite: (**a**) load-deflection curve, (**b**) AE—acoustic emission effects, (**c**) 2D acoustic spectrum (frequency amplitude depending on sound intensity) [11].

3. Results

Figure 3 presents the testing area for the four-point bending test with the AE acoustic emission measurements. Subsequent pictures show the characteristic stages of the ESD concrete failure process.

Figure 3b indicates a crack occurring at the f_{cr} point, Figure 3c shows the multicracking (micro- and macrocracks), Figure 3d shows the progressing crack propagation, and Figure 3e shows the sample after the completed test.

The load-deflection curve of the ESD composite and matrix (concrete without the dispersed reinforcement) is presented in Figure 4a. Above the curves, there are the results of the AE measurement with characteristic failure process events.

The ESD effects in the quasi-brittle composite were described with the use of the formula defining any points on the load-deflection curve f_x (load; deflection; absorbed energy). This formula enables the description and assessment of the ESD effects in the elastic range, strengthening, deflection control, and propagation areas. The matrix is characterized by f_{max} (Table 1).

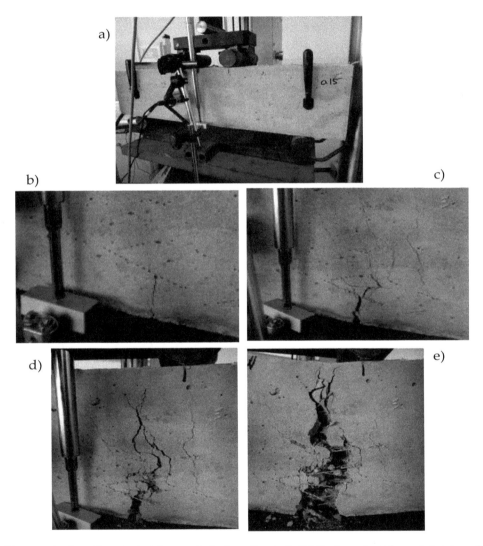

Figure 3. Four-point bending test: (**a**) sample before the test, (**b**) first crack at f_{cr} point, (**c**) multicracking (micro- and macrocracks), (**d**) destruction - propagation process, (**e**) view after the test.

Table 1. Mechanical properties of the matrix (concrete without reinforcement) and ESD composite.

Composite		Load F [N]	Deflection ε [mm]	Work W [kJ]	Ratio	Load	Deflection	Work
matrix	f_{max}	32.9	1.42	23.0	-	-	-	-
ESD	f_{cr}	38.9	1.87	36.3	$A_{E/Ematrix}$	1.2	1.3	1.6
	f_{tb}	49.7	3.46	106.7	$A_{S/E}$	0.3	0.9	1.9
	f_d	38.9	4.06	133.3	$A_{D/E}$	-	0.4	0.7

For the ESD composite, the following results were achieved: f_{cr}, f_{max}, f_d, (Table 1, Figure 4). Comparing the elastic range area of the ESD composite and the matrix, an x-time improvement was achieved for load, deflection, and absorbed energy $A_E/A_{Ematrix}$. $A_{S/E}$ and $A_{D/E}$ are the comparison of the strengthening A_S and deflection control A_D areas to the elastic range A_E. The amount of absorbed energy in the strengthening area was considerably larger than in the deflection control area. The failure process propagation range following the deflection control area was not important in the ESD composites and was omitted.

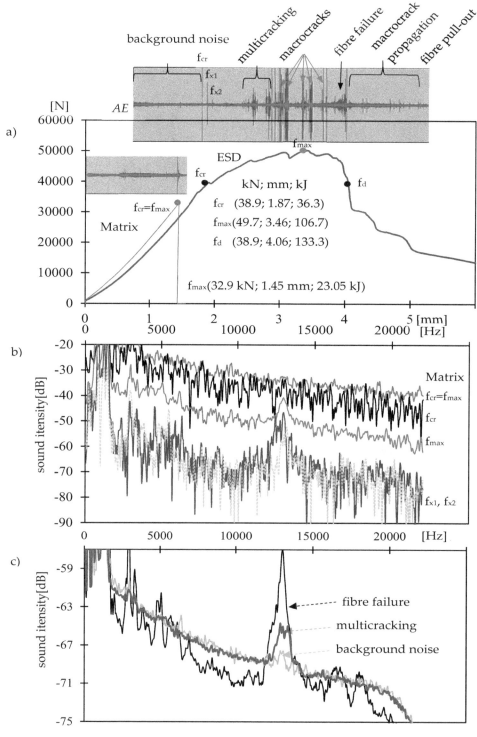

Figure 4. Matrix and ESD composite: (**a**) load-deflection curve, (**b**) 2D spectra of the matrix and ESD composite, (**c**) 2D spectrum of the ESD composite.

Two-dimensional (2D) spectra of the matrix and ESD composite in the frequency range of 0–22 kHz are presented in Figure 4. Figure 4b shows the matrix spectrum for a crack $f_{cr} = f_{max}$, additional spectra of the ESD sample for the first crack, f_{cr}, and subsequent cracks, f_{x1}, f_{x2}, and f_{max}. Figure 4c presents the ESD concrete spectra compared to the spectra of background noise, multicracking, and the fiber failure process. Three-dimensional (3D) spectra of the ESD composite are presented in Figures 5 and 6.

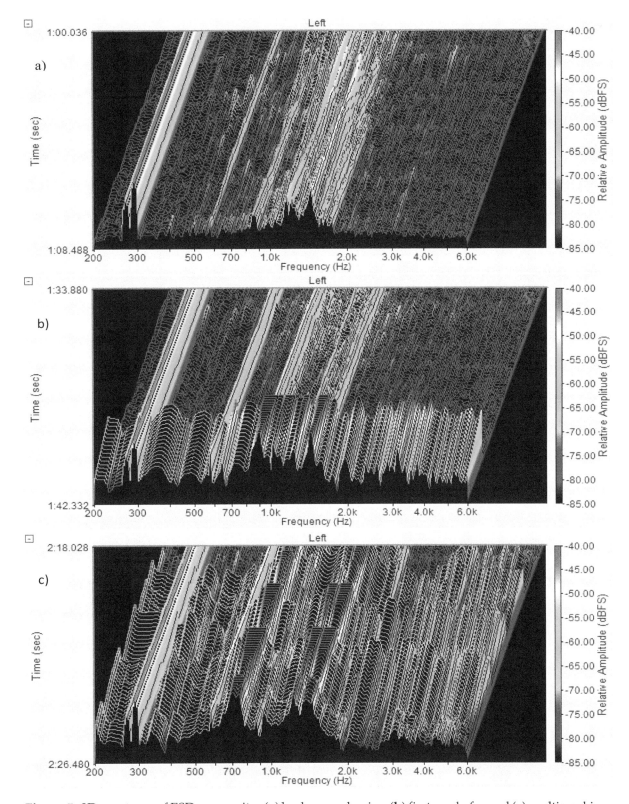

Figure 5. 3D spectrum of ESD composite: (**a**) background noise; (**b**) first crack, f_{cr}; and (**c**) multicracking.

The spectrum frequency range was limited to 200–6000 Hz as the greatest changes were observed within this frequency range in the sound spectra connected with the failure process. Figure 5a shows the background noise spectrum for the ESD composite recorded during the test, Figure 5b presents the first crack, f_{cr}, and Figure 5c shows the multicracking. Figure 6a displays the macrocrack spectrum, Figure 6b shows the reinforcement destruction, and Figure 6c presents the fiber pull-out effect.

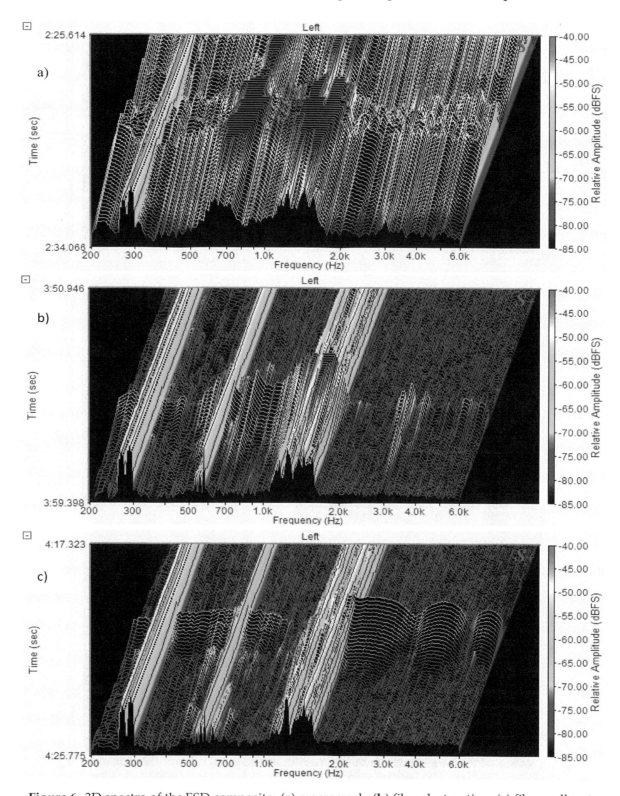

Figure 6. 3D spectra of the ESD composite: (**a**) macrocrack, (**b**) fiber destruction, (**c**) fiber pull-out.

4. Discussion

The existing provisions in the ASTM 1018 standard concerning the identification of characteristic points LOP, MOR, and ASTM indices I_5, I_{10}, I_{15} [35–37] have been extended by adding the possibility of describing any area or point f_x (load, deflection, energy). The introduced f_{cr}, f_{max}, and f_d points enable a precise description of the elastic range, strengthening, deflection control, and propagation areas as well as their comparison with one another with respect to the same sample or different samples.

The obtained results indicate that the ESD composite achieved the best effects in the strengthening area As. The improvement of properties in the elastic range AE was not good enough, which resulted from a low elasticity module of the fibers, causing a more significant deflection in this area. The deflection control range in the A_D area could be improved by increasing the fiber-matrix bond of the dispersed reinforcement. That effect may be achieved by increasing the strength of the composite or modifying the surface and geometry of the fibers.

2D and 3D spectra in the lowest frequency range did not record well (due to the head's measurement range from 0.2 to 20 kHz) and were not taken into account in the interpretation of the results (2D spectrum 0–0.2 kHz). The 3D spectrum frequency range was limited to 200–6000 Hz. Within that frequency range, the greatest changes were observed in the sound spectra connected with the failure process.

Concrete without reinforcement (matrix) is characterized by a catastrophic destruction process. The appearing crack causes a destruction—breaking in halves—of the sample. The sound spectrum (within the range from −20 to −40 dB) corresponding to that process was positioned the highest when compared to other spectra characterizing various destruction processes, as shown in Figure 4a and is characterized by a small range of amplitudes. The background noise spectrum of the matrix has not been presented here, but was similar to the background noise spectrum of the ESD composite.

The ESD composite in the elastic range showed 2D and 3D spectra of background noise located low and within the range of the lowest amplitudes, as seen in Figures 4c and 5a. The sound corresponding to f_{cr} was characterized by significant sound intensity, and the corresponding spectrum was located high, immediately below curve f_{cr} for the matrix, and the range of amplitudes was much larger (Figures 4a and 5b).

Subsequent cracks, f_{x1} and f_{x2}, were situated at the level of background noise spectrum, but with the greatest amplitude range (Figure 4b).

The sound intensity of the multicracking was similar to background noise spectrum, but with a slightly greater range of amplitudes and significant amplitudes in a narrow frequency range (Figures 4c and 5c).

Macrocracks showed the highest sound intensity. What is worth noting is the fact that the corresponding spectra were not characterized by the greatest amplitude range. The spectra were located the highest (Figure 4b).

The fiber pull-out process was characterized by a small range of amplitudes with a wide range of wavelengths (Figure 6c).

The sound spectrum corresponding to fiber failure was positioned low. The lower position of that spectrum when compared to that of the background noise may result from the manner of determining that spectrum. The background noise spectrum was determined with respect to a long period of time before the first crack and refers to a number of background noises in that period, whereas the fiber failure spectrum refers to a single signal. The sound spectrum for fiber failure was characterized by great amplitudes with a strong spike at 12–15 kHz. The average sound intensities of the fiber failure and the background noise were on a comparable level. The analysis of the background noise spectrum for a single sound in a short period of time resulted in a slight decrease in the sound intensity, but the amplitudes did not change significantly.

Frequency spike 12–15 kHz is an interesting, recurring correlation that may be used in the future for the identification of the failure process (Figure 4b,c). It is worth noting that that spike did not occur in the case of a catastrophic fracture f_{cr} (in a sample without reinforcement) in a short period of time. Frequency spike 12–15 kHz occurs in the case of defects generating acoustic effects that last for a longer period of time such as fiber failure, fiber pull out, and micro- and macrocracks that are blocked (stop propagating) or propagate slowly.

The conducted tests confirmed the possibility of identifying the failure process in traditional and ESD cement composites. The analysis of data showed that the 3D spectrum provided better general information for the identification of the failure process at each stage of the process, whereas the 2D spectrum enabled a more precise characterization of each of the sound spectra (sound intensity, amplitudes, frequency range) and their correlation with each of the failure processes. The simultaneous occurrence of failure processes makes their identification difficult. In order to differentiate them accurately, it is necessary to separately identify the sounds that do not overlap.

Summarizing the conducted tests, it can be stated that the analysis of 2D and 3D spectra is a good method of controlling the failure processes in the ESD cement composites. It increases the safety in the use of construction elements and enables correct decision-making in whether and how they should be repaired.

5. Conclusions

The research has allowed for the following conclusions to be formulated conclusions:

(1)   The 3D sound spectrum is a good tool for the observation and identification of failure processes in cement composites.

(2)   It has been noticed that the 2D spectrum enables a more precise identification and description of the sound spectra (sound intensity, amplitudes, frequency range) corresponding to different failure processes.

(3)   It has been suggested that for the analysis and identification of failure processes, both the 2D and 3D spectra should be used at the same time in a wide frequency range.

(4)   The development of ESD cement composites and the identification of failure processes with the use of AE and 2D and 3D spectra enables the control of failure processes (particularly useful in seismic areas or during natural disasters) or decisions of whether and how damaged cement composites should be repaired.

References

1.   Brandt, A.M. Fiber reinforced cement-based (FRC) composites after over 40 years of development in building and civil engineering. *Compos. Struct.* **2008**, *86*, 3–9, ISSN: 0263-8223. [CrossRef]
2.   Kucharska, L.; Brandt, A.M. Pitch-based carbon fibre reinforced cement composites. In *Materials for the New Millenium, Proceedings of the Materials Engineering Conference ASCE, Washington, DC, USA, 10–14 November 1996*; Chong, K.P., Ed.; American Society of Civil Engineers: New York, NY, USA, 1996; Volume 1, pp. 1271–1280.
3.   Yuyama, S.; Ohtsu, M. Acoustic Emission evaluation in concrete. In *Acoustic Emission-Beyond the Millennium*; Kishi, T., Ohtsu, M., Yuyama, S., Eds.; Elsevier: Amsterdam, The Netherlands; Tokyo, Japan, 2000; pp. 187–213.
4.   Landis, E.; Ballion, L. Experiments to relate acoustic energy to fracture energy of concrete. *J. Eng. Mech.* **2002**, *128*, 698–702. [CrossRef]
5.   Ouyang, C.S.; Landis, E.; Shh, S.P. Damage assessment in concrete using quantitative acoustic emission. *J. Eng. Mech.* **1991**, *117*, 2681–2698. [CrossRef]
6.   Ranchowski, Z.; Jóźwiak-Niedźwiecka, D.; Brandt, A.M.; Dębowski, T. Application of acoustic emission method to determine critical stress in fibre reinforced mortar beams. *Arch. Acoust.* **2012**, *37*, 261–268. [CrossRef]
7.   Kim, B.; Weiss, W.J. Using acoustic emission to quantify damage in restrained fiber-reinforced cement mortars. *Cem. Concr. Res.* **2003**, *33*, 207–214. [CrossRef]

8.   Shieldsa, Y.; Garboczib, E.; Weissc, J.; Farnam, Y. Freeze-thaw crack determination in cementitious materials using 3DX-ray computed tomography and acoustic emission. *Cem. Concr. Comp.* **2018**, *89*, 120–129. [CrossRef]

9.   Hoła, J. Acoustic Emission investigation of failure of high-strength concrete. *Arch. Acoust.* **1999**, *24*, 233–244.

10. Logoń, D. Monitoring of microcracking effect and crack propagation in cement composites (HPFRC) using the acoustic emission (AE). In Proceedings of the The 7th Youth Symposium on Experimental Solid Mechanics (YSESM '08), Wojcieszyce, Poland, 14–17 May 2008.

11. Logoń, D. The application of acoustic emission to diagnose the destruction process in FSD cement composites. In *Proceedings of the International Symposium on Brittle Matrix Composites (BMC-11), Warsaw, Poland, 28–30 September 2015*; Brandt, A.M., Ed.; Institute of Fundamental Technological Research: Warsaw, Poland, 2015; pp. 299–308.

12. Schabowicz, K.; Gorzelańczyk, T.; Szymków, M. Identification of the degree of degradation of fibre-cement boards exposed to fire by means of the acoustic emission method and artificial neural networks. *Materials* **2019**, *12*, 656. [CrossRef] [PubMed]

13. Chen, B.; Juanyu Liu, J. Experimental study on AE characteristics of free-point-bending concrete beams. *Cem. Concr. Res.* **2004**, *34*, 391–397. [CrossRef]

14. Reinhardt, H.W.; Weiler, B.; Grosse, C. Nondestructive testing of steel fibre reinforced concrete. In Proceedings of the Brittle Matrix Composites (BMC-6), Warsaw, Poland, 9–11 October 2000; pp. 17–32.

15. Granger, S.; Pijaudier, G.; Loukili, A.; Marlot, D.; Lenain, J.C. Monitoring of cracking and healing in an ultra high performance cementitious material using the time reversal technique. *Cem. Concr. Res.* **2009**, *39*, 296–302. [CrossRef]

16. Ohtsu, M. The history and development of acoustic emission in concrete engineering. *Mag. Concr. Res.* **1996**, *48*, 321–330. [CrossRef]

17. Ranachowski, Z.; Schabowicz, K. The contribution of fibre reinforcement system to the overall toughness of cellulose fibre concrete panels. *Constr. Build. Mater.* **2017**, *156*, 1028–1034. [CrossRef]

18. Logoń, D. FSD cement composites as a substitute for continuous reinforcement. In *Proceedings of the Eleventh International Symposium on Brittle Matrix Composites (BMC-11), Warsaw, Poland, 28–30 September 2015*; Brandt, A.M., Ed.; Institute of Fundamental Technological Research: Warsaw, Poland, 2015; pp. 251–260.

19. Parmar, D. *Non-Destructive Bridge Testing With Advanced Micro-II Digital AE system. Final Report*; Hampton University, Eastern Seaboard Intermodal Transportation Applications Center (ESITAC): Hampton, VA, USA, 2011.

20. Ono, K.; Gołaski, L.; Gębski, P. Diagnostic of reinforced concrete bridges by acoustic emission. *J. Acoust. Emiss.* **2002**, *20*, 83–98.

21. Paul, S.C.; Pirskawetz, S.; Zijl, G.P.A.G.; Schmidt, W. Acoustic emission for characterising the crack propagation in strain-hardening cement-based composites (SHCC). *Cem. Concr. Res.* **2015**, *69*, 19–24. [CrossRef]

22. Watanab, K.; Niwa, J.; Iwanami, M.; Yokota, H. Localized failure of concrete in compression identified by AE method. *Constr. Build. Mater.* **2004**, *18*, 189–196. [CrossRef]

23. Soulioti, D.; Barkoula, N.M.; Paipetis, A.; Matikas, T.E.; Shiotani, T.; Aggelis, D.G. Acoustic emission behavior of steel fibre reinforced concrete under bending. *Constr. Build. Mater.* **2009**, *23*, 3532–3536. [CrossRef]

24. Šimonová, H.; Topolář, L.; Schmid, P.; Keršner, Z.; Rovnaník, P. Effect of carbon nanotubes in metakaolin-based geopolymer mortars on fracture toughness parameters and acoustic emission signals. In Proceedings of the International Symposium on Brittle Matrix Composites (BMC 11), Warsaw, Poland, 28–30 September 2015; pp. 261–288.

25. Shahidan, S.; Rhys Pulin, R.; Bunnori, N.M.; Holford, K.M. Damage classification in reinforced concrete beam by acoustic emission signal analysis. *Constr. Build. Mater.* **2013**, *45*, 78–86. [CrossRef]

26. Aggelis, D.G.; Mpalaskas, A.C.; Matikas, T.E. Investigation of different modes in cement-based materials by acoustic emission. *Cem. Concr. Res.* **2013**, *48*, 1–8. [CrossRef]

27. Elaqra, H.; Godin, N.; Peix, G.; R'Mili, M.; Fantozzi, G. Damage evolution analysis in mortar, during compressive loading using acoustic emission and X-ray tomography: Effects of the sand/cement ratio. *Cem. Concr. Res.* **2007**, *37*, 703–713. [CrossRef]

28. Shiotani, T.; Li, Z.; Yuyama, S.; Ohtsu, M. Application of the AE Improved b-Value to Quantitative Evaluation of Fracture Process in Concrete Materials. *J. Acoust. Emiss.* **2004**, *19*, 118–133.

29. Ohtsu, M. Determination of crack orientation by acoustic emission. *Mater. Eval.* **1987**, *45*, 1070–1075.

30. Ohno, K.; Ohtsu, M. Crack classification in concrete based on acoustic emission. *Constr. Build. Mater.* **2010**, *24*, 2339–2346. [CrossRef]

31. Ohtsu, M. Elastic wave methods for NDE in concrete based on generalized theory of acoustic emission. *Constr. Build. Mater.* **2016**, *122*, 845–855. [CrossRef]

32. Van Steen, C.; Verstrynge, E.; Wevers, M.; Vandewalle, L. Assessing the bond behaviour of corroded smooth and ribbed rebars with acoustic emission monitoring. *Cem. Concr. Res.* **2019**, *120*, 176–186. [CrossRef]

33. Tsangouri, E.; Michels, L.; El Kadi, M.; Tysmans, T.G.; Aggelis, D. A fundamental investigation of textile reinforced cementitious composites tensile response by Acoustic Emission. *Cem. Concr. Res.* **2019**, *123*, 105776. [CrossRef]

34. Kumar Das, A.; Suthar, D.; Leung, C.K.Y. Machine learning based crack mode classification from unlabeled acoustic emission waveform features. *Cem. Concr. Res.* **2019**, *121*, 42–57.

35. ASTM 1018. *Standard Test Method for Flexural Toughness and First Crack Strength of Fiber–Reinforced Concrete*; ASTM 1018: West Conshohocken, PA, USA, 1992; Volume 04.02.

36. EN 14651. Test method for metallic fibre concrete. In *Measuring the Flexural Tensile Strength (Limit of Proportionality (LOP), Residual)*; European Committee for Standardization (CEN): Brussels, Belgium, 2005.

37. Japan Concrete Institute Standard (JCI). *Method of Test for Bending Moment-curvature Curve of Fiber-reinforced Cementitious Composites, S-003-2007*; Japanese Concrete Institute Standard Committee: Tokyo, Japan, 2007.

8

Non-Destructive Testing of a Sport Tribune under Synchronized Crowd-Induced Excitation using Vibration Analysis

Karol Grębowski [1,*], **Magdalena Rucka** [2]🆔 **and Krzysztof Wilde** [2]

[1]   Department of Technical Fundamentals of Architectural Design, Faculty of Architecture, Gdansk University of Technology, Narutowicza 11/12, 80-233 Gdansk, Poland

[2]   Department of Mechanics of Materials and Structures, Faculty of Civil and Environmental Engineering, Gdansk University of Technology, Narutowicza 11/12, 80-233 Gdansk, Poland

*   Correspondence: karol.grebowski@pg.edu.pl

Abstract: This paper presents the concept of repairing the stand of a motorbike speedway stadium. The synchronized dancing of fans cheering during a meeting brought the stand into excessive resonance. The main goal of this research was to propose a method for the structural tuning of stadium stands. Non-destructive testing by vibration methods was conducted on a selected stand segment, the structure of which recurred on the remaining stadium segments. Through experiments, we determined the vibration forms throughout the stand, taking into account the dynamic impact of fans. Numerical analyses were performed on the 3-D finite element method (FEM) stadium model to identify the dynamic jump load function. The results obtained on the basis of sensitivity tests using the finite element method allowed the tuning of the stadium structure to successfully meet the requirements of the serviceability limit state.

Keywords: non-destructive testing; reinforced concrete grandstand stadium; vibration analysis; crowd-induced excitation; structural tuning

1. Introduction

Stadiums are sport objects that require a particular level of attention to ensure human security. During the last century, many tragedies have been caused mainly by negligence at the stage of designing the structure. Nowadays, many structural projects are sought to be optimized. The results are slender, more effective and economical objects; however, they are often sensitive to dynamic impacts.

Since 1902, more than 20 construction disasters of stadium stands have taken place around the world [1,2]. These range from the stand collapse at Ibrox Park stadium in Glasgow (Scotland) on April 5, 1902 during the Scotland vs. England match, which resulted in 26 dead and 550 injured supporters, to the tragedy that took place on November 26, 2007 at the Salvador stadium (Brazil) during the Bahia vs. Vila Nova match, which resulted in eight deaths and 150 injured due to the collapse of the stand.

Strong vibrations caused by moving people mainly occur in structures with low rigidity and mass. Stadiums are an example of this type of structure. High dynamic loads occur, caused by jumping fans cheering their teams. This type of load is often omitted during the structure design, because procedures in current standards or guidelines regarding the jump load are limited.

In recent years, many studies have been devoted to the dynamic impact of stadium structures. Development in sensor technology, through the supply of adapted vibration transducers (low-level vibration, high-level/shock vibration, near-static sensitivity, etc.) in conjunction with the dissemination of digital processing algorithms, has led to a notable increase in the study of vibrations caused by crowd-induced excitation in structures [3]. Modal identification results can be used as part of the vibration reduction methodologies [4]. In [5], it was found that crowd occupation can significantly alter the modal properties of a stadium, and that the changes vary according to crowd configuration. A large number of publications on structural response and stadium vibration resistance due to crowd-induced loads have shown that the dynamic response of stadiums depends not only on the basic mechanical characteristics, i.e., mass, stiffness and damping, but also on the nature of the load, which can become complex in the case of jumping supporters (e.g., [6]). In order to limit the vibrations of the structure, one of the countermeasures is frequency tuning a structure [7,8]. A comprehensive analysis of the literature on dynamic performance tests of existing stadium structures was conducted in [9]. It was concluded that the available knowledge on this subject is not yet sufficiently advanced and thus jump load is not currently included in most design standards. Therefore, the dynamic identification based on non-destructive testing methods is increasingly popular and effective in civil engineering research. Several literature studies have shown how non-destructive testing and structural health monitoring can be efficiently used to assess the durability of reinforced concrete structures [8–15].

An example of stand failure due to the dynamic impact of fans is the case of the Swiss Krono Arena motorbike speedway stadium in Zielona Góra (Poland). The new stand was built in 2009–2010. In July 2010, it was put into service. The stand is a reinforced concrete structure that is used in a very specific way. Speedway meetings are a sport discipline that arouse great emotions and interest among supporters. One of the basic forms of team cheering at the Swiss Krono Arena is the so-called "Labado dance". The dance is considered as almost a hymn among speedway fans. In this dance, fans put their hands on their neighbor's shoulders and in the rhythm of the animator's drumming, they jump simultaneously to the song "(...) we dance labado, labado, (...)". In 2010 it was noticed that the fans' dance causes an increase in the vibrations of the structure, which led to a discussion on the safety of using the new stand. Research on the assessment of the harmfulness of vibrations was carried out in 2010–2011 by a team from University of Zielona Góra, and in the years 2011–2012 by a team from Gdańsk University of Technology. The Labado dance appeared to be dangerous for the stand's structure, because by jumping fans caused a vertical periodic force, which when synchronized movements of a large number of people could lead to resonant vibrations, possibly ending with the collapse of the structure. A large number of people generating regular dynamic excitations could be able to destroy any bridge or reinforced concrete stadium stand, which is why this phenomenon is considered vandalism.

This paper presents non-destructive testing of a sport facility subjected to dynamic interactions in the form of jumping fans. Experimental tests were conducted in the field on the Swiss Krono Arena speedway stadium in Zielona Góra. Later on, numerical simulations were performed to determine the dynamic characteristics of the structure. Validation was carried out by comparing the numerical model with the results obtained in the field. After the validation of the numerical model, the identification of the jump load dynamic function was determined based on laboratory tests. Lastly, the stadium stand tuning model was re-tested in situ after structure strengthening.

2. Theoretical Background of the Experimental Modal Analysis

Modal analysis is a method of determining dynamic properties of structures (i.e., natural frequencies, modal shapes and damping coefficients) under vibrational excitation. The analysis involves the registration of vibrational signals and the application of different signal processing techniques. In general, two types of modal analysis can be distinguished: operational modal analysis (OMA) [16–19] and experimental modal analysis (EMA) [20–24]. Environmental excitations (e.g., wind, sea waves, micro-seismic vibrations, traffic loads, etc.) are used in the OMA technique, and only the structure's response is recorded. In the case of EMA, both excitation and response are measured. The most common approach in EMA is the impulse test, which is based on the excitation of the structure using a modal hammer and the harmonic test, in which electromechanical or piezoelectric actuators are usually applied (Figure 1).

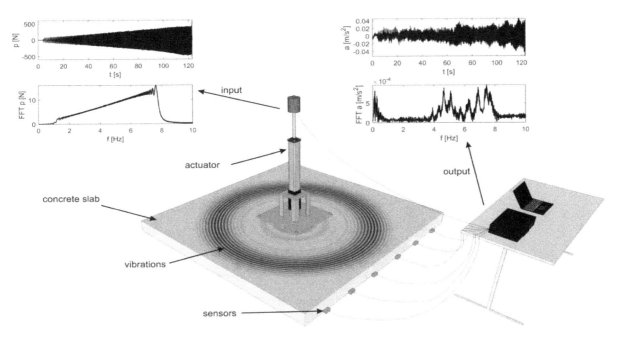

Figure 1. Scheme of the experimental modal analysis technique.

First, the input force $p(t)$ and output displacement $u(t)$ time signals are measured in EMA, and then they are transformed into the frequency domain by Fourier transform, resulting in $P(\omega)$ and $U(\omega)$ signals (Figure 1). The next step is the calculation of the frequency response function (FRF) in all measured points, which is a ratio of the j-th output and the k-th input:

$$H_{jk}(\omega) = \frac{U_j(\omega)}{P_j(\omega)}. \tag{1}$$

In the practical application of EMA, acceleration signals are often measured instead of displacement signals, as they are more convenient in analytical works. Then, the acceleration function can be defined as:

$$A_{jk}(\omega) = \frac{\ddot{U}_j(\omega)}{P_j(\omega)}. \tag{2}$$

Both the receptance and acceleration functions describe the same dynamic properties of the system and the relationship between them is given by:

$$A_{jk}(\omega) = -\omega^2 H_{jk}(\omega). \tag{3}$$

Based on the frequency response functions (FRFs), resonance frequencies, mode shapes and damping coefficients can be identified (e.g., [25,26]).

3. Experimental Dynamic Identification of the Grandstand

3.1. Description of the Structure

The Swiss Krono Arena motorbike speedway stadium is located in Zielona Góra (Poland). The grandstand was erected in 2009–2010 and its geometry is a circular sector with a center angle of 150° (Figure 2).

The stand consists of a reinforced concrete structure in which prefabricated under-seat beams rest on the main beams of reinforced concrete girders. The main girder is composed of two parts and it is supported on three reinforced concrete columns. The first part of the main girder (BR1) is an inclined simply supported beam, located in the bottom part of the stand. The second part of the girder (BR2), with a variable cross-sectional height, also works as a simply supported beam, and it is finished with a cantilever beam located in the upper part of the stand.

(a)   (b)

Figure 2. Swiss Krono Arena: (**a**) the view on the grandstand, (**b**) the location of the grandstand on the land development plan.

Both girders are connected to each other by an articulated joint. The main beams are based on columns via elastomeric pads. Steel anchors prevent horizontal displacement of the main beams against columns. The columns and the monolithic wall supporting the main girder are located on reinforced concrete strip foundations. The walls of additional objects are made of silicate brick, while the stairs structure is made as a prefabricated reinforced concrete element (Figure 3).

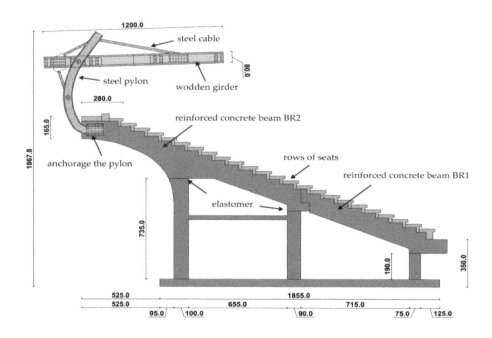

Figure 3. Cross-section of the stand (dimensions in cm).

The roof structure over the stand consists of three elements: steel columns, which are attached to the cantilever ends of the main load-bearing girder; the roof, which consists of wooden girders; and purlins, to which the covering trapezoidal sheeting and cables are attached. The cables connect the wooden roof structure with steel columns. The steel supporting columns of the roof structure are braced with steel braces. Roof bracings are also used in the wooden roof structure.

3.2. Test Procedure and Vibration Measurement

The measurements were conducted on a selected section of the stand, as shown in Figure 4. The experimental program included a sweep sine test and two types of people-induced vibrations, namely single jumps and synchronous Labado dancing.

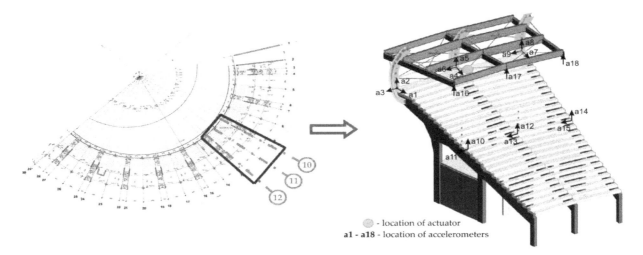

- location of actuator
a1 - a18 - location of accelerometers

Figure 4. The stand section selected for field tests.

Triaxial piezoelectric accelerometers 356B18 (PCB Piezotronics, Inc., Depew, NY, USA) were used for the measurement of vibrations. They were attached to both the concrete part of the stand as well as to the roof structure. Vibrations signals were registered at nine points (Figure 4), in one, two and three directions. In total, 18 acceleration signals were acquired (a1 to a18). Data acquisition and signal conditioning were performed by the LMS SCADAS portable system (Siemens, Leuven, Belgium). The sampling frequency was set as 256 Hz.

In the first stage, the experimental modal analysis was conducted. The harmonic load was excited by means of the electromechanical actuator shown in Figure 5a. The excitation signal, created by an arbitrary signal generator, was a sweep sine of a smoothly adjustable frequency from 1 Hz to 8 Hz (Figure 5b).

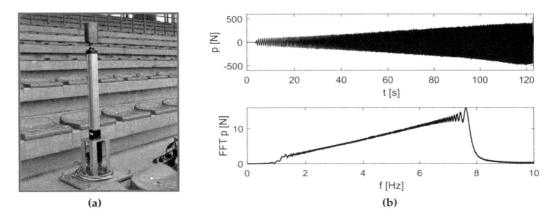

(a)   (b)

Figure 5. Experimental setup: (a) electromechanical actuator, (b) excitation signal in the time and frequency domains.

Dynamic parameters were determined based on the frequency response functions, according to procedure described in Section 2. The imaginary parts of the FRFs for signals registered on the girder and on the roof are given in Figure 6. Several peaks can be distinguished. The experiment allowed the identification of three natural frequencies in the range from 2 to 5 Hz (Figure 6). Their values were: 3.36 Hz, 3.92 Hz and 4.68 Hz. Experimentally determined mode shapes are presented in Figure 7. The directions and values of displacements for particular degrees of freedom are shown by means of vectors and numerical values. The shapes of modes of the girder and roof are similar, wherein the amplitude for the roof is much larger. The dominant vibration types for each identified form are vertical mode shapes. The main displacements of the girder and roof occurred as a result of vertical motion.

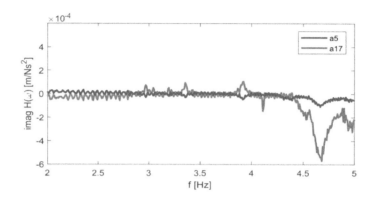

Figure 6. Imaginary parts of the frequency response function (FRF) for acceleration signals a5 (girder) and a17 (roof).

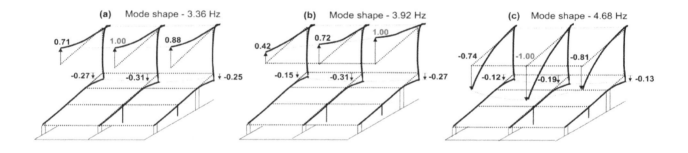

Figure 7. Identified mode shapes and natural vibrations: (**a**) 3.36 Hz, (**b**) 3.92 Hz, (**c**) 4.68 Hz.

The next step included the non-destructive evaluation of the stand throughout the analysis of the forced vibrations. The impact of the real dynamic forces was implemented thanks to the fans who, at the request of the club authorities, came to participate in the measurements. About 400–450 people participated in the research. For the purpose of the research, the fans presented two forms of active support: the so-called Scotland-type jumps and the Labado dance. By making the Scotland-type jumps, the fans performed single jumps to a drum beat. They started very slowly and finished with asynchronous jumps. In the Labado dance, almost all fans performed synchronous jumps with a constant frequency.

During the research, one of the most important tasks was to find the most unfavorable load combination. Different combinations of fans' arrangement on stadium stand rows were tested. One of the most unfavorable combinations, resulting in the greatest vibrations of the cantilever part of the stand, was selected for the purpose of the tests. Figure 8 presents the most unfavorable setting in which the dancing fans are located at the bottom of the stand (beam BR1) and on the upper part of the stand (four rows at the cantilever part of beam BR2).

(a)   (b)

Figure 8. Dynamic tests with fans: (**a**) experimental setup, (**b**) scheme showing the arrangement of people during tests (orange color—rows indicating the most unfavorable fan positions).

The results of the in situ tests are shown in Figures 9 and 10 for the Scotland-type jumps and the Labado dance, respectively. In the case of the Scotland-type jumps, many jump events are visible, with decreasing time between individual jumps. In the Fourier transformation results, two wide peaks are visible, the first around 2.5 Hz and the second around 4.5 Hz. The Labado dance, in which jumps were more regular, resulted in narrower peaks concentrated at 2.2 and 4.4 Hz.

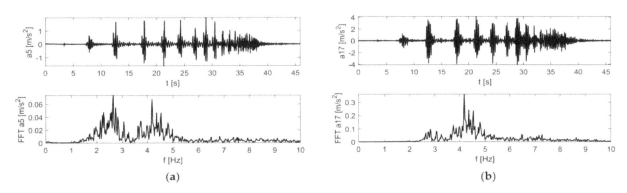

(a)   (b)

Figure 9. Accelerations registered during dynamic tests with fans (Scotland-type jumps) at: (**a**) girder, (**b**) roof.

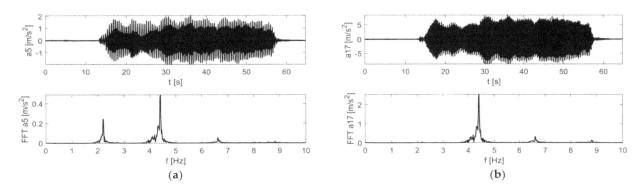

(a)   (b)

Figure 10. Accelerations registered during dynamic tests with fans (Labado dance) at: (**a**) girder, (**b**) roof.

Based on the obtained results, it was found that the frequency of vibrations of the end of the cantilever girder was equal to 2.2 Hz, coinciding with the frequency of the exciting force, which was caused by jumping supporters. The end of the roof structure vibrated at 4.4 Hz, which means that the roof vibrated twice as fast as the end of the girder.

4. Finite Element Method Modeling

4.1. Identification of the Jump Load Function

Experimental tests were conducted in the laboratory of the Gdańsk University of Technology in order to identify the jump load function. The test object was a composite plate with dimensions of 190 cm × 7.5 cm × 40 cm (Figure 11a). The plate was made of two sheets of poplar plywood of a thickness of 2 cm. Between them, C 20 class wood with a 3.5 cm thickness was inserted. Material parameters adopted for testing were: bending strength 20 MPa, stretching along fibers 12 MPa, compression along fibers 19 MPa, modulus of elasticity along fibers 10 GPa, average density 330 kg/m^3. The plate was placed on supports using elastomer pads with dimensions of 7 cm × 40 cm × 0.5 cm.

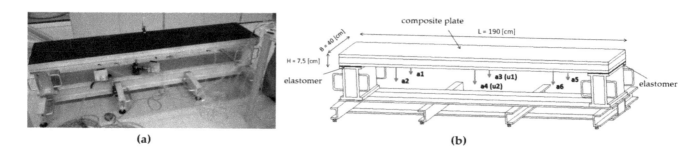

Figure 11. Experimental setup: (**a**) general view, (**b**) geometry of the tested plate and the location of points for measurement of accelerations (a1 to a5) and displacements (u1, u2).

The plate was subjected to dynamic tests. The acceleration measurements (a1 to a6) were performed at six points by means of the PCB accelerometers model 356A16, while displacement measurements were taken at two points (u1 to u2) by means of optoNCDT 1302 laser sensors, as shown in Figure 11b. The LMS SCADAS vibration measurement system was used to record the time histories. The research agenda included measurements of free vibrations as well as measurements of forced vibrations in the form of synchronized jumps performed by one, two and three people (Figure 12).

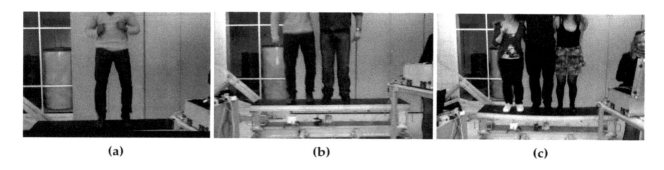

Figure 12. Implementation of excited vibrations in the form of rhythmical jumps of (**a**) one person, (**b**) two people, (**c**) three people.

The first natural frequency of the plate was identified using a typical impact test. Its value was 26 Hz. Figure 13 present the results of vibrations in the form of accelerations and displacements recorded during the jumps of three people. Apart from the jump frequency, further components of harmonically excited vibrations are visible in the Fourier transform diagrams. Eight harmonic components are visible on the Fourier transforms diagrams on acceleration, while the Fourier transforms of displacement signals show four harmonic components.

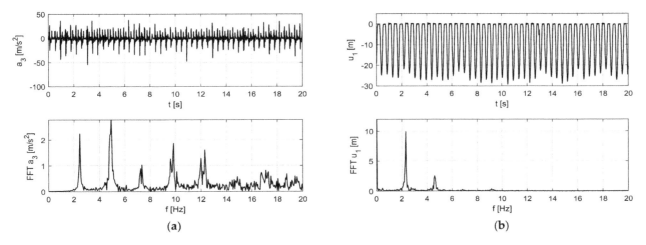

Figure 13. Results of vibrations in the form of (**a**) accelerations and (**b**) displacements recorded during the jumps of three people.

The dynamic impact function during the jumping was described by means of an impulse shaped in half of the sinus function period form. The loading cycle for the Labado dance includes the flight phase and the contact phase. On the basis of the measurement results, the duration of the periodic force full period was identified as 0.45005 s, which is composed of: the phase of load contact with the plate equal to 0.27003 s and the flight phase equal to 0.18002 s. The shape of a single dance cycle is shown in Figure 14a. The in situ tests indicated the jump frequency to be 2.2 Hz. The amplitude of the load function was determined as the average value of the weight of fans participating in the experiment, and was equal to approximately 0.78 kN.

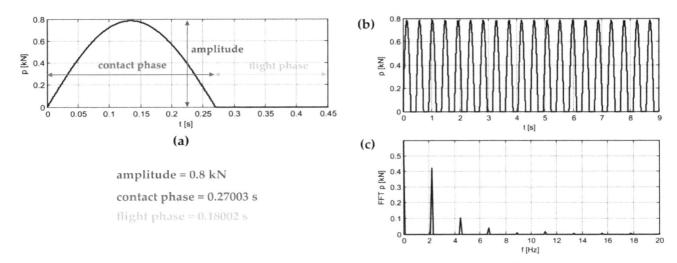

amplitude = 0.8 kN

contact phase = 0.27003 s

flight phase = 0.18002 s

Figure 14. Labado function: (**a**) one jump, (**b**) 20 jumps, (**c**) Fourier transformation of the Labado function.

Figure 14b,c shows excitation over time during 20 rhythmic jumps and the corresponding Fourier transform. The Fourier transform graph proves that, in addition to the 2.2 Hz jump frequency, further harmonics are present in the spectrum. The occurrence of higher harmonics is characteristic of this periodic signal. Higher harmonic components are also present on transforms from the experimental data.

In order to validate the laboratory tests results, numerical simulations for the composite plate were performed. The calculated plate length was set as 183 cm. It was assumed that the plate works in a simply supported scheme. In the discretization process, the slab was divided into 504 solid elements in accordance with the results obtained during the convergence division analysis. Numerical calculations were performed for excitation caused by one, two and three people (Figure 15).

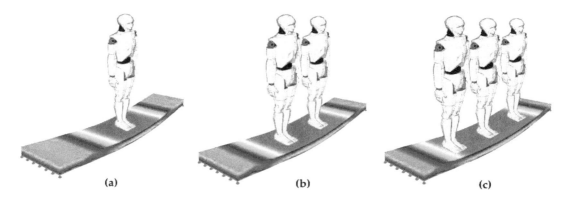

Figure 15. Numerical simulations of the implementation of excited vibrations in the form of rhythmic jumps performed by: (**a**) one person, (**b**) two people, (**c**) three people.

Figure 16 presents a comparison of the numerical and experimental results in the form of displacement in time, registered at the middle of the span, and their Fourier transforms for the cases of one and three people jumping. In the flight phase, plate vibrations occurred at a frequency of 26 Hz for both the experimental and numerical results. Slight phase shifts between the experimental and numerical graphs can be observed. The reason for the shifts lies in the fact that the real jumps were not perfectly repetitive. A very good experiment consistency of the numerical simulations was obtained, which proves the correctness of the assumed load model for the Labado dance. The excitation frequency was approximately 2.2 Hz during both the experimental and numerical tests. Later in the article, the obtained Labado dance load model is used to simulate vibrations and structure tuning of the stand at the Swiss Krono Arena speedway stadium.

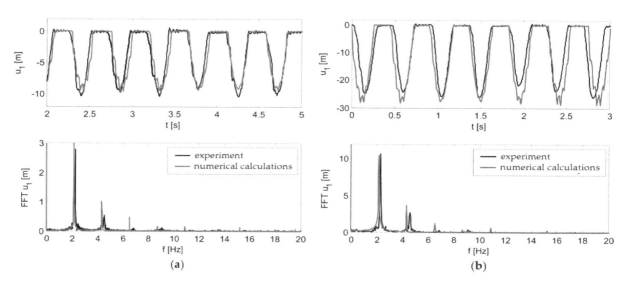

Figure 16. Numerical model validation results for jumps performed by: (**a**) one person, (**b**) three people.

4.2. Numerical Analysis

The creation of the 3-D FEM model of the stand at the Swiss Krono Arena stadium was implemented in two stages with the use of the commercial FEM software SOFiSTiK (SOFiSTiK AG, Oberschleißheim, Germany). In the first stage, models of the main structural parts of the stand were performed using solid elements (Figure 17b), while in the second stage they were performed with the use of beam elements (Figure 17a). The construction of beam models was verified based on the results of numerical analyses carried out on solid models, due to which a very high consistency was obtained between the results of the static and dynamic tests. The FEM calculation model of the entire prefabricated 3-D stadium contains approximately 3008 beam elements and 2001 nodes, in accordance with the results collected during the convergence analysis. Substitute cross-sections were applied in the beam model, in which the cross-section of concrete and steel was replaced with a representative cross-section of a homogeneous material. The under-seat beams of the stand were modeled as beam elements. The columns were articulately connected with the girders, and an articulated joint can also be found at the connection of the two main girders. Glued laminated timber girders were modeled as beam elements, and steel bracings and bar elements model the truss elements. In models that consider various stand strengthening methods, the tensile elements are modeled using tensile elements (cable) taking into account the prestressing force. Non-structural elements of significant weight (e.g., non-structural steel columns) were modeled as concentrated masses applied at the place of mounting. The material parameters taken for the purpose of the numerical analysis were adopted in accordance with the construction and structural design.

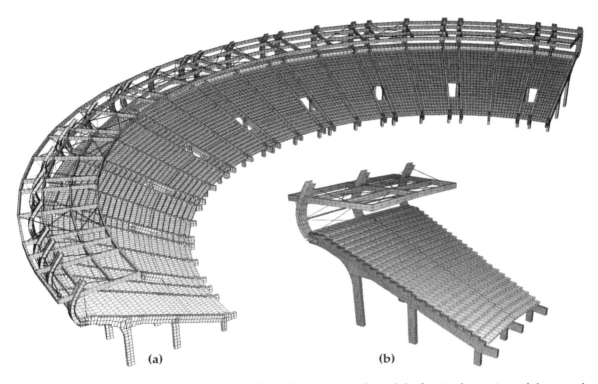

(a)　　**(b)**

Figure 17. (**a**) Numerical model of the stadium, (**b**) numerical model of a single section of the stand.

The eigenvalues were calculated at the beginning. Thirty eigenvalues of the system were calculated. The total mass contributions in the dynamic response were more than 90% in directions: X (longitudinal), Y (transverse) and Z (vertical), which complied with the standard recommendations. The first three eigenvalues and their corresponding natural vibration frequencies determined on the basis of modal analysis are presented below (Figure 18).

The determined basic natural vibration frequency of the stand construction with a roof and without strengthening was used to simulate forced oscillations and to determine the degree of stiffness reduction of the concrete cross-section due to scratches. For the purpose of the analysis, the structure was loaded with its own weight and an evenly distributed load on the stands of 8.0 kN/m^2, in accordance with PN-EN-1991-1-1 [27]. It was assumed that this load changes harmonically according the to the relation $p(t) = p\sin(\omega t)$, where the amplitude $p = 8.0$ kN/m^2 and the circular frequency $\omega = 15.71$ rad/s (which corresponds to the frequency $f = 2.3$ Hz). It has been assumed that the load of such nature lasts $t = 60$ s, the same as in the experiment.

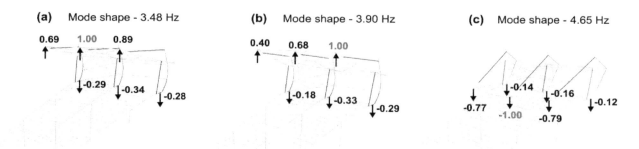

Figure 18. Mode shapes and frequencies of natural vibrations of a single stand section obtained from numerical simulations: (**a**) 3.48 Hz, (**b**) 3.90 Hz, (**c**) 4.65 Hz.

Figure 19 shows the stresses in the BR2 girder. Figure 19b shows equivalent von Mises stresses, Figure 19c presents the main tensile stresses and Figure 19d demonstrates the normal horizontal stresses in the girder plane (in the global coordinate system). The main tensile stress in the upper part of the BR2 beam was greater than the concrete tensile strength, i.e., $\sigma_1 = 8\,\text{MPa} > \sigma_{cr} = f_{ctm} = 3.2\,\text{MPa}$, which means that the reinforced concrete section becomes scratched.

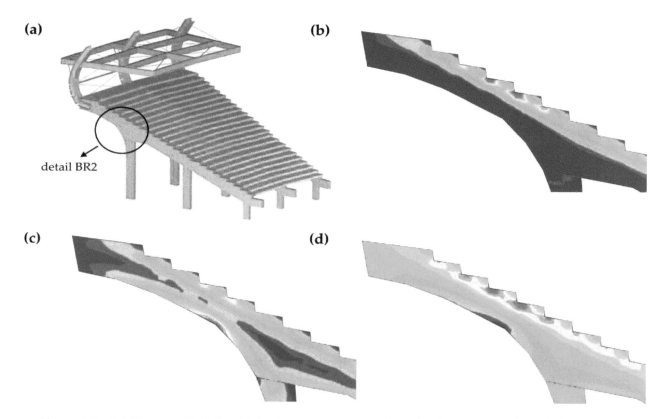

Figure 19. (**a**) Placement of the highest stress concentrations in the main girder of the stadium stand. (**b**) von Mises map of stresses in the BR2-reinforced concrete girder area (red color—maximum stress equal to 10 MPa). (**c**) Map of the main tensile stresses within the BR2-reinforced concrete girder (red color—maximum tensile stress equal to 8 MPa). (**d**) Map of the horizontal normal stress s11 within the BR2-reinforced concrete girder (red color—maximum tensile stress equal to 8 MPa, blue color—maximum compressive stress equal to 8 MPa).

The decrease in the stiffness of the scratched girder was equal to about 50%. The decrease of the stiffness of the reinforced concrete section after scratching was determined on the basis of the obtained stress distribution in the prefabricated beam supporting the stands. During the numerical simulation, the basic natural vibration frequencies of the structure were determined, taking into account the stiffness reduction of the reinforced concrete cross-section due to scratching.

In the further stage of the numerical analysis, the stadium structure was loaded with jumping fans. Each of the fans was replaced by a single concentrated force with a load value of 0.8 kN, as this load corresponded to the average human weight. A jump function determined during experimental research was assigned to every concentrated force. The duration time was equal to 40 s. The forces were set in the most unfavorable load combination, i.e., at the bottom of the stand (chairs on the whole BR1 beam width) and on the upper part of the stand (four rows above the cantilever section). Simulations of stand vibrations for the excitation form of non-synchronized Scotland-type jumps were carried out at the beginning (Figure 20). Afterwards, an analysis for the form of the synchronized Labado dance was performed (Figure 21).

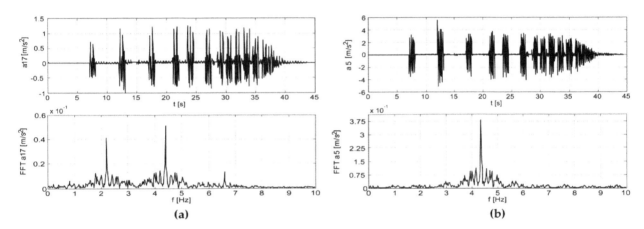

Figure 20. Results of numerical simulations from the excitation of vibrations by jumping fans (Scotland-type jumps): (**a**) girder, (**b**) roof.

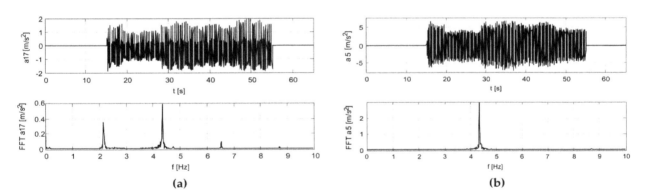

Figure 21. Results of numerical simulations from the excited vibrations caused by jumping fans (Labado dance): (**a**) girder, (**b**) roof.

On the basis of the results obtained during non-destructive tests and numerical simulations, it was concluded that as a result of dynamic interactions, the phenomenon of higher harmonic resonance occurred on the stand and the roof structure. The vibrations frequency at the end of the roof was equal to approximately 4.7 Hz, which coincides with the second harmonic of the excitation force, i.e., 4.4 Hz, caused by jumping fans. The form and dynamic characteristics of the Labado dance can be considered as vandalism. A large number of jumping fans with their synchronized movements drives the tribune structure's vibrations, and their regular dance could cause the collapse of the entire object. The results of non-destructive tests and numerical simulations showed a significant excess of the serviceability limit state.

5. Structural Tuning

An original method of stand structural tuning was proposed in response to the excess of the serviceability limit state. This method assumed the implementation of wall bracings stiffening the steel roofing columns, as well as roof bracings on the wooden roof structure (Figures 22 and 23). The main goal of the concept of bracing of all columns and roof girders was to significantly limit the free end of the wooden roof vibration amplitudes, and hence reduce the concrete cantilever displacements. The bracing cross-sections were the same as those of the bracings already installed in the stand structure. The bracings were made from Macalloy bar rods with a yield point of 520 MPa and tensile strength of 690 MPa.

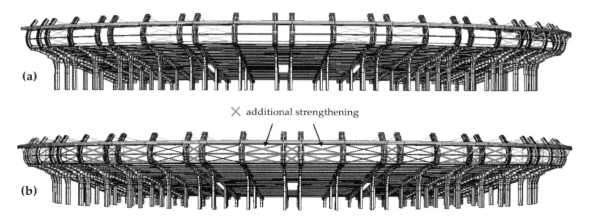

Figure 22. Additional wall bracings (**a**) before structural tuning, (**b**) after structural tuning.

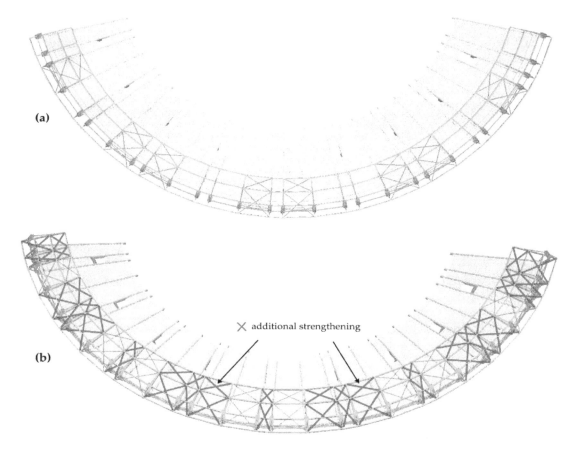

Figure 23. Additional roof bracings (**a**) before structural tuning, (**b**) after structural tuning.

The stand FEM model with additional roof bracings and vertical bracings was implemented in SOFiSTiK software. The beam stand model was made on the basis of the object design documentation (Figure 24). The basic model parameters were determined on the basis of detailed models with the use of solid elements, in which the location of reinforcement bars and details of beams support, such as elastomer pads and anchors, were taken into account. The models include the scratching of concrete elements by reducing the value of the concrete modulus of elasticity. The FEM model parameters were updated on the basis of the obtained measurement data.

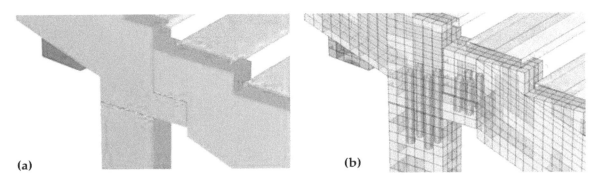

(a)   **(b)**

Figure 24. Detailed view of the stand's load-bearing beams: (**a**) support, (**b**) connection.

Figure 25 presents the numerical results obtained during the calculations for the load induced by excitation fans, for the purposes of post-strengthening the stadium structure. The exemplary first form of natural vibration frequencies determined numerically is shown in Figure 26. The dynamic analysis proved that in the range of 2.9 Hz to 5 Hz, there are at least 15 natural vibration frequencies. Such a large number of natural vibration frequencies is ordinary for a spatial structure made of repetitive elements. The strengthening target was to improve the spatial structure performance and reduce natural vibration frequencies within the range of 4.3–4.7 Hz, in order to eliminate the phenomenon of the beat and characteristic of the excitation of two very close natural vibration frequencies.

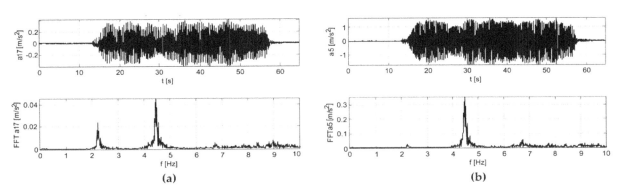

(a)   **(b)**

Figure 25. The results of numerical simulations after stand structural tuning from vibrational excitation by the jumping supporters: (**a**) girder, (**b**) roof.

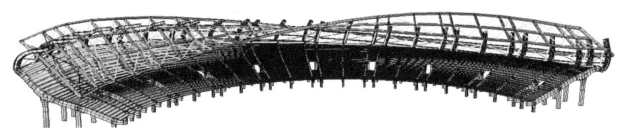

Figure 26. The first numerical form of natural vibrations of the stand after structural tuning.

Figure 27 shows the visualization of vibrations induced by synchronized dancing of the crowd in the stands, located in rows 22–25 throughout the stadium. The applied dynamic load was a periodic function consisting of a contact phase (sinusoidal impulse) and a flight phase (no load). The load function of a jumping crowd is typically not included in standard regulations. The integration of motion equations was carried out using the Newmark–Wilson method. Figure 27a shows a plane view of the relations of the object degrees of freedom vibration amplitudes in the case when the roof and vertical bracings were fixed in accordance with the design, i.e., every two fields. Figure 27b presents the nature of vibrations for bracings installed in each roofing field. The bracings installation resulted in the roof structure stiffening and in the change of its vibrational spatial character. The calculated amplitude of vibrations from the fans' synchronized dance on the whole structure in rows 22–25 was equal to 3.6 mm at the end of the concrete cantilever, and 5 mm at the free end of the roof. The strengthening method caused a slight reduction in the vibrations of the concrete cantilevers and a significant reduction in the vibration amplitudes of the free end of the roof. A requirement for the correct bracing performance during vibrations was the presence of an initially prestressing force. However, the reduction of concrete cantilevers vibration amplitudes was problematic, because they took over the main part of the dynamic forces generated by fans dancing directly on them.

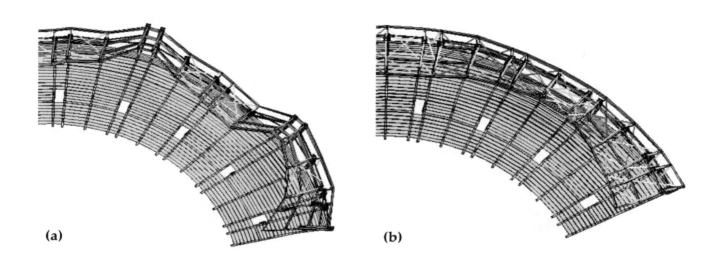

(a)   (b)

Figure 27. Top view of the stand's roof loaded with synchronous jumps of people showing the vibration amplitudes (**a**) before structural tuning, (**b**) after structural tuning.

Repeated in situ tests were carried out in order to verify the correctness of the original concept of stadium stand tuning. The results of measurements made after structural tuning are shown in Figures 28 and 29. It can be seen that that frequencies identified during crowd-induced vibrations are similar to the measurements before tuning (see Figures 9 and 10), because the same nature of the jumping load. It is important, however, that the natural frequencies of the structure were shifted after structural tuning (see Figure 28), decreasing resonance significantly. The individual resonance zones moved towards the higher frequencies.

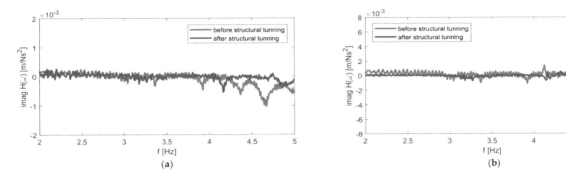

Figure 28. Experimental transition functions for the end of the concrete cantilever **(a)** before and **(b)** after stand structural tuning.

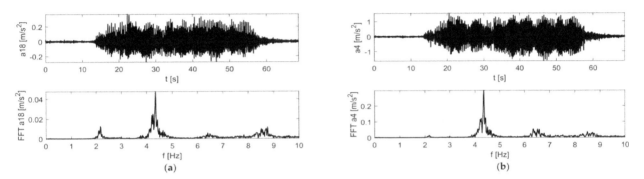

Figure 29. Results of in situ tests after the stand structural tuning from vibration excitation with the jumping supporters: **(a)** girder, **(b)** roof.

6. Conclusions

In this study, we adopted the vibration testing method for non-destructive diagnostics of a stadium stand. Experimental and numerical studies were conducted on the real object. The reason for undertaking the investigation was excessive stand vibrations, in particular vertical oscillations of the free end of the roof. On the basis of the conducted vibration tests, the following conclusions of the technical condition of the stadium stand were formulated:

(1) The key natural vibration frequency of 4.68 Hz coincided with the frequency of the higher harmonics of the excitation force (approximately 4.4 Hz).

(2) The reinforced concrete cross-section of the BR2 main girder was scratched. The calculated decrease in the stiffness after scratching was equal to about 50%. The inclusion of scratches reduced the calculated initial natural vibration frequencies of the structure by 5%–10% in comparison with the unscratched cross-section.

(3) The dynamic and synchronic nature of the Labado dance is accidental, but can constitute vandalism. Fans jumping to the rhythm of the animator's regular drumming results in increasing stand vibrations, causing the serviceability limit state to be exceeded. The stand vibrations can lead to material fatigue and, as a result, to its complete destruction.

(4) The proposal to solve the problem of overlapping frequencies causing resonance by changing the animator's drumming rhythm and using it for the needs of dynamic impact elimination in favor of the static impact of people on the stand structure was not accepted by club authorities nor by fans. This objection was explained by the long-term Ladabo dance tradition at the Swiss Krono Arena stadium. Any changes in the drumming rhythm would result in the loss of the cheering spirit among fans used to dancing the Labado.

(5) The proposed original concept of stadium structural tuning significantly reduced the amplitude of the stand structure vibrations. Individual resonance zones were moved towards higher frequencies after tuning, which allowed for safe use of the stadium.

(6)   There is a high probability of the occurrence of double natural vibrations frequencies or frequencies with close values in the case of spatial constructions with large dimensions and repeated construction segments. Therein, the phenomenon of the beat will appear, which may cause an increase in tribune elements displacement. The post-tuning FEM model of a structure does not indicate the occurrence of such a case. However, if the actual post-tuning stand structure had double frequencies in a range of up to 5 Hz, additional structure strengthening might be necessary.

(7)   The implementation of the architectural structure form resulted in a static scheme in which a roof structure cantilever steel column was mounted at the end of the stand's reinforced concrete cantilever. The girders working as cantilevers were fixed at the cantilever steel column end. This results in fragility under dynamic load applied to the cantilever part of the concrete beams. Sports facilities must ensure the freedom of supporters gathered in the stands and all possible dances and ways to cheer their favorite teams must be included in the design and implementation process.

(8)   The non-destructive tests and numerical simulations conducted in this paper are innovative and pioneering. Currently, no standard has been developed that allows for the determination of performance conditions and methods for testing the resistance of stadium stands to jump loads. The normative documents also lack a ready-to-use jump load function that could be used when designing a structure.

Finally, it can be concluded that the use of the vibration-based method as a non-destructive testing technique for massive structures such as stadiums enabled the creation of a jump function in order to study dance phenomena on the stands. Therefore, it was possible to prevent failures of and damage to the stadium structure, thus ensuring safety for fans supporting their teams. At the design stage, with the use of the created function, the dance influence as a negative factor affecting the stands can be taken into account, allowing the efficient elimination of potential errors and mistakes.

Author Contributions: Conceptualization and Methodology, K.W., M.R. and K.G.; Experimental Investigations, M.R. and K.W.; FEM Calculations, K.G.; Formal Analysis, K.G., M.R. and K.W.; Visualization, K.G. and M.R.; Writing—Original Draft Preparation, K.G.; Writing—Review and Editing, M.R.; Supervision, K.W.

Acknowledgments: The authors would like to thank the Swiss Krono Arena for providing access to the structure and supplementary data.

References

1.   Melrose, A.; Hampton, P.; Manu, P. Safety at sports stadia. *Procedia Eng.* **2011**, *14*, 2205–2211. [CrossRef]
2.   Soomaroo, L.; Murray, V. Disasters at Mass Gatherings: Lessons from History. *PLOS Curr. Disasters* **2012**. [CrossRef] [PubMed]
3.   Proença, J.M.; Branco, F. Case studies of vibrations in structures. *Rev. Eur. génie Civ.* **2007**, *9*, 159–186. [CrossRef]
4.   Sachse, R.; Pavic, A.; Reynolds, P. Parametric study of modal properties of damped two-degree-of-freedom crowd–structure dynamic systems. *J. Sound Vib.* **2004**, *274*, 461–480. [CrossRef]
5.   Reynolds, P.; Pavic, A. Vibration Performance of a Large Cantilever Grandstand during an International Football Match. *J. Perform. Constr. Facil.* **2006**, *20*, 202–212. [CrossRef]
6.   Yang, Y. Comparison of bouncing loads provided by three different human structure interaction models. In Proceedings of the 2010 International Conference on Mechanic Automation and Control Engineering, Wuhan, China, 26–28 June 2010; pp. 804–807.
7.   Bachmann, H. Case Studies of Structures with Man-Induced Vibrations. *J. Struct. Eng.* **1992**, *118*, 631–647. [CrossRef]
8.   Santos, F.; Cismaşiu, C.; Cismaşiu, I.; Bedon, C. Dynamic Characterisation and Finite Element Updating of a RC Stadium Grandstand. *Building* **2018**, *8*, 141. [CrossRef]
9.   Jones, C.; Reynolds, P.; Pavic, A. Vibration serviceability of stadia structures subjected to dynamic crowd loads: A literature review. *J. Sound Vib.* **2011**, *330*, 1531–1566. [CrossRef]
10. Gul, M.; Catbas, F.N. A Review of Structural Health Monitoring of a Football Stadium for Human Comfort and Structural Performance. *Struct. Congr.* **2013**, 2445–2454.

11. Ren, L.; Yuan, C.-L.; Li, H.-N.; Yi, T.-H. Structural Health Monitoring System Developed for Dalian Stadium. *Int. J. Struct. Stab. Dyn.* **2015**, *16*, 1640018. [CrossRef]

12. Di Lorenzo, E.; Manzato, S.; Peeters, B.; Marulo, F.; Desmet, W. Structural Health Monitoring strategies based on the estimation of modal parameters. *Procedia Eng.* **2017**, *199*, 3182–3187. [CrossRef]

13. Spencer, B.F.; Ruiz-Sandoval, M.E.; Kurata, N.; Ruiz-Sandoval, M.E. Smart sensing technology: opportunities and challenges. *Struct. Control. Heal. Monit.* **2004**, *11*, 349–368. [CrossRef]

14. Bedon, C.; Bergamo, E.; Izzi, M.; Noè, S. Prototyping and Validation of MEMS Accelerometers for Structural Health Monitoring—The Case Study of the Pietratagliata Cable-Stayed Bridge. *J. Sens. Actuator Networks* **2018**, *7*, 30. [CrossRef]

15. Hoła, J.; Bień, J.; Sadowski, Ł.; Schabowicz, K. Non-destructive and semi-destructive diagnostics of concrete structures in assessment of their durability. *Bull. Pol. Acad. Sci. Tech. Sci.* **2015**, *63*, 87–96. [CrossRef]

16. Brandt, A. A signal processing framework for operational modal analysis in time and frequency domain. *Mech. Syst. Signal Process.* **2019**, *115*, 380–393. [CrossRef]

17. Chen, G.-W.; Omenzetter, P.; Beskhyroun, S. Operational modal analysis of an eleven-span concrete bridge subjected to weak ambient excitations. *Eng. Struct.* **2017**, *151*, 839–860. [CrossRef]

18. Torres, W.; Almazán, J.L.; Sandoval, C.; Boroschek, R. Operational modal analysis and FE model updating of the Metropolitan Cathedral of Santiago, Chile. *Eng. Struct.* **2017**, *143*, 169–188. [CrossRef]

19. Idehara, S.J.; Júnior, M.D. Modal analysis of structures under non-stationary excitation. *Eng. Struct.* **2015**, *99*, 56–62. [CrossRef]

20. Jannifar, A.; Zubir, M.; Kazi, S. Development of a new driving impact system to be used in experimental modal analysis (EMA) under operational condition. *Sens. Actuators A Phys.* **2017**, *263*, 398–414. [CrossRef]

21. Prashant, S.W.; Chougule, V.; Mitra, A.C. Investigation on Modal Parameters of Rectangular Cantilever Beam Using Experimental Modal Analysis. *Mater. Today Proc.* **2015**, *2*, 2121–2130. [CrossRef]

22. Hu, S.-L.J.; Yang, W.-L.; Liu, F.-S.; Li, H.-J. Fundamental comparison of time-domain experimental modal analysis methods based on high- and first-order matrix models. *J. Sound Vib.* **2014**, *333*, 6869–6884. [CrossRef]

23. Clemente, P.; Marulo, F.; Lecce, L.; Bifulco, A. Experimental modal analysis of the Garigliano cable-stayed bridge. *Soil Dyn. Earthq. Eng.* **1998**, *17*, 485–493. [CrossRef]

24. Lin, R. Identification of modal parameters of unmeasured modes using multiple FRF modal analysis method. *Mech. Syst. Signal Process.* **2011**, *25*, 151–162. [CrossRef]

25. Rucka, M.; Wilde, K. Application of continuous wavelet transform in vibration based damage detection method for beams and plates. *J. Sound Vib.* **2006**, *297*, 536–550. [CrossRef]

26. Chroscielewski, J.; Miśkiewicz, M.; Pyrzowski, Ł.; Rucka, M.; Sobczyk, B.; Wilde, K. Modal properties identification of a novel sandwich footbridge—Comparison of measured dynamic response and FEA. *Compos. Part B Eng.* **2018**, *151*, 245–255. [CrossRef]

27. EN C. 1-1: *Eurocode 1: Actions on Structures–Part 1-1: General Actions–Densities, Self-Weight, Imposed Loads for Buildings*; European Committee for Standardization: Brussels, Belgium, 2002.

9

Assessment of the Condition of Wharf Timber Sheet Wall Material by Means of Selected Non-Destructive Methods

Tomasz Nowak⬤, Anna Karolak *⬤, Maciej Sobótka⬤ and Marek Wyjadłowski⬤

Wroclaw University of Science and Technology, Wybrzeze Wyspianskiego 27, 50-370 Wroclaw, Poland; tomasz.nowak@pwr.edu.pl (T.N.); maciej.sobotka@pwr.edu.pl (M.S.); marek.wyjadlowski@pwr.edu.pl (M.W.)
* Correspondence: anna.karolak@pwr.edu.pl

Abstract: This paper presents an assessment of the condition of wood coming from a wharf timber sheet wall after 70 years of service in a (sea) water environment. Samples taken from the structure's different zones, i.e., the zone impacted by waves and characterised by variable water-air conditions, the zone immersed in water and the zone embedded in the ground, were subjected to non-destructive or semi-destructive tests. Also, the basic parameters of the material, such as its density and moisture content, were determined. Moreover, the ultrasonic, stress wave and drilling resistance methods were used. Then, an X-ray microtomographic analysis was carried out. The results provided information about the structure of the material on the micro and macroscale, and the condition of the material was assessed on their basis. Also, correlations between the particular parameters were determined. Moreover, the methods themselves were evaluated with regard to their usefulness for the in situ testing of timber and to estimate, on this basis, the mechanical parameters needed for the static load analysis of the whole structure.

Keywords: timber structures; non-destructive methods; ultrasonic wave; stress wave; drilling resistance; X-ray micro-computed tomography

1. Introduction

Timber as a universal building material has been used in many kinds of structures in, e.g., geotechnical engineering. In this field, one of the principal uses of structural timber is as timber pile foundations in low bearing capacity building lands situated in river deltas and beds and on peat bogs, which often were important locations because of their strategic position. As a foundation method, timber piles have been used almost everywhere in Europe in, i.a., the Scandinavian countries, the countries of Eastern and Middle Europe, such as Poland, the Baltic countries and Russia, as well as Germany, Great Britain and the Netherlands, where the foundation of historical buildings is most fully documented. Also, in the south of Europe, for instance in Venice, deep foundation timber piles may be found in nearly all the historical buildings erected since the 12th century [1].

Wood is a heterogeneous, hygroscopic, cellular and anisotropic material. Its mechanical properties depend on many factors, such as density and water content, which means that the creation of a constitutive model of wood poses a great challenge [2].

Wood biodegrades over time. Under the impact of external factors, wooden members undergo chemical and physical changes. Wood can be regarded as a durable material when it is completely immersed in water, and so protected against decomposition caused by aerobic fungi.

In nature, five basic chemical processes occur which reconvert a wooden material into carbon dioxide and water: oxidation, hydrolysis, dehydration, reduction and free radical reactions [3].

Table 1 shows major degradation pathways and the chemistries involved in the pathways [3].

Table 1. Major degradation pathways and chemistries.

Biological Degradation	Fungi, bacteria, insects, termites, enzymatic reactions, oxidation, hydrolysis, reduction, free radical reactions
Chemical Reactions	Oxidation, hydrolysis, reduction, free radical reactions
Mechanical Degradation	Chewing dust, wind, hail, snow, sand, stress, crack, fracture, abrasion, compression
Thermal Degradation	Lightning, fire, sun
Water Degradation	Rain, sea, ice, acid rain, dew
Water Interactions	Swelling, shrinking, freezing, cracking, erosion
Weather Degradation	Ultraviolet radiation, water, heat

All the above factors can be significant when testing wood samples taken from a retaining structure since the latter is exposed to the air, water and soil environment. Therefore, wood analyses should be based on more than one research technique to gain a deeper insight into the change of wood parameters over time in different environmental conditions [3].

In the considered case, since the structure was in service in the water environment and in saturated soils, the hazards can be divided into the following three main groups:

• a low or variable water level,
• excessive loading,
• decomposition of wood in the water environment.

If timber members are above the groundwater table, access to oxygen makes the activation of wood decomposing fungi possible. The decomposition rate is determined by the time during which the timber members are above groundwaters and the member's length situated above the groundwater table [4]. It is estimated that the maximum rate of decomposition caused by fungi attacking water-saturated wood amounts to approximately 10 mm/year [1]. However, the long-lasting service of timber structures below the groundwater table does not prevent decomposition [5], and examinations have shown some of the pile foundations in Venice to be in extremely bad condition.

This paper presents research methods which enable one to determine the condition and some parameters of a material which has been in service in the water-soil environment. For the best results, wood should be tested on different levels of detail, i.e., macro, submacro and microlevels. On the macrolevel, acoustic methods, drilling resistance method or laboratory tests of basic material parameters, such as density and moisture content, are used. On the microlevel the cell wall of the wood is tested and different elements, such as hardwood, sapwood and annual rings, are identified [6], using, e.g., X-ray micro-computed tomography.

The considered timber sheet wall was made of tongue-and-groove jointed timber piles. The history of this structure is not well known because of its previous military use. The timber sheet wall had been in service for about 70 years. The wharf is in the Swina straight connecting the Szczecin Lagoon with the Baltic Sea. In the Swina strait, fresh water (fully or partially) mixes with seawater due to the stratification. The salinity of the Swina strait ranges from 1‰ to 8‰. The average salinity of the Baltic Sea amounts to about 7‰, generally ranging from 2 to 12‰. It can be assumed that the water environmental conditions correspond to low salinity seawater.

After the timber sheet wall had been dug out and dismantled, its members were closely examined with regard to their original and current cross-sectional dimensions and to the quality of the wood. In the photograph (Figure 1a,b) of the dismantled members of the wall, one can see pile surfaces which were in service in diverse environmental conditions: completely embedded in the ground, stayed in water and stayed in the variable water-air environment. One can notice that the timber embedded in the ground, under the groundwater table, has preserved its constant volume. Bacteria destroy the cellulose very slowly, while the lignin remains constant, and water replaces the large cellulose molecules. The original waterfront layout has been reconstructed, see Figure 1c.

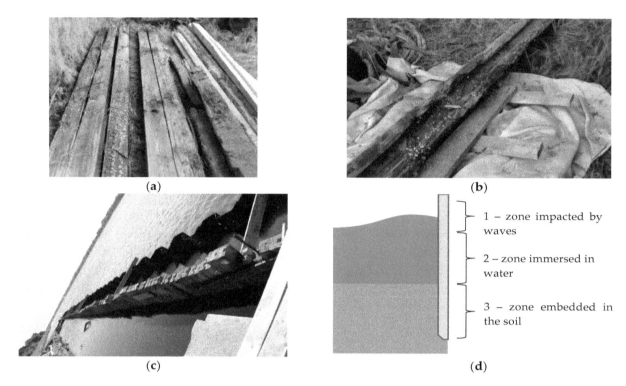

Figure 1. (**a,b**) View of wharf timber sheet pile wall from which samples were taken for testing (the zone impacted by waves is marked with red ellipse); (**c**) view of rebuilt wharf timber sheet pile wall; (**d**) scheme of the global structure with the location of the specimens.

The main objective of the work is to develop a methodology for testing of wooden structural members using non-destructive techniques. This is aimed at obtaining information which is necessary to assess the technical condition of the material in wooden members and to conduct a global structure analysis. In particular, to carry out such analysis, it is required to estimate the values of mechanical parameters and to assess possible zones of destruction. The detailed aim is to compare the quality of wood subjected to various environmental conditions (Figure 1d).

2. Selected Non-destructive Methods for Wood Assessment

2.1. Brief Survey of Methods

Material tests for wood can be divided into three groups: destructive tests (DT), semi-destructive tests (SDT) [7] and non-destructive/quasi-non-destructive tests (NDT) [8]. Unlike destructive tests, the tests belonging to the latter two groups do not affect or only slightly affect the properties of the tested sample, whereby the parameters of a wooden member can be determined with no detriment to its value. Their undeniable advantage is also the mobility of the testing equipment, whereby tests can be carried out in situ when it is not possible to take samples for laboratory tests (as in the case of heritage assets). Among the non-destructive and semi-destructive methods one can distinguish global testing methods (e.g., ultrasonic and stress wave techniques) and local testing methods (e.g., the drilling resistance method).

In order to acquire detailed data on the values of the physical and mechanical parameters of wood both non-destructive and destructive methods should be used. If the results yielded by the two testing methods are found to correlate, the data acquired in this way are fully sufficient for further static load analyses of the structural members or the whole building structure. Nevertheless, even using only non-destructive methods (as in the case of, e.g., heritage assets), one can obtain some information about the properties of the tested member's material, assess the technical condition of the structure or acquire some data helpful in evaluating this condition or in the design of possible repairs or upgrades. Thanks to the use of non-destructive methods one can also detect internal damage or flaws in the wood [9].

Among the non-destructive and quasi-non-destructive testing methods used to assess and diagnose timber structures, the most common are the ones presented in Table 2, and also described in detail in, i.a., [7,8,10,11]. The non-destructive and quasi-non-destructive methods can be divided into two groups: global testing methods (e.g., visual evaluation and ultrasonic and stress wave techniques) and local testing methods (e.g., the drill resistance method, the core drilling method and the hardness test method).

Table 2. Selected methods available for assessing timber in building structure.

Organoleptic Methods	Acoustic Methods	Quasi-Non-destructive (Semi-Destructive) Methods	Radiographic Methods	Other Methods
Visual evaluation Acoustic evaluation Fragrance evaluation	Stress waves Ultrasonic technique Acoustic emission	Drilling resistance Core drilling Screw withdrawal Hardness tests Needle penetration Pin pushing Tension microspecimens	X-rays Gamma rays	Computed tomography Ground penetrating radar Near infrared spectrometry

2.2. Acoustic Methods

2.2.1. Idea of Acoustic Test

Using acoustic testing methods, such as the ultrasonic and stress wave techniques, one can evaluate the properties of wood by analysing the velocity of wave propagation in the tested material. The methods can be used to estimate selected mechanical properties (e.g., the modulus of elasticity) of a material and to detect its internal structural discontinuities.

The basic parameter used in the acoustic methods is sound wave propagation velocity (V), defined as follows:

$$V = L/T,   (1)$$

where L is the distance (between two measuring points) covered by the excited sound wave, and T is the time needed to cover this distance.

Knowing the velocity of wave propagation and the wood density (ρ), one can determine the dynamic modulus of elasticity (MOE_{dyn}), which can be interrelated with the static modulus of elasticity (MOE_{stat}) [10]. The dynamic modulus of elasticity is calculated from the formula:

$$MOE_{dyn} = V^2 \times \rho,   (2)$$

where Vis the velocity of sound wave propagation, and ρ is the density of the tested element.

The velocity of sound wave propagation largely depends on the structure of the material. In the case of wood, it depends on the grain direction being several times higher (usually 3–5 times higher) along than across the grain [10,12].

According to [12], for wood with no significant structural flaws the velocity of sound wave propagation amounts to 3500–5000 m/s along the grain and to 1000–1500 m/s across the grain. Other values than the above ones may indicate internal discontinuities in the material structure. The lower values of the velocity across the grain are due to the internal structure of this material (on its way the wave encounters more cell walls, whereby the time in which it covers the distance increases, whereas in the longitudinal direction there are fewer barriers or they do not occur, whereby the velocity is higher).

2.2.2. Description of Testing Methods and Devices

Several kinds of devices are used for testing by means of ultrasonic or stress wave methods. In this case study, two of them were used and the test results are presented in Section 3.

The Fakopp Microsecond Timer (Figure 2a) uses the stress wave technique. The test consists of exciting a stress wave with a single strike of a special hammer. The device probes are driven directly into the tested sample. There is no need to drill holes as in the case of other devices (e.g., Sylvatest Trio). The device measures the time of wave propagation between the two probes (the receiving probe and the transmitting probe).

(b)

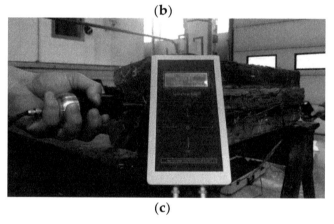

(a)   (c)

Figure 2. Devices for testing by acoustic method: (**a**) Fakopp Microsecond Timer using stress wave; (**b**) Sylvatest Trio device using ultrasonic wave; (**c**) Sylvatest Trio device during test.

In the case of the ultrasonic technique, the measurement can be performed in two ways: directly and indirectly. The first way consists of transmitting a signal from the transmitting probe to the receiving probe, with the probes placed on the opposite sides of the tested sample. As regard the second way, there is no need to place probes on the opposite sides of the sample because the signal is registered as reflected (the echo method). Owing to this, the range of the applicability of this test widens since only a unilateral access is required (which is useful when testing, e.g., historical monuments in situ).

Sylvatest Trio (Figure 2b), manufactured by the firm CBS-CBT, is another device which one can use to non-destructively evaluate the properties of wood [13]. The device measures the time needed for an ultrasonic wave to pass between transmitting-receiving probes placed against the tested sample, and the energy of this wave. In order to carry out the test the tips of the probes should be inserted into previously drilled holes each 5 mm in diameter and 10 mm deep. One should bear in mind that because of the high sensitivity of the device other mechanical waves excited near the test site can affect the test results. Also, the material's moisture content and internal stresses can significantly influence the results.

In order to obtain exhaustive results, it is recommended, for both methods, to perform a large number of measurements in different points and directions.

2.2.3. Correlation between Physical and Mechanical Properties of Wood and Results Yielded by Acoustic Methods

In many studies (e.g., [14–20]) based on acoustic methods attempts were made to assess the effectiveness of the methods and to find a correlation between the physical and mechanical properties of wood and the parameter values obtained from measurements.

According to the above studies, there is a strong correlation between the dynamic modulus of elasticity (MOE_{dyn}) and the static modulus of elasticity (MOE_{stat}). According to [18], for sound wood free of flaws the determination coefficient for the static and dynamic modulus of elasticity amounts to 0.96.

Also, comparative analyses of the effectiveness of the Fakopp Microsecond Timer (Fakopp Enterprise Bt., Agfalva, Hungary) and the Sylvatest Trio device (-CBS-CBT, Choisy-le-Roi, France) were carried out. They showed the two devices to be highly effective [18,19] and confirmed the correlation between the value of MOE_{dyn} and that of MOE_{stat} [20].

As part of other investigations, the decrease in the value of the velocity of the ultrasonic wave and the stress wave was analysed. According to [12], a reduction in the velocity by about 30% can correspond to a 50% fall in the load bearing capacity, while a reduction in the level of velocity by more than 50% can indicate considerable damage and the loss of load bearing capacity by the tested element. According to the results of the above research the relative decrease in the value of the velocity of wave propagation between two measuring points (ΔV_{rel}) describes the degree of damage to the material. The value of ΔV_{rel} is defined by Equation (3):

$$\Delta V_{rel} = [(V_{ref} - V_{mes})/V_{ref}] \cdot 100\%, \tag{3}$$

where ΔV_{rel} is the relative decrease in velocity, V_{ref} is the reference velocity (value of the velocity for a sound wood, taken from tests or literature), and V_{mes} is the measured velocity.

The relation between the relative decrease in velocity and the degree of damage is shown in Table 3 [21].

Table 3. Relation between relative velocity decrease and degree of damage.

Relative Velocity Decrease [%]	Degree of Damage [%]
0–10	no destruction
10–20	10
20–30	20
30–40	30
40–50	40
≥50	≥50

2.3. Drilling Resistance Method

2.3.1. Description of Test

One of the semi-destructive (SDT) methods is the drilling resistance method.

After the test a small borehole, below 3.0 mm in diameter, (not larger than the exit hole of most of the woodworm) remains in the sample material, but with no detriment to the properties of the element, whereby the test can be regarded as semi-destructive [7].

The test consists of measuring the energy needed to drill the resistance drilling device's metal needle into the material. The test makes it possible to detect structural discontinuities, damage, knots and other flaws and also to estimate the density and strength of the material [22]. The device measures the drilling resistance of a drill with a diameter of 1.5–3.0 mm and a length of 300–500 mm, rotating at a constant speed of about 1500 rpm (Figure 3).

Figure 3. IML RESI PD-400S device used in tests.

Drilling resistance is closely connected with the difference in density between the zones of early and late wood [22], the structure of the annual rings [23,24], changes in wood density caused by, i.a., biological decomposition, and the drilling angle [25]. The device registers the measurement results at every 0.1 mm, in the form of drilling resistance-depth graphs. The peaks in the graph correspond to the high resistance and high density of the material while the declines represent its low resistance and low density. The flatline in the diagram indicates places where the material does not show any drilling resistance, which means that the material is completely decomposed. During drilling the measurement in the entry and exit zones is disturbed because of the time needed for the drill to assume the proper position and rotational speed. Consequently, the graph in these zones usually has the form of a smoothly rising or declining curve.

Using the device one can detect structural flaws and discontinuities in timber elements without adversely affecting their useful properties (see, e.g., [14,26,27]).

2.3.2. Correlation between Physical and Mechanical Properties of Wood and Drilling Resistance Results

Attempts have been made to correlate drilling resistance results with strength test results in order to estimate the mechanical parameters of wood in the structure (e.g., [27–33]). Diagrams of relative resistance (RA) versus drilling depth (H) make it possible to evaluate the parameters of wood through the correlation between the average value of the resistance measure (RM) parameter and the density, strength and the modulus of elasticity of the wood. The value of RM can be calculated from formula 4 [25]:

$$\text{RM} = \frac{\int_0^H \text{RA} \cdot dh}{H},\tag{4}$$

where $\int_0^H \text{RA} \cdot dh$ is the area under the drilling resistance graph, and H is the drilling depth.

Attempts have also been made to correlate the resistance measure with different material parameters (density, longitudinal modulus of elasticity, transverse modulus of elasticity, longitudinal compressive strength, transverse compressive strength and bending strength) for different wood species, new wood and old wood. The results of some of the endeavours presented high determination coefficients amounting to 0.78 for the transverse compressive strength and to 0.67 for the modulus of elasticity [16] as well as to 0.64 for the modulus of elasticity and the longitudinal compressive strength [25]. As regards density, the determination coefficients of 0.71 [25], 0.75 for Pine, 0.74 for Spruce, 0.65 for Fir [28], 0.70 [29], 0.80 [30] or even 0.88 [31] were obtained. However, some researchers [32,34,35] did not obtain such a good correlation.

In general, tests performed on non-decayed, defect free, small sized laboratory specimens provide high values of correlation coefficients. On the other hand, results of the onsite tests of full-sized elements must be analysed with greater caution due to the possible presence of defects. In this context the drilling resistance method should be perceived to be a qualitative method.

The RM parameter value is influenced by many factors, such as the tree species, the condition of the wood and its moisture content and the drilling direction [32]. The results should be treated as not a quantitative, but qualitative assessment and the resistance drilling method test can be a complement to other tests or the starting point for a preliminary inspection of timber members or the location of damage inside the cross section.

2.4. X-Ray Micro-Computed Tomography

X-ray micro-computed tomography (Skyscan 1172, Bruker, Kontich, Belgium) is a state-of-the-art non-destructive technique for visualizing the inner structure of the tested object [36,37]. In essence, the tests consists of mathematically reconstructing the three-dimensional microstructure of the tested material on the basis of a series of high-resolution X-ray pictures. The scanning consists of recording a series of projections taken during the slow rotation of the sample placed on the scanner's rotary fixture [38]. The Bruker SkyScan 1172 microtomograph used in the tests and a view of a sample placed in the scanning chamber are shown in Figure 4 below.

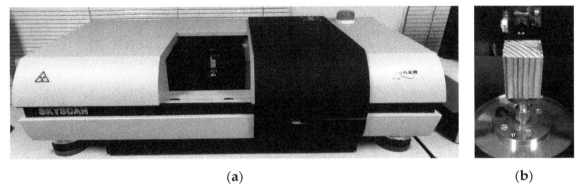

(a)   (b)

Figure 4. (a) Bruker SkyScan 1172 device; (b) sample of timber mounted on stage inside scanning chamber.

A single projection taken at a set sample rotation angle shows (in greyscale) the distribution (registered by the detector) of the intensity of the X-ray radiation emitted by the source and attenuated by passing through the sample. After a series of projections is recorded, the mathematical reconstruction of the tested material is carried out. The fact that according to the Lambert-Beer law, radiation absorption depends on the material's attenuation coefficient and on the thickness of the layer which the radiation must penetrate, is exploited for this purpose. The most commonly used algorithms are based on back projection, e.g., the Feldkamp algorithm [39] used in the present study. The result of the reconstruction is a series of images representing the cross sections of the examined object. The images show (in greyscale) the distributions of the attenuation coefficient, i.e., a characteristic of the material. The cross sections, arranged one above the other in space, make up a three-dimensional image of the internal microstructure of the examined object. Using image analysis techniques, quantitative and qualitative analyses of the material's microstructure can be carried out on the basis of the reconstructed images [40,41].

Since the absorption coefficient depends mainly on density, the greyscale of the resulting image can be treated as a monotonic density function. Consequently, the correlation between the level of brightness of the pixels (voxels) and the density of the tested material can be determined. For example, in [41,42] the linear correlation between density and the brightness level of the resultant tomographic image was used to determine the density of wood.

When testing heterogenous materials (composites), a morphometric analysis is carried out [38,43]. Its aim is to quantitatively characterise the morphology of a given component of the composite, particularly by determining the shape parameters and form of the geometrical objects constituting the area (in space) occupied by the considered component. Such an analysis is carried out on binary images obtained from segmentation.

3. Materials and Methods

In this case study, non-destructive and semi-destructive testing methods, i.e. the ultrasonic and stress wave techniques and the drilling resistance technique, were used to estimate the parameters of samples taken from the structure. Also c.a. 20 mm × 20 mm × 30 mm samples were cut out from the same structural members and subjected to scanning in the microtomograph in order to augment the non-destructive test results. The direct result of the scanning is a 3D image of the microstructure of the tested material. The level of brightness in the images is approximately proportional to the local density of the tested material. Owing to this a semi-quantitative comparison of wood density for the different zones of the timber sheet wall could be made.

Using the testing methods and devices described in Section 2, a series of tests were carried out on samples taken from the wharf timber sheet wall. The samples dimensions were about 18 cm × 20 cm cross-section and 60 cm length (Figure 5). The samples come from different locations in the wharf structure, which means that in the course of their service they were submerged to different levels and exposed to the variable impact of water. The material of the samples is pinewood (*Pinus sylvestris L*). The tests were carried out in a laboratory at Wroclaw University of Science and Technology.

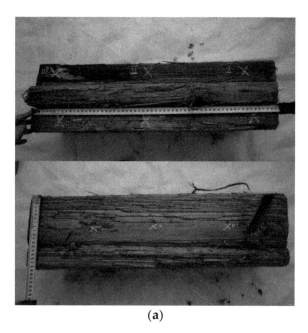 

(a)   (b)

Figure 5. Samples and their dimensions: (**a**) sample 1 cross section 18 cm × 20 cm and length 60 cm, (**b**) sample 3 cross-section 18 cm × 20 cm and length 60 cm with cut.

The acoustic tests and the stress wave tests were carried out using respectively the Sylvatest Trio device and the Fakopp Microsecond Timer.

A new generation device IML RESI PD400 with a drill length of about 400 mm was used for drilling resistance testing. The device can register both drilling resistance and the feed force at every 0.1 mm. Five-millimetre deep entry and exit zones were assumed when calculating mean drilling resistance RM from formula (5). The zones were not taken into account in the calculations. The places where drilling resistance (RA) amounted to less than 5% and where the under-five-per-cent values in the diagram extended for minimum 5 mm were regarded as zones with flaws.

Moreover, the moisture content in the samples was determined using an FMW moisture meter (of the resistance type with a hammer probe).

Density was determined using 20 mm × 20 mm × 400 mm flawless samples prepared from the tested timber samples (16, 12 and 12 samples from each of the member, altogether 40 samples) with a moisture content of 18%. In accordance with the standard procedure [44], density was calculated from formula (5):

$$\rho = \rho(u)\cdot[1 - 0.005\,(u - u_{ref})], \qquad (5)$$

where ρ is density, u is the sample's moisture content during testing, and u_{ref} is the reference moisture content = 12%.

Also, the correction due to the size of the sample is needed; the value should be divided by 1.05 [44].

Density determinations and the drilling resistance tests were carried out for the wood moisture content of about 18%. Acoustic tests were carried out three times for chosen different sample moisture content levels: about 30% (direct after taking samples from the structure), 24–28% and 17–18% to examine the effect of moisture content on the measurement results.

Also, a series of scans of the small samples were performed using the Bruker SkyScan 1172 microtomograph.

The total number of three samples were scanned using X-ray micro-computed tomography (see Figure 6 below):

- "1" a portion of timber from the zone impacted by waves—a sample with c.a. 20 mm × 20 mm × 30 mm dimensions,
- "2" a portion of timber from the zone submerged in water)—a sample with c.a., 20 mm × 20 mm × 30 mm dimensions,
- "3" a portion of timber from the zone sunk in the ground—a sample with c.a., 20 mm × 20 mm × 30 mm dimensions.

Figure 6. Samples prepared for scanning: "1", "2" and "3" (from left to right).

The samples were scanned in the Bruker SkyScan 1172 device (Figure 4a). The same set of scanning parameters was used for each of the samples in order to ensure identical scanning conditions. The selected major scanning parameters are summarised in Table 4 below.

Table 4. Scanning parameters.

Parameter	Value
Source Voltage	59 kV
Source Current	167 μA
Projection image size	2000 × 1333 pix
Image Pixel Size	13,56 μm
Filter	Al foil
Exposure	750 ms
Rotation Step	0.24°
Frame Averaging	ON (6)
Random Movement	ON (10)
Use 360 Rotation	YES

4. Results and Discussion

4.1. Density and Moisture Content

Density was determined for the 40 flawless small samples and the results are presented in Table 5.

Table 5. Determined densities.

Sample	Number of Measurements	Density ρ				
		Mean Value from Tests [kg/m^3]	Mean value Calculated according to [44] [kg/m^3]	Range [kg/m^3]	Standard Deviation [kg/m^3]	Coefficient of Variation [%]
1	16	511.5	472.5	449.6–547.0	27.2	5.3
2	12	514.4	475.2	484.4–549.1	25.8	5.0
3	12	623.0	575.5	598.8–671.1	20.5	3.3
summary	40	545.8	504.2	449.6–671.1	58.5	10.8

On the basis of the measurements performed by means of a resistance-type moisture meter the moisture content in the samples was determined to amount to 18 ± 1%. In addition, testing by acoustic methods was carried out for two more different moisture content levels, i.e., about 30% and 24–28%.

4.2. Drilling Resistance

Drilling resistance tests were carried out on 3 samples with a moisture content of 18 ± 1%. Twenty measurements were performed on each of the samples. An exemplary drilling resistance curve and a feed force curve are shown in Figure 7. The depth to which the biological corrosion of the wood extends (3 mm in this case) can be easily read off the diagram. The drilling resistance test results for the particular samples are presented in Tables 6 and 7.

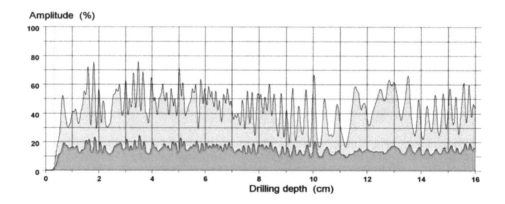

Figure 7. Exemplary drilling resistance curve (green) and feed force curve (grey).

Table 6. Mean drilling resistance tests results.

Sample	Number of Measurements	Drilling Resistance RM [%]			
		Mean	Range	Standard Deviation	Coefficient of Variation
1	20	16.5	14.9–18.3	1.2	7.4
2	20	16.3	13.5–19.2	2.3	14.1
3	20	16.9	15.2–20.1	1.6	9.5
summary	60	16.6	13.5–20.1	1.8	10.6

Table 7. Mean feed force test results.

Sample	Number of Measurements	Feed Force *FM* [%]			
		Mean	Range	Standard Deviation	Coefficient of Variation
1	20	47.4	42.3–53.4	3.5	7.3
2	20	47.7	35.9–61.7	9.8	20.6
3	20	55.3	44.3–70.6	7.7	13.9
summary	60	50.1	35.9–70.6	8.2	16.4

Only 60 measurements were carried out as part of the laboratory tests, but the number of samples in in situ tests is usually not larger because of the not fully non-destructive character of the testing method. Since the test has a pointwise character and wood is a heterogenous material, it is necessary to perform numerous measurements to assess its condition and density. Therefore, one cannot responsibly evaluate wood on the basis of single measurements.

Many factors have a bearing on drilling resistance, e.g., moisture content, drill sharpness, drilling angle and direction and battery charge status [45]. Moreover, wood flaws, such as knots (resulting in very high drilling resistance) and damaged zones (zero or close to zero drilling resistance) affect the RM value, which was taken into account in the analysis. The places with knots were neglected in drilling resistance and feed force calculations.

Despite the quite good correlation (Figure 8) between drilling resistance and feed force ($R^2 = 0.8114$), no correlation between these quantities and density was found (Figures 9 and 10), which casts doubt on the correlativity between them [32,46]. The results obtained using the drilling resistance method can be used to estimate the depth of wood damage in static load analyses to reduce the cross sections of the members.

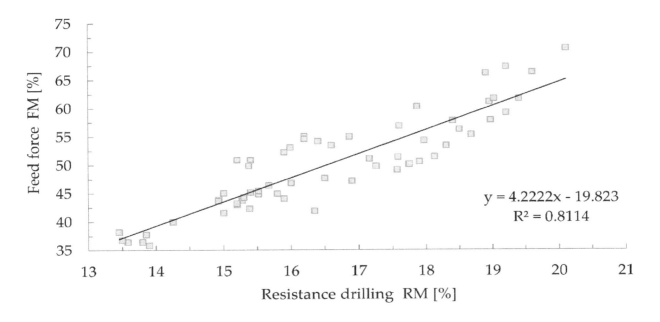

$$y = 4.2222x - 19.823$$
$$R^2 = 0.8114$$

Figure 8. Correlation between drilling resistance and feed force.

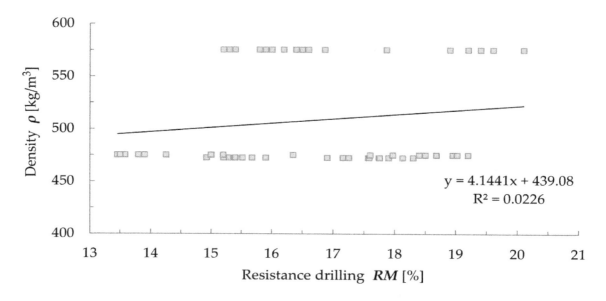

Figure 9. Correlation between drilling resistance and density.

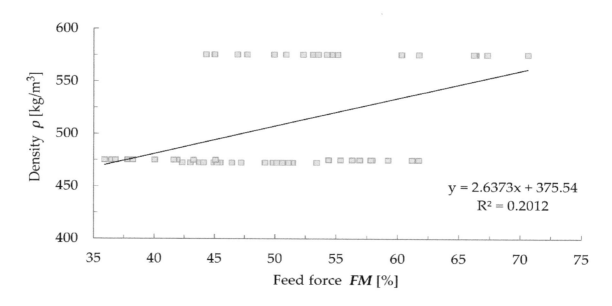

Figure 10. Correlation between feed force and density.

4.3. Stress and Ultrasonic Waves

The propagation times of the stress wave and the ultrasonic wave and the length of the distance covered by the waves in both the directions (along and across) relative to the grain were registered and used to calculate the velocity of wave propagation in the material. Then the dynamic moduli of elasticity were calculated using the densities measured for the particular samples (sample 1—472.5 kg/m^3, sample 2—475.2 kg/m^3, sample 3—575.5 kg/m^3). The results are shown in Tables 8 and 9. Also the dynamic elasticity modulus values parallel and perpendicular to the grain, yielded by the two methods were correlated for selected samples. The results are presented in Figure 11.

Table 8. Fakopp Microsecond Timer test results: velocity of stress wave propagation and dynamic modulus of elasticity depending on moisture content and direction relative to grain.

Sample	Direction Relative to Grain	V [m/s]			MOE_{dyn} [GPa]		
		Moisture Content			Moisture Content		
		~30%	24–28%	~18%	~30%	24–28%	~18%
1	parallel	4872.8	5376.6	5644.8	11.22	13.66	15.06
	perpendicular	1153.7	1439.9	1443.8	0.63	0.98	0.98
2	parallel	-	-	-	-	-	-
	perpendicular	1218.5	1372.8	1270.6	0.71	0.90	0.77
3	parallel	-	-	-	-	-	-
	perpendicular	1406.1	1447.7	1665.2	1.14	1.21	1.60

Table 9. Sylvatest Trio test results: velocity of ultrasonic wave propagation and dynamic modulus of elasticity depending on moisture content and direction relative to grain.

Sample	Direction Relative to Grain	V [m/s]			MOE_{dyn} [GPa]		
		Moisture Content			Moisture Content		
		~30%	24–28%	~18%	~30%	24–28%	~18%
1	parallel	5128.9	5855.4	6035.2	12.43	16.20	17.21
	perpendicular	1118.4	1311.5	1356.9	0.59	0.81	1.04
2	parallel	-	-	-	-	-	-
	perpendicular	1030.2	1311.4	1481.1	0.50	0.82	1.04
3	parallel	-	-	-	-	-	-
	perpendicular	1199.8	1251.1	1422.3	0.68	0.90	1.04

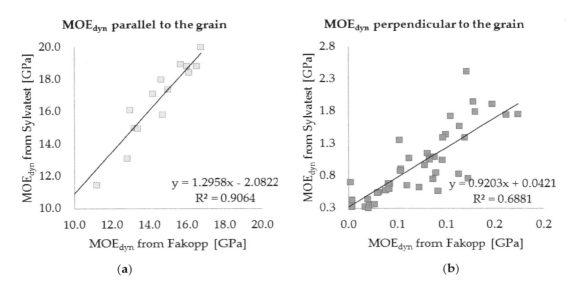

Figure 11. Correlation between dynamic modulus of elasticity determined using ultrasonic wave method (Sylvatest) and stress wave method (Fakopp) for all the samples for all moisture content levels for both directions relative to grain: **(a)** parallel; **(b)** perpendicular.

As one can see above, the results yielded by the two measuring methods using the Sylvatest Trio device and the Fakopp Microsecond Timer are similar. For the selected samples the correlation coefficient along and across the grain amounts to respectively $R^2 = 0.9064$ and $R^2 = 0.6881$. This result can be regarded as satisfactory and it indicates the two methods can be used complementarily to estimate the material parameters of wood.

4.4. X-Ray Micro-Computed Tomography

3D images of the samples were reconstructed on the basis of a series of X-ray projections, using the Feldkamp algorithm in the NRecon software (version 1.7.1.0) by Bruker. Selected major reconstruction parameters are summarised in Table 10 below.

Table 10. Reconstruction parameters.

Parameter	Value
Pixel Size	13.53217 μm
Smoothing	2 pix
Ring Artefact Correction	19
Beam Hardening Correction	41%
Minimum for CS to Image Conversion	0.000
Maximum for CS to Image Conversion	0.030

Exemplary projections for all of the tested samples are shown in Figure 12.

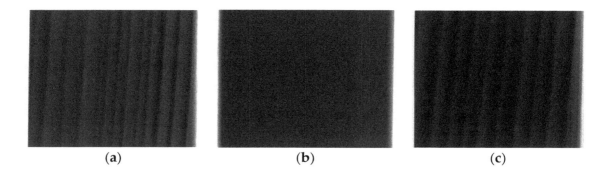

(a)   (b)   (c)

Figure 12. Exemplary projections for samples: (a) "1"; (b) "2"; (c) "3" (scale 200%).

Exemplary cross sections of the samples, obtained by reconstructing the 3D sample model are shown in Figures 13 and 14, at a scale of respectively 200% and 800% (an enlarged fragment of the image).

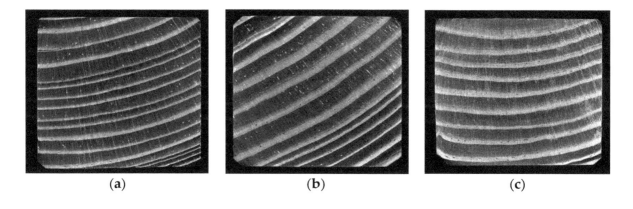

(a)   (b)   (c)

Figure 13. Exemplary cross sections of samples: (a) "1"; (b) "2"; (c) "3" (scale 200%).

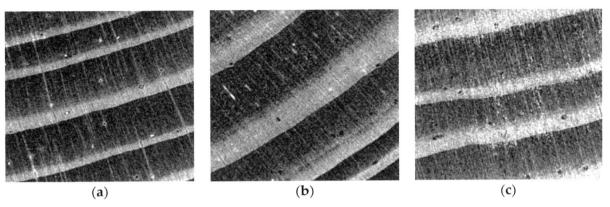

Figure 14. Exemplary cross section (scale 800%): (**a**) sample "1"; (**b**) sample "3"; (**c**) sample "2".

Figure 15 shows the rendering of the 3D model of the samples.

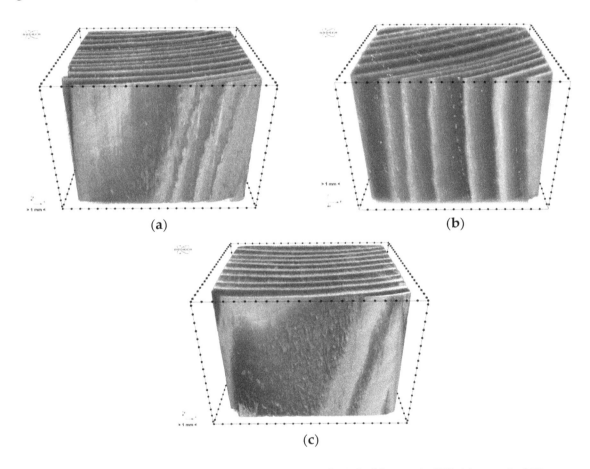

Figure 15. Reconstruction—3D view: (**a**) sample "1", (**b**) sample "2", (**c**) sample "3".

The cell structure is practically invisible due to the adopted scanning resolution. Only the early and late wood with local flaws (small microcrack in sample "1") and higher-density inclusions (samples "1" and "2") can be distinguished. In the images obtained from scanning, sample "3" is generally brighter than the other two, which unambiguously indicates its higher density. This is particularly visible in the cross sections (Figures 13 and 14). It also appears from the reconstruction that the late growth rings, which are denser (brighter in the imaging results), are thicker and occupy more material volume in samples "2" and "3", whereas in sample "1" the volume fraction of late growth rings is clearly smaller. Moreover, small highly dense (probably mineral) inclusions are noticeable in samples "1" and "2".

A cubic area of $(1200 \text{ vox})^3$ completely contained within the volume of the tested material, i.e., the so-called volume of interest (VOI), was selected in order to quantitatively characterise the above observations. The selection of VOI is shown in Figure 16 below.

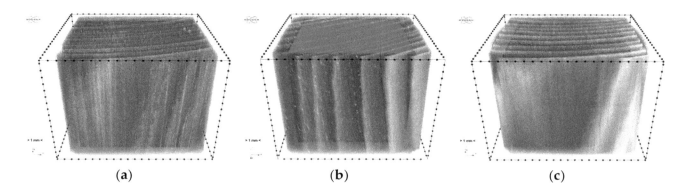

Figure 16. Selection of VOI: (a) sample "1"; (b) sample "2"; (c) sample "3".

As mentioned, there is a correlation between greyscale and density. Thanks to the use of such a correlation in the linear form as in [41,42], the spatial distributions of local density in the analysed samples (Figure 17) and the statistical distributions (histograms) of density in the VOI of the particular samples (Figure 18) were determined. The correlation coefficients were determined by comparing the mean density of a given sample with the average grey level in VOI, and the density of the air with the average grey level of area outside the sample (visible in the images obtained from X-ray micro-computed tomography). The coefficient of proportionality of this correlation, determined independently for each of the three samples, amounted to respectively 5.884, 5.808 and 5.906. The distributions presented below were obtained using the mean value of this coefficient, i.e., 5.866.

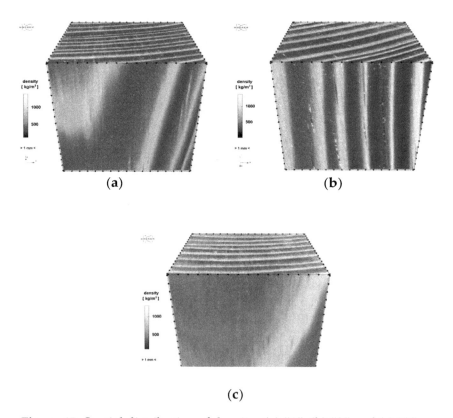

Figure 17. Spatial distribution of density: (a) "1"; (b) "2" and (c) "3".

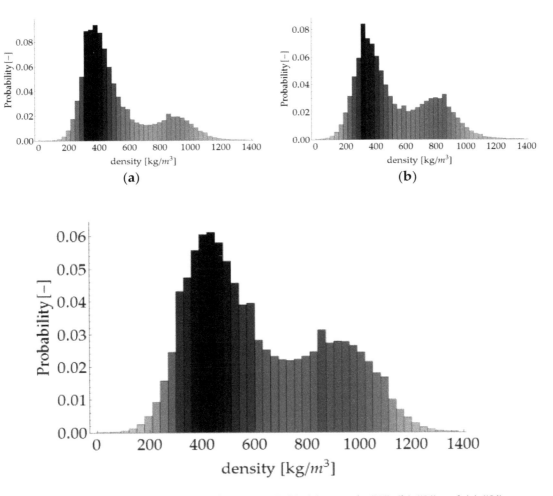

Figure 18. Histogram of local density in VOI: (**a**) sample "1"; (**b**) "2" and (**c**) "3".

As one can see, the above histograms represent bimodal probability distributions. This means that there occur two principal components. In the case of this analysis, these are the early and late growth rings. Segmentation, i.e. partitioning the image into segments occupied by the particular components, was performed using the thresholding preceded by a smoothing filter. The threshold value of brightness corresponds to the grey level value at which the image histogram reaches a minimum (sample "1": 135, "2": 122, "3": 149). A morphometric analysis was carried out for the two image components. In particular, the volume fraction of the component and the spatial distribution of its local thickness [47] were determined and then its mean thickness was calculated. The calculated values are shown in Table 11 and the thickness distributions are presented in Figures 19 and 20. In order to avoid the error connected with determining thickness at the boundary of VOI, the area of the latter was limited to the volume of $(1000 \text{ vox})^3$ by applying dilation with a radius of 100 voxels.

Table 11. Summary of analysis of micro-computed tomography images.

Parameter	Value in Sample:		
	"1"	"2"	"3"
Mean density [kg/m^3]	510	519	619
Mean density of early wood [kg/m^3]	410	367	463
Mean density of late wood [kg/m^3]	912	814	934
Volume fraction of early wood [%]	80.1	66.2	66.9
Volume fraction of late wood [%]	19.8	33.6	33.1
Mean thickness of early growth [mm]	1.29	1.34	1.26
Mean thickness of late growth [mm]	0.35	0.70	0.56

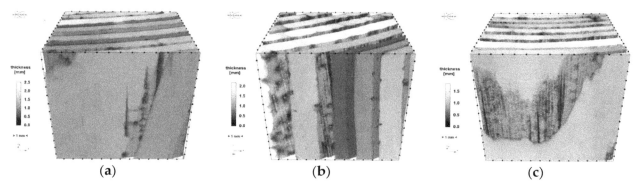

Figure 19. Distribution of local thickness of early growth rings: (**a**) sample "1"; (**b**) sample "2"; (**c**) sample "3".

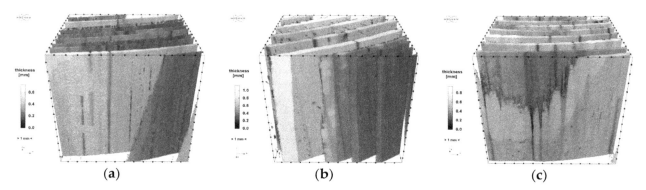

Figure 20. Distribution of local thickness of later growth rings: (**a**) sample "1"; (**b**) sample "2"; (**c**) sample "3".

4.5. Comparative Analysis for Samples from Different Zones

Results obtained for samples from different zones: the zone impacted by waves (sample 1), the zone immersed in water (sample 2) and the zone embedded in the ground (sample 3) are summarised in the Table 12.

Table 12. Selected results for different samples.

Testing Method	Parameter	Value in Sample:		
		"1"	"2"	"3"
Laboratory test	Mean density (at moisture content 18%) [kg/m^3]	511.5	514.4	623.0
Acoustic method (Fakopp)	Mean MOE$_{dyn}$ perpendicular to the grain [GPa]	0.98	0.77	1.6
Resistance drilling method	Mean RM [%]	16.5	16.3	16.9
	Mean FM [%]	47.4	47.7	55.3

The obtained results confirm the engineering "intuition" about the impact of the environment on the degradation level of the material. The best condition of wood was observed for sample embedded in the soil (sample 3). The highest values of density were obtained for this sample. The same applies to the modulus of elasticity and the FM parameter, which are positively correlated with density. At the same time differences between parameters for samples from the zone impacted by waves (sample 1) and the zone immersed in water (sample 2) are not significant. Moreover, in the whole member there were no changes observed in the material within the surface layer exposed to direct environmental impact in the contact zone. Eventually, despite the noticeable differences between the different zones, it can be stated that in the tested timber members, after 70 years of operation, no significant destruction, reducing the safety of use, was found. In that sense, the condition of a wharf timber sheet wall's material may be described as fairly good.

5. Conclusions

For a reliable assessment of the technical condition of timber structures, the use of non-destructive examinations is recommended, in addition to visual evaluation. Still, there are no comprehensive studies in this area, which present correlations enabling estimation of the mechanical parameters of wood and the degree of destruction, although some attempts of predicting these parameters or instance by regression analysis were made (among others [48]).

None of the currently known non-destructive methods used to assess the condition of timber members does not allow for an unambiguous estimation of the strength characteristics of wood. This is not possible even when using the X-ray method [41], which enables relatively accurate measurement of wood density. It results from the material inhomogeneous internal structure, including different defects, f. ex. knots, which have a significant impact on the strength parameters of wood.

In the case of the resistance drilling tests, the obtained coefficients of determination between drilling resistance and density ($R^2 = 0.0226$) and between the feed force and density ($R^2 = 0.2012$) do not indicate any correlation between the results. Drilling resistance testing should be treated not as a quantitative, but qualitative assessment. The results obtained by means of this method can be used to estimate the depth of material damage in static load analyses to reduce the cross sections of the analysed members [14,49].

The acoustic testing (using the Sylvatest Trio device and the Fakopp Microsecond Timer) provided data on the value of the dynamic modulus of elasticity which can be correlated with the static modulus of elasticity. The latter is the basic mechanical parameter needed to carry out a global structural analysis. The obtained coefficients of determination between the values of MOE_{dyn} yielded by the two measuring methods ($R^2 = 0.9064$ along the grain and $R^2 = 0.6881$ across the grain) for the selected samples are satisfactory, showing the results to be reliable. The velocity of acoustic wave propagation (and so the modulus of elasticity) clearly decreases as the moisture content in the wood increases. The acoustic testing methods can be regarded as useful for estimating material stiffness parameters (Young's modulus), but they require further research in order to develop correlations comprising wood moisture content. Currently, research is underway to correlate the dynamic modulus of elasticity with the static modulus of elasticity. On the basis of the determined value of modulus of elasticity, it is possible to estimate mechanical parameters of material, for example according to standard procedure [50].

One should bear in mind that the two methods supply information about the local state of the material. In order to determine the global parameters, one should perform the largest possible number of measurements, which is not always possible, especially in the case of in situ testing. Therefore, in order to obtain the most accurate data on the tested member or structure, it is recommended to combine several testing methods.

Despite their non-destructive character, in most cases the testing methods require samples to be taken to determine the density of the material.

Thanks to the use of X-ray micro-computed tomography, the internal microstructure of the wood could be imaged. The results of the laboratory density measurements were used as input data for determining the correlation between the greyscale of the tomography results and the local density of the wood. Consequently, it became possible, for example, to estimate the density of the early and late wood. It should be noted that the correlation was determined independently for each of the three tested samples and very good agreement was obtained. The values of the coefficient of proportionality do not differ by more than 2%. This means that such a correlation can be a tool for the precise evaluation of the local density of the tested material on the microscale. As a result of the morphomorphic analysis based on the scanning results the volume fractions and morphology of the particular wood components, i.e. the early and late growth rings, could be determined. The data acquired from the analysis in the microtomograph can be useful in micromechanical modelling aimed at estimating the effective parameters of the material on the basis of microscale information.

Summing up, resistance drilling tests enable to determine the depth of material decayed zones, acoustic methods provide estimation of mechanical parameters. The application of the X-ray microtomography allows detailed insight to be gained into the microstructure of the material in a different scale of observation. In particular, it makes it possible to determine the occurrence of microdefects and to determine the parameters (density) of wood constituents, i.e. early and late growths. It must be pointed out that the applied methods are not equivalent, but rather, they are complementary.

The paper presents the methodology for comprehensive wood testing in structural members using the described research methods. The set of results obtained from these methods makes it possible to assess the material, and consequently, to perform a global analysis of the structure. In particular, it is possible to estimate the value of mechanical parameters, whereas the qualitative evaluation makes it possible to determine the extent of possible material destruction and the location of any defects.

Author Contributions: Conceptualization, T.N., M.W.; methodology, A.K., T.N., M.S., M.W.; software, M.S.; validation, A.K., T.N., M.S., M.W.; formal analysis, A.K., T.N., M.S., M.W.; investigation, A.K., T.N., M.S.; resources, A.K., T.N., M.S., M.W.; data curation, A.K., T.N., M.S., M.W.; writing—original draft preparation, A.K., T.N., M.S., M.W.; writing—review and editing, A.K.; visualization, M.S.; supervision, T.N., M.W.; project administration, A.K., M.W.; funding acquisition, T.N., M.S.

Acknowledgments: We thank Filip Patalas and Krzysztof Wujczyk for their help in preparing the samples.

References

1.   Francesca, C.; Paolo Simonini, P.; Lionello, A. Long-term mechanical behavior of wooden pile foundation in Venice. In Proceedings of the 2nd International Symposium on Geotechnical Engineering for the Preservation of Monuments and Historic Sites, Napoli, Italy, 30–31 May 2013. [CrossRef]

2.   Klaassen, K.W.M.; Creemers, J.G.M. Wooden foundation piles and its underestimated relevance for cultural heritage. *J. Cult. Herit.* **2012**, *135*, 123–128. [CrossRef]

3.   Nilson, T.; Rowell, R. Historical wood—Structure and properties. *J. Cult. Herit.* **2012**, *135*, 55–59. [CrossRef]

4.   Klaassen, K.W.M. Life Expectation of Wooden Foundations—A Non-Destructive Approach. In Proceedings of the International Symposium Non-Destructive Testing in Civil Engineering (NDT-CE), Berlin, Germany, 15–17 September 2015; pp. 775–779.

5.   Klaassen, R.K.M. Bacterial decay in wooden foundation piles—Pattern and causes: A study of historical foundation piles the Netherlands. *Int. Biodeterior. Biodegrad.* **2008**, *61*, 45–60. [CrossRef]

6.   Panshin, A.J.; de Zeeuw, C. *Textbook of Wood Technology. Structure, Identification, Properties, and Uses of the Commercial Woods in the United States and Canada*, 4th ed.; Mc Graw-Hill Book Company: New York, NY, USA, 1980.

7.   Tannert, T.; Anthony, R.; Kasal, B.; Kloiber, M.; Piazza, M.; Riggio, M.; Rinn, F.; Widmann, R.; Yamaguchi, N. In situ assessment of structural timber using semi-destructive techniques. *Mater. Struct.* **2014**, *47*, 767–785. [CrossRef]

8.   Riggio, M.; Anthony, R.; Augelli, F.; Kasal, B.; Lechner, T.; Muller, W.; Tannert, T. In situ assessment of structural timber using non-destructive techniques. *Mater. Struct.* **2014**, *47*, 749–766. [CrossRef]

9.   Dolwin, J.A.; Lonsdale, D.; Barnet, J. Detection of decay in trees. *Arboric. J.* **1999**, *23*, 139–149. [CrossRef]

10.   Kasal, B.; Lear, G.; Tannert, T. Stress waves. In *In Situ Assessment of Structural Timber. RILEM State-of-the-Art Reports*; Kasal, B., Tannert, T., Eds.; Springer: Dordrecht, The Netherlands, 2010; Volume 7, pp. 5–24, ISBN 978-94-007-0559-3.

11.   Dackermann, U.; Crews, K.; Kasal, B.; Li, J.; Riggio, M.; Rinn, F.; Tannert, T. In situ assessment of structural timber using stress-wave measurements. *Mater. Struct.* **2014**, *47*, 787–803. [CrossRef]

12.   Wang, X.; Divos, F.; Pilon, C.; Brashaw, B.K.; Ross, R.J.; Pellerin, R.F. *Assessment of Decay in Standing Timber Using Stress Wave Timing Nondestructive Evaluation Tools: A Guide for Use and Interpretation*; Gen. Tech. Rep. FPL-GTR-147; US Department of Agriculture, Forest Service, Forest Products Laboratory: Madison, WI, USA, 2004. [CrossRef]

13.   Sandoz, J.L. Grading of construction timber by ultrasound. *Wood Sci. Technol.* **1989**, *23*, 95–108. [CrossRef]

14. Lechner, T.; Nowak, T.; Kliger, R. In situ assessment of the timber floor structure of the Skansen Lejonet fortification, Sweden. *Constr. Build. Mater.* **2014**, *58*, 85–93. [CrossRef]

15. García, M.C.; Seco, J.F.G.; Prieto, E.H. Improving the prediction of strength and rigidity of structural timber by combining ultrasound techniques with visual grading parameters. *Mater. Constr.* **2007**, *57*, 49–59. [CrossRef]

16. Lourenço, P.B.; Feio, A.O.; Machado, J.S. Chestnut wood in compression perpendicular to the grain: Non-destructive correlations for test results in new and old wood. *Constr. Build. Mater.* **2007**, *21*, 1617–1627. [CrossRef]

17. Ilharco, T.; Lechner, T.; Nowak, T. Assessment of timber floors by means of non-destructive testing methods. *Constr. Build. Mater.* **2015**, *101*, 1206–1214. [CrossRef]

18. Íñiguez, G.; Martínez, R.; Bobadilla, I.; Arriaga, F.; Esteban, M. Mechanical properties assessment of structural coniferous timber by means of parallel and perpendicular to the grain wave velocity. In Proceedings of the 16th International Symposium on Nondestructive Testing of Wood, Beijing, China, 11–13 May 2009.

19. Esteban, M.; Arriaga, F.; Íñiguez, G.; Bobadilla, I. Structural assessment and reinforcement of ancient timber trusses. In Proceedings of the International Conference on Structures & Architecture, Guimarães, Portugal, 21–23 July 2010.

20. Nowak, T.; Hamrol-Bielecka, K.; Jasieńko, J. Experimental testing of glued laminated timber members using ultrasonic and stress wave techniques. In Proceedings of the International Conference on Structural Health Assessment of Timber Structures, SHATIS '15, Wroclaw, Poland, 9–11 September 2015; pp. 523–533.

21. Fakopp Enterprise Microsecond Timer. Available online: http://www.fakopp.com/site/microsecond-timer (accessed on 20 November 2018).

22. Rinn, F. Practical application of micro-resistance drilling for timber inspection. *Holztechnologie* **2013**, *54*, 32–38.

23. Hiroshima, T. Applying age-based mortality analysis to a natural forest stand in Japan. *J. For. Res.* **2014**, *19*, 379–387. [CrossRef]

24. Wang, S.Y.; Chiu, C.M.; Lin, C.J. Application of the drilling resistance method for annual ring characteristics: Evaluation of Taiwania (Taiwania cryptomerioides) trees grown with different thinning and pruning treatments. *J. Wood Sci.* **2003**, *49*, 116–124. [CrossRef]

25. Feio, A.O.; Machado, J.S.; Lourenço, P.B. Compressive behavior and NDT correlations for chestnut wood (Castanea sativa Mill.). In Proceedings of the 4th International Seminar on Structural Analysis of Historical Constructions, Padova, Italy, 10–13 November 2004; pp. 369–375.

26. Jasieńko, J.; Nowak, T.; Bednarz, Ł. Baroque structural ceiling over the Leopoldinum Auditorium in Wrocław University: Tests, conservation, and a strengthening concept. *Int. J. Archit. Herit.* **2014**, *8*, 269–289. [CrossRef]

27. Branco, J.M.; Piazza, M.; Cruz, P.J. Structural analysis of two King-post timber trusses: Non-destructive evaluation and load-carrying tests. *Constr. Build. Mater.* **2010**, *24*, 371–383. [CrossRef]

28. Kloiber, M.; Tippner, J.; Hrivnák, J. Mechanical properties of wood examined by semi-destructive devices. *Mater. Struct.* **2014**, *47*, 199–212. [CrossRef]

29. Morales-Conde, M.J.; Rodríguez-Liñán, C.; Saporiti-Machado, J. Predicting the density of structural timber members in service. The combine use of wood cores and drill resistance data. *Mater. Constr.* **2014**, *64*, 1–11. [CrossRef]

30. Acuña, L.; Basterra, L.A.; Casado, M.M.; López, G.; Ramón-Cueto, G.; Relea, E.; Martínez, C.; González, A. Application of resistograph to obtain the density and to differentiate wood species. *Mater. Constr.* **2011**, *61*, 451–464. [CrossRef]

31. Tseng, Y.J.; Hsu, M.F. Evaluating the mechanical properties of wooden components using drill resistance method. In Proceedings of the 10th World Conference on Timber Engineering, Miyazaki, Japan, 2–5 June 2008; pp. 303–310.

32. Nowak, T.; Jasieńko, J.; Hamrol-Bielecka, K. In situ assessment of structural timber using the resistance drilling method–evaluation of usefulness. *Constr. Build. Mater.* **2016**, *102*, 403–415. [CrossRef]

33. Jasieńko, J.; Nowak, T.; Hamrol, K. Selected methods of diagnosis of historic timber structures–principles and possibilities of assessment. *Adv. Mater. Res.* **2013**, *778*, 225–232. [CrossRef]

34. Piazza, M.; Riggio, M. Visual strength-grading and NDT of timber in traditional structures. *J. Build. Apprais.* **2008**, *3*, 267–296. [CrossRef]

35. Sousa, H. Methodologies for Safety Assessment of Existing Timber Structures. Ph.D. Thesis, Department of Civil Engineering, University of Minho, Minho, Portugal, 2013.

36. Salvo, L.; Cloetens, P.; Maire, E.; Zabler, S.; Blandin, J.J.; Buffière, J.Y.; Ludwig, W.; Boller, E.; Bellet, D.; Josserond, C. X-ray micro-tomography an attractive characterisation technique in materials science. *Nucl. Instrum. Methods Phys. Res. Sect. B Beam Interact. Mater. At.* **2003**, *200*, 273–286. [CrossRef]

37. Schabowicz, K.; Jóźwiak-Niedźwiedzka, D.; Ranachowski, Z.; Kudela, S.; Dvorak, T. Microstructural characterization of cellulose fibres in reinforced cement boards. *Arch. Civ. Mech. Eng.* **2018**, *18*, 1068–1078. [CrossRef]

38. Cała, M.; Cyran, K.; Kawa, M.; Kolano, M.; Łydżba, D.; Pachnicz, M.; Rajczakowska, M.; Różański, A.; Sobótka, M.; Stefaniuk, D.; et al. Identification of Microstructural Properties of Shale by combined Use of X-Ray Micro-CT and Nanoindentation Tests. *Procedia Eng.* **2017**, *191*, 735–743. [CrossRef]

39. Feldkamp, L.A.; Davis, L.C.; Kress, J.W. Practical cone-beam algorithm. *J. Opt. Soc. Am. A* **1984**, *1*, 612–619. [CrossRef]

40. Elliott, J.C.; Dover, S.D. X-ray microtomography. *J. Microsc.* **1982**, *126*, 211–213. [CrossRef]

41. Lechner, T.; Sandin, Y.; Kliger, R. Assessment of density in timber using X-ray equipment. *Int. J. Archit. Herit.* **2013**, *7*, 416–433. [CrossRef]

42. Lazarescu, C.; Watanabe, K.; Avramidis, S. Density and moisture profile evolution during timber drying by CT scanning measurements. *Dry. Technol.* **2010**, *28*, 460–467. [CrossRef]

43. Rajczakowska, M.; Stefaniuk, D.; Łydżba, D. Microstructure Characterization by Means of X-Ray Micro-CT and Nanoindentation Measurements. *Studia Geotech. Mech.* **2015**, *37*, 75–84. [CrossRef]

44. *PN-EN 384:2016-10—Structural Timber. Determination of Characteristic Values of Mechanical Properties and Density*; Polish Committee for Standarization: Warsaw, Poland, 2016.

45. Kraft, U.; Pribbernow, D. *Handbuch der Holzprüfung. Anleitungen und Beispiele*; Verlag Bau+Technik GmbH: Düsseldorf, Germany, 2006.

46. Feio, A.O.; Lourenço, P.B.; Machado, J.S. Non-destructive evaluation of the mechanical behavior of chestnut wood in tension and compression parallel to grain. *Int. J. Archit. Herit.* **2007**, *1*, 272–292. [CrossRef]

47. Hildebrand, T.; Rüegsegger, P. A new method for the model-independent assessment of thickness in three-dimensional images. *J. Microsc.* **1997**, *185*, 67–75. [CrossRef]

48. Sousa, H.S.; Branco, J.M.; Machado, J.S.; Lourenço, P.B. Predicting mechanical properties of timber elements by regression analysis considering multicollinearity of non-destructive test results. In Proceedings of the International Conference on Structural Health Assessment of Timber Structures, SHATIS '17, Istanbul, Turkey, 20–22 September 2017; pp. 485–493.

49. Cuartero, J.; Cabaleiro, M.; Sousa, H.S.; Branco, J.M. Tridimensional parametric model for prediction of structural safety of existing timber roofs using laser scanner and drilling resistance tests. *Eng. Struct.* **2019**, *185*, 58–67. [CrossRef]

50. *PN-EN 338:2016-06—Structural Timber. Strength Classes*; Polish Committee for Standarization: Warsaw, Poland, 2016.

10

Viscoelastic Parameters of Asphalt Mixtures Identified in Static and Dynamic Tests

Piotr Mackiewicz *[ID] and **Antoni Szydło**

Faculty of Civil Engineering, Wrocław University of Science and Technology, 50-370 Wrocław, Poland
* Correspondence: piotr.mackiewicz@pwr.edu.pl

Abstract: We present two methods used in the identification of viscoelastic parameters of asphalt mixtures used in pavements. The static creep test and the dynamic test, with a frequency of 10 Hz, were carried out based on the four-point bending beam (4BP). In the method identifying viscoelastic parameters for the Brugers' model, we included the course of a creeping curve (for static creep) and fatigue hysteresis (for dynamic test). It was shown that these parameters depend significantly on the load time, method used, and temperature and asphalt content. A similar variation of parameters depending on temperature was found for the two tests, but different absolute values were obtained. Additionally, the share of viscous deformations in relation to total deformations is presented, on the basis of back calculations and finite element methods. We obtained a significant contribution of viscous deformations (about 93% for the static test and 25% for the dynamic test) for the temperature 25 °C. The received rheological parameters from both methods appeared to be sensitive to a change in asphalt content, which means that these methods can be used to design an optimal asphalt mixture composition—e.g., due to the permanent deformation of pavement. We also found that the parameters should be determined using the creep curve for the static analyses with persistent load, whereas in the case of the dynamic studies, the hysteresis is more appropriate. The 4BP static creep and dynamic tests are sufficient methods for determining the rheological parameters for materials designed for flexible pavements. In the 4BP dynamic test, we determined relationships between damping and viscosity coefficients, showing material variability depending on the test temperature.

Keywords: viscoelastic parameters; creep test; fatigue tests; asphalt mixtures; Burgers model; four point bending beam

1. Introduction

The selection of correct material parameters is very important, both in engineering practice and scientific study. The determination of reliable material properties is also essential in further structural analyses. The appropriate material parameters and the model enable the use of efficient numerical methods, and determine the state of stresses and deformations in the construction model. It is especially important for asphalt mixtures used as the main material in vulnerable road pavements. Such mixtures are thermo-rheological, changing their properties under thermal conditions and load time. In various conditions, both under static and dynamic loads, they reveal their rheological characteristics. These properties are much more important in the description of the material in higher temperatures than in lower ones, in which linear–elastic models are sufficient to model the material parameters. The asphalt layers in the road pavements show both elastic and viscous features. The elastic properties dominate at the lower temperatures, and are responsible for irreversible deformations of the asphalt pavement, whereas the viscous features are typical of the higher temperatures. Therefore, proper identification of the rheological parameters of asphalt mixtures based on the results of laboratory tests is not easy. The typical static tests in which these parameters are defined include the static creep, testing under the constant load when the cylindrical specimens are compressed and the beams are bent [1]. Dynamic tests are analogous, testing with compressed cylindrical specimens [2] and bend fatigue beams.

Different rheological parameters can be obtained in various mounting schemes and load conditions, characterized by duration and frequency. Therefore, the choice of proper research method is important. This method is used to determine these parameters and the models describing the behaviour of the structure. In the case of road pavements, both static (parking lots, crossroads, etc.) and repetitive loads with short-term impact are analysed. As mentioned earlier, the asphalt mixtures become viscous over time in high temperatures for a long-term static load, whereas the accumulation of permanent deformations resulting in permanent deformation (i.e., ruts) occurs under dynamic loading. However, changing the viscoelastic dissipative energy is also important in fatigue tests. This change significantly affects the fatigue destruction of the material. This publication analyses the behaviour of the asphalt mixture under the static and dynamic loading for a four-point bending beam (4BP).

2. Identification of Rheological Parameters

Many rheological models are used in the common road practice. As already mentioned, the asphalt mixtures expose their rheological properties at high temperatures. The viscoelastic models are used to describe these properties. The viscoelasticity theory is increasingly being used in the analysis of asphalt pavements, due to its good description of flow and deformation of road materials.

According to Reiner and Ward [3], the first papers about rheology come from the thirties of the previous century. However, this discipline has intensively developed since the 1950s [4], and deals with materials and constructions of buildings as well as road pavements. In the fundamental work edited by Reiner and Ward [3], there is a chapter devoted to the rheology of materials and asphalt pavements written by Van der Poela [5]. Regarding asphalt mixtures, one of the first works by Monismith et al. [6] deserves special attention. The authors found that asphalt pavement mixtures also exhibit linear viscoelastic properties at very low deformations. By studying the creep of the asphalt pavement mixtures, Vakili [7] draws the same conclusions. Goodrich [8] studied asphalt mixtures with mineral fillers, as well as the large aggregate under oscillations with small amplitudes, and found again that these materials show linear viscoelastic features at very small deformations.

In theoretical considerations, Kisiel and Lysik [9], Nowacki [10], and Jakowluk [11] contributed significantly to development the rheology in construction. The use of rheological models in the description of asphalt mixtures can be found, among others, in the works [12–15]. The identified rheological parameters were also studied under different static [16–19] and dynamic load conditions [6,13,20]. However, no comprehensive comparisons have been made to the four-point study of static and dynamic conditions, although the 4BP is commonly used. There are also no comparisons to other various laboratory studies.

Currently, there are many analytic methods [21–23] and numerical models, including micromechanical models [24] and anisotropic models [25,26], in which the material parameters of asphalt mixtures are used in the assessment of the behaviour of flexible pavement.

Currently, due to the high availability of software for numerical calculations, no attention is paid to the selection of the appropriate research method, the application of the appropriate model, or the use of valid parameters in the models of surfaces. Moreover, the entire creep curve is not included with the load curve in the determination of parameters. Both simple and complex rheological models were analysed. For example, the complex constitutive models, with and without damage, can be found in [27–31]. Other studies have addressed advanced pavement structural models with and without dynamic effects [32–34].

It has been found that the Burgers model, among many other viscoelastic models, reliably describes asphalt concrete behaviour [12,17,35,36]. The model diagram along with its parameters is shown in Figure 1.

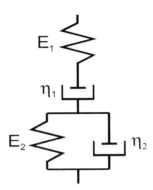

Figure 1. Burgers rheological model. E_1: immediate elastic modulus in the Burgers model (Pa); E_2: delayed elastic modulus in the Burgers model (Pa); η_1: viscosity coefficient in the Burgers model (Pa·s), η_2: viscosity coefficient of elastic delay in the Burgers model (Pa·s).

The study of static creeping was performed under the 4BP bending conditions. The creep curve in the Burgers model has its graphic interpretation, shown in Figure 2. Parameters can be determined by immediate deformations, maximum deformations (elastic moduli), and the rate of deformations (viscosity coefficients). However, such interpretation is not very accurate, because large errors can occur when the immediate deformations are registered during elastic recurrence. Therefore, we proposed to determine these parameters using numerical methods, taking into account the overall creep curve at loading and the curve at unloading.

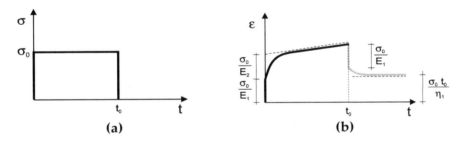

Figure 2. The creep curve in the Burgers model. The graphical finding of the parameters is presented. (**a**)—stress vs. time curve, (**b**)—strain vs. time curve.

A conjugated gradient method was used to approximate the laboratory creep curve using the theoretical curve. This is an effective method for solving optimization problems. The minimum of function was determined from each point in a given search direction. The rheological parameters of the Burgers model were the sought variables: E_1, E_2, η_1, η_2 (see Figure 1). The target function is described by Equation (1):

$$\Delta p = \sqrt{\frac{\sum\limits_{i=1}^{i=l} (\varepsilon_{ti} - \varepsilon_{li})^2}{l}} \cdot 100\% \tag{1}$$

where ε_{li} is the deformation measured on the sample, ε_{ti} is the theoretical deformation calculated for the model, and l is the number of measured points.

The theoretical deformations were determined from the equations of the Burgers model, starting from the differential constitutive relationship between stress σ and deformation ε:

$$\sigma + a\dot{\sigma} + b\ddot{\sigma} = c\dot{\varepsilon} + d\ddot{\varepsilon} \tag{2}$$

$$\sigma + \left(\frac{\eta_1}{E_1} + \frac{\eta_1}{E_2} + \frac{\eta_2}{E_2}\right) \cdot \dot{\sigma} + \frac{\eta_1\eta_2}{E_1 E_2}\ddot{\sigma} = \eta_1\dot{\varepsilon} + \frac{\eta_1\eta_2}{E_2}\ddot{\varepsilon} \tag{3}$$

After the solution of these equations, the relationship between the deformation $\varepsilon(t)$ and the time t was obtained Equations (4) and (5):

$$\text{for the load } t < t_0 \quad \varepsilon(t) = \sigma_0 \left[\frac{1}{E_1} + \frac{t}{\eta_1} + \frac{1}{E_2} \left(1 - e^{\frac{-tE_2}{\eta_2}} \right) \right], \tag{4}$$

$$\text{or the unload } t > t_0 \quad \varepsilon(t) = \sigma_0 \left[\frac{t_0}{\eta_1} - \frac{1}{E_2} e^{\frac{-tE_2}{\eta_2}} \left(1 - e^{\frac{t_0 E_2}{\eta_2}} \right) \right] \tag{5}$$

The identification of rheological parameters can effectively contribute to the optimization of mixture composition also under fatigue conditions, when there exists energy dissipation due to microcracks. The procedure for determining the rheological parameters under repetitive stress conditions was performed. In this test, the parameters were determined at the 10 Hz load frequency. This frequency was applied according to the European Standard EN 12697-24:2012 [37], in order to evaluate the fatigue characteristics of asphalt mixtures. The identification of the parameters was performed by the selection of parameters in the Burgers model for hysteresis, describing the relationship between stress σ and deformation ε (Figure 3), and using the conjugate gradient method to minimize the function in Equation (1).

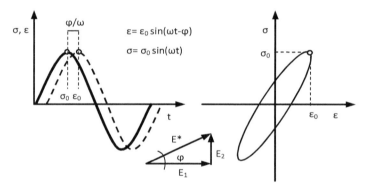

Figure 3. Fatigue hysteresis used for the identification of parameters. ϕ: phase angle (°); ω: angular frequency = $2\pi f$ (1/s); t: time (s); σ_0: the amplitude of stress (MPa); ε_0: the amplitude of deformation (-); E^*: composite modulus (MPa); E_1: the real element of the composite modulus (MPa); E_2: the imaginary element of composite modulus (MPa).

The controlled amplitude of displacement and the time of its delay in relation to the acting force were recorded directly on the basis of the fatigue strength test, using the variable force. Based on the basic dependencies for the 4BP beam (Equations (6)–(8)), it is possible to determine the required stress value σ, deformation ε, and phase angle ϕ in any cycle and at any time of the load, as follows:

$$\sigma = \frac{3Pa}{bh^2} \tag{6}$$

$$\varepsilon = \frac{12\Delta h}{3L^2 - 4a^2} \tag{7}$$

$$\varphi = 360 fs \tag{8}$$

where P is the force (N); b and h are the beam width and height, respectively (m); a is the distance between the support and the force (m), and $a = L/2$; Δ is the displacement (m); L is the spacing of the supports (m); f is the frequency (Hz), and $f = \omega/2\pi$; and s is the delay time between the force P and the displacement Δ (s).

According to Figure 3, it is possible to determine the complex stiffness modulus E^* and phase angle φ between deformation and stress:

$$E* = \frac{\sigma_0 \sin(\omega t)}{\varepsilon_0 \sin(\omega t - \varphi)} \qquad (9)$$

$$tg\varphi = \frac{E_2}{E_1} \qquad (10)$$

Additional conditions for the agreement between the phase angle and the composite modulus found in the test and model were introduced in the search criteria for the most accurate matching of the laboratory results, with the hysteresis determined by the Burgers model. The dependence of the phase angle and the complex modulus on the parameters in the Burgers model are described by Equations (11) and (12):

$$E* = \omega \left[\frac{c^2 + (d\omega)^2}{(b\omega^2 - 1)^2 + (a\omega)^2} \right]^{1/2} \qquad (11)$$

$$tg\varphi = \frac{ad\omega^2 - (b\omega^2 - 1)c}{(b\omega^2 - 1)d\omega + ac\omega} \qquad (12)$$

Using Equation (2), for the cyclic symmetrical sinusoidal deformation $\varepsilon = \varepsilon_0 \sin(\omega t - \phi)$, we obtain the relationship

$$\sigma + a\dot{\sigma} + b\ddot{\sigma} = -\varepsilon_0 \omega [d\omega \sin(\omega t) - c \cos(\omega t)] \qquad (13)$$

The variables a, b, c, and d present in the constitutive Equation (13) are described by Equations (14)–(17):

$$a = \frac{\eta_1}{E_1} + \frac{\eta_1}{E_2} + \frac{\eta_2}{E_2} \qquad (14)$$

$$b = \frac{\eta_1 \eta_2}{E_1 E_2} \qquad (15)$$

$$c = \eta_1 \qquad (16)$$

$$d = \frac{\eta_1 \eta_2}{E_2} \qquad (17)$$

3. Materials and Methods

The static creeping test with 4BP bending was performed on the NAT (Nottingham Asphalt Tester, University of Nottingham, Nottingham, UK) device, which enables the efficient testing of many asphalt mixtures under various mounting and loading patterns. This device is characterized by the good reproducibility of results. The technical conditions for the study were adopted according to manual (Cooper Research Technology [38]). The following conditions were applied: a constant load with 0.30 MPa (15% of the bending strength at 25 °C), a load time of 1800 s, and an unload time of 510 s. In order to determine rheological parameters, we applied four temperatures: −5 °C, 0 °C, 10 °C, and 25 °C. The dimensions of the samples were as follows: the width was 60 mm, the height was 50 mm, and the length was 384 mm. In Figure 4, a schematic diagram of the static 4BP creep test is shown.

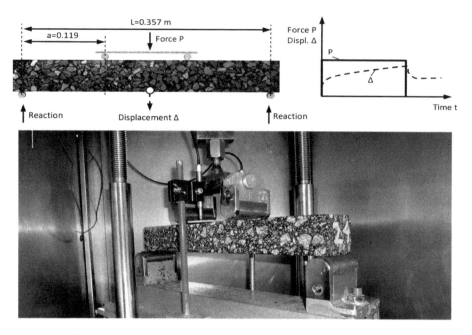

Figure 4. The scheme of the static four-point bending beam (4BP) creep test.

The 4BP dynamic test consists of the cyclic bending of the beam supported in four points, as shown in Figure 5 (accordance with EN 12697-24:2012 [37]). Due to common research practice, the study was conducted under the sinusoidal kinematic constraints, with the controlled deformation. The amplitude of deformation was 100×10^{-6}. This method allows us to compare the received results to the known fatigue criteria in the design practice [35,39]. The dimensions of the bending beams and the temperature conditions were assumed as for the static testing. The basic parameter that was determined during the test was the fatigue hysteresis, which depends on the number of load cycles. The tests with the fixed peak-to-peak strain allowed us to record the change in stresses in relation to the load cycles.

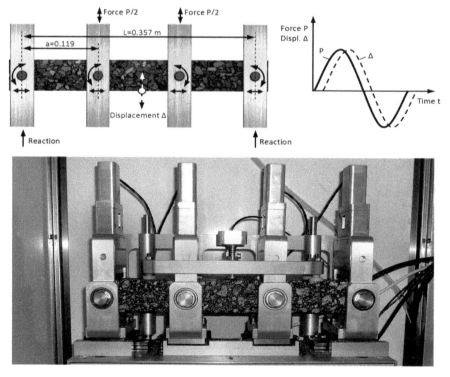

Figure 5. The scheme of the dynamic 4BP test.

The research was carried out with Cooper Research Technology Ltd. Beam-Flex, on typical asphalt mixture commonly used in building road pavements, which was laid on the binding surfaces AC16W with asphalt 35/50. Mixtures with different asphalt content were analyzed—i.e., 4.0%, 4.5%, and 5.3%. The formulas of the mixtures were previously designed in accordance with the current technical requirements.

4. Results of the Parameters' Identification

Based on the presented procedure for the identification of the viscoelastic parameters and the tests performed under various temperature conditions and asphalt content, we derived the parameters of the Burgers model for the creep study at static (Table 1) and dynamic (Table 2) loading.

Table 1. Rheological parameters of the Burger's model obtained for the creep 4BP study.

Temperature	−5 °C	0 °C	10 °C	25 °C
	asphalt content: 4.0%			
E_1 (MPa)	5470	3619	1498	607
η_1 (MPa·s)	1,497,854	769,973	308,374	121,832
E_2 (MPa)	4311	3250	2235	1213
η_2 (MPa·s)	1,881,548	821,849	34,011	28,569
	asphalt content: 4.5%			
E_1 (MPa)	5733	3902	1762	742
η_1 (MPa·s)	1,570,013	830,165	362,676	148,891
E_2 (MPa)	4519	3504	2628	1483
η_2 (MPa·s)	1,972,192	886,096	40,000	34,914
	asphalt content: 5.3%			
E_1 (MPa)	5135	3474	1601	485
η_1 (MPa·s)	1,406,134	739,012	329,545	97,405
E_2 (MPa)	4047	3119	2388	970
η_2 (MPa·s)	1,766,333	788,801	36,346	22,841

Table 2. Rheological parameters of the Burger's model obtained for the dynamic 4BP study.

Temperature	−5 °C	0 °C	10 °C	25 °C
	asphalt content: 4.0%			
E_1 (MPa)	26,494	20,164	11,703	8001
η_1 (MPa·s)	23,454	17,520	9441	4327
E_2 (MPa)	31,980	23,912	12,865	5797
η_2 (MPa·s)	5165	3399	870	132
	asphalt content: 4.5%			
E_1 (MPa)	27,770	21,740	13,764	9778
η_1 (MPa·s)	24,584	18,890	11,103	5288
E_2 (MPa)	33,521	25,781	15,130	7084
η_2 (MPa·s)	5414	3665	1023	161
	asphalt content: 5.3%			
E_1 (MPa)	24,871	19,353	12,507	6397
η_1 (MPa·s)	22,018	16,816	10,089	3459
E_2 (MPa)	30,022	22,950	13,748	4634
η_2 (MPa·s)	4849	3263	930	105

For a mixture with asphalt content 4.5%, the results of study and the approximation of curves using the Burgers creep model in the static test for various temperatures are shown in Figure 6, whereas the results of the dynamic test, as well as the approximation of curves σ–ε using the Burgers model, are presented in Figures 7–10.

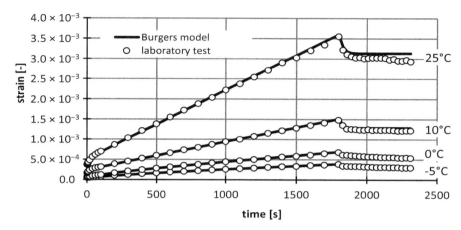

Figure 6. The results of laboratory study and the approximation of creep curves using the Burgers model in the static test (asphalt mixtures with an asphalt content of 4.5%).

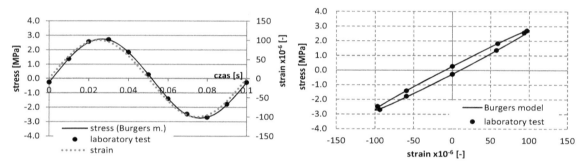

Figure 7. The results of laboratory study and the approximation of curves σ–ε using the Burgers model in the dynamic test at the temperature −5 °C (asphalt mixtures with an asphalt content of 4.5%).

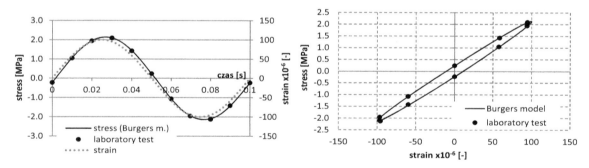

Figure 8. The results of laboratory study and the approximation of curves σ–ε using the Burgers model in the dynamic test at the temperature 0 °C (asphalt mixtures with an asphalt content of 4.5%).

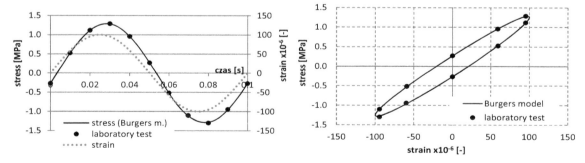

Figure 9. The results of laboratory study and the approximation of curves σ–ε using the Burgers model in the dynamic test at the temperature 10 °C (asphalt mixtures with an asphalt content of 4.5%).

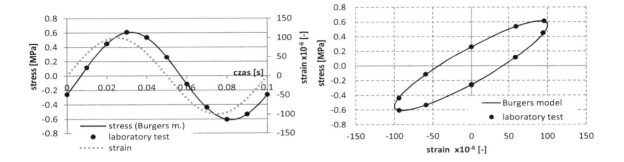

Figure 10. The results of laboratory study and the approximation of curves σ–ε using the Burgers model in the dynamic test at the temperature 25 °C (asphalt mixtures with an asphalt content of 4.5%).

It may be noticed that the obtained parameters differ. Moduli of instant elasticity E_1 obtained in the dynamic 4BP test are about 5 to 13 times larger than those received in the static test. Similarly, moduli of delayed elasticity E_2 are about 5 to 7 times greater in the dynamic than in the static tests. On the other hand, viscosity coefficients η_1 and η_2 are about two and three times smaller in the dynamic test than in the static test, respectively. This results from the very short time of variable loading, and consequently, of the short time of the material deformation response. A comprehensive comparison of changes in parameter values for different temperatures and asphalt content is shown in Figure 11 (a static test) and Figure 12 (a dynamic test).

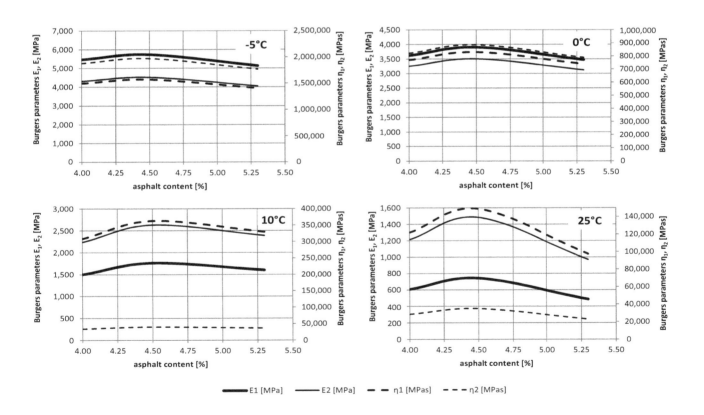

Figure 11. The relationship of Burgers parameters on the temperature in the static test.

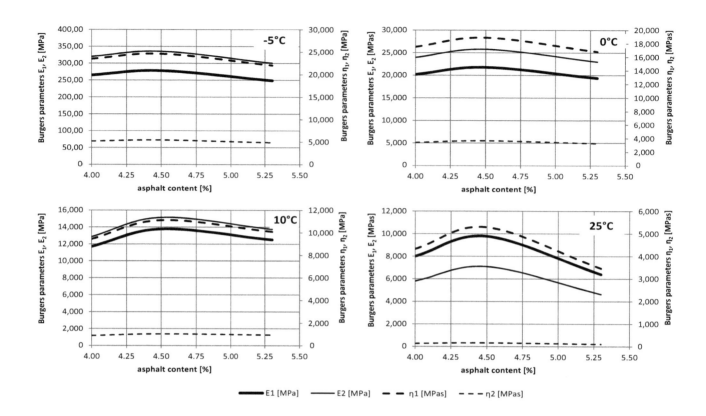

Figure 12. The relationship of Burgers parameters on the temperature in the dynamic test.

For the temperature and asphalt content range analysed, changes in the parameters were observed. For lower temperatures, a smaller change in parameters is observed depending on the asphalt content. For temperature 25 °C, the largest parameter values (except for η_2) were obtained for the optimal asphalt content 4.5%. For other temperatures, there are less pronounced extremes associated with the composition of the mixture.

Moreover, it is worth noting as the load time increased in the creep static test, smaller values of the elastic parameters E_1 and E_2 were obtained, whereas the viscosity parameters η_1 and η_2 were higher. In the dynamic test, the response of material influenced by the short-variable load was more elastic. However, the rheological features were visible over the entire range of analysed temperatures. It is worth noting that the values of parameters, mainly related to the viscosities η_1 and η_2, were correlated with the phase angle ϕ. Its value increases with increasing temperature. The angle change in the low temperature range is practically linear (Figure 13). At higher temperatures and higher angles, the material will have a larger contribution of viscous rather than elastic characters, while the lower the angle, the more elastic the material. Moreover, at higher temperatures, there is a greater variation in the angle value depending on the asphalt content in the mixture. For 5.3% asphalt content, the highest values of the phase angle ϕ were obtained.

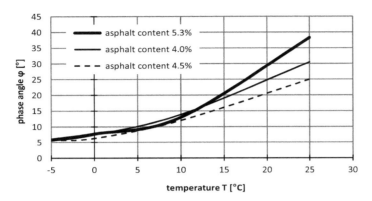

Figure 13. The relationship between the phase angle and temperature.

5. Numerical Verification of Rheological Parameters

For a selected mixture with the optimum asphalt content of 4.5%, the numerical verification of rheological parameters was performed using the finite element method. Three-dimensional static and dynamic models were developed. They included the appropriate sample geometry and load conditions that are consistent with the previously described test procedure (Figure 14). Previously, we analysed the division of the model into finite elements. The dimensions of the model were in agreement with those of the laboratory tested sample. To build the model, we used 410,000 eight-node volume elements. In the middle part of the beam, the density of the element grid was greater. Such discretization allowed for the convergence of results for displacements and deformations. The calculation of the model was carried out in the SOLIDWORKS-COSMOS/M software, ver. 2010, Structural Research and Analysis Corporation, Santa Monica, CA, USA.

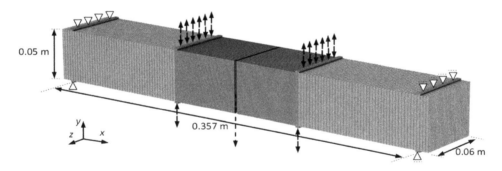

Figure 14. The relationship between the phase angle and temperature.

Rheological parameters were appropriately applied for the static and dynamic testing, according to the date presented in Table 1; Table 2. In the dynamic study, we assumed the density to be 2400 kg/m^3, with appropriate damping parameters associated with the dynamic analysis included in the static testing. The dynamic problem of discretization in Finite Element Method (FEM) is described by the classical equation [40]

$$[M]\{\ddot{u}(t)\} + [C]\{\dot{u}(t)\} + [K]\{u(t)\} = \{F(t)\} \tag{18}$$

where: $[M]$ is the matrix mass, $[C]$ is the damping matrix, $[K]$ is the stiffness matrix, $\{F(t)\}$ is the load vector variable in time t, $\{\ddot{u}(t)\}$ is the acceleration vector in the time t, $\{\dot{u}(t)\}$ is the velocity vector in time t, and $\{u(t)\}$ is the displacement vector in time t.

Selecting appropriate damping for the material is an important issue in the dynamic model. This is a complex phenomenon that involves the dissipation of energy through a variety of mechanisms, such as internal friction, cyclic thermal effects, microscopic material deformation, and micro- and macrocracks. The damping process consists of damping material, structural damping, and viscous damping associates with energy dissipation. To realistically simulate the material behavior under a short-term load, the damping factor is included as an important parameter. This is difficult to model, but the existing damping models are available in numerical calculation programs. The damping models can depend on the frequency or viscosity. The Rayleigh's damping model is quite often used in the structural dynamic analysis. To include damping effects, the damping coefficients α and β should be calculated. They are present in Rayleigh's damping matrix [40]:

$$[C] = \alpha[M] + \beta[K] \tag{19}$$

The damping coefficients are related to the angular frequency, in the form of Rayleigh's damping coefficient:

$$\xi = \frac{\alpha}{2\omega} + \frac{\beta\omega}{2} \tag{20}$$

At present, no effective experimental methods have been developed to identify damping parameters for asphalt mixtures. There is no simple correlation between the damping and static or dynamic deflections. In practice, it is possible to use FEM, and it is enough to associate only the damping with the rigidity of the system [41]. The damping was also applied for this beam model, but with only the stiffness obtaining satisfactory results. The damping parameter β was identified in the model using iterative back-calculations maintaining the agreement between the deformations from the laboratory studies. The following damping parameters β were obtained: 0.002 1/s for −5 °C, 0.0025 1/s for 0 °C, 0.004 1/s for 10 °C, and 0.005 1/s for 25 °C. Correlation between damping parameters and viscosity coefficients was determined (Figure 15). We have described the relationships between viscosity parameters and the damping coefficient using linear regression functions. High correlation coefficients (close to 1) were obtained for this function. As the value of viscosity parameters increases, the value of damping coefficients decreases.

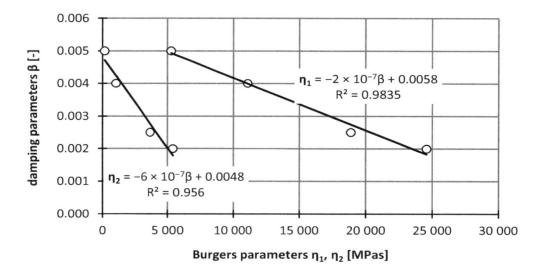

Figure 15. The relationship between the phase angle and the damping parameters (asphalt mixtures with an asphalt content of 4.5%).

Figure 16 shows the results of deformation for the static creep tests for different temperatures, and Figure 17 presents the results for the fatigue dynamic test.

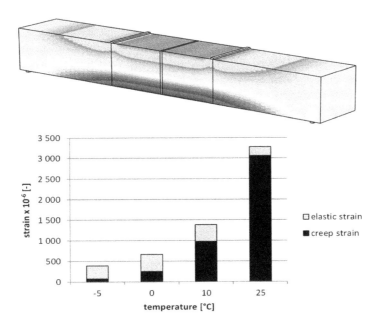

Figure 16. The deformation results for the static creep test (asphalt mixtures with an asphalt content of 4.5%).

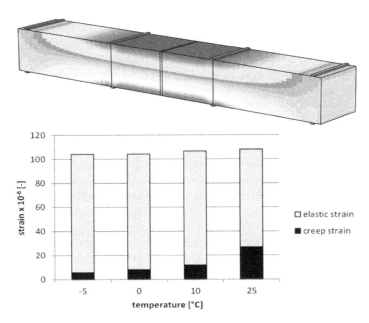

Figure 17. The deformation results for the fatigue dynamic test (asphalt mixtures with an asphalt content of 4.5%).

The numerical analysis allows us to indicate the proportion of elastic and viscous deformations depending on the test temperature. It has been found that with increasing temperature, the proportion of viscous deformations also increases, and in a dynamic temperature test is 6% at −5 °C, 8% at 0 °C, 11% at 10 °C, and 25% at 25 °C. The results for the creeping test are 19% at −5 °C, 37% at 0 °C, 71% at 10 °C, and 93% at 25 °C. The differentiation of the contribution of weak deformations results from different time intervals of the load in both studies. It is worth noting, however, that in the case of the static creep testing, the increase in temperature results in a significant nonlinear growth in the value of viscous deformations.

6. Conclusions

The different values of viscoelastic parameters in a Burgers model were determined. The variability of the parameters in temperature was obtained for the static and dynamic tests.

1. These parameters depended significantly on the duration of the load. Therefore, appropriate parameters should be chosen depending on the load time when the behaviour of asphalt mixtures in the pavement is modelled.
2. For the static long-term load tests, the parameters should be derived from creep curves, and for dynamic tests, they should be determined from the hysteresis.
3. It was found that the use of the Burgers viscoelastic model is justified for dynamic loads with the frequency of 10 Hz. For higher frequencies and at lower temperatures, the determination of the parameters may be of lesser importance, because the material has parameters similar to the elastic model, due to its low phase angle.
4. The creep test using static and dynamic 4BP loading is an effective method for determining rheological parameters under the assumed load time, the number of cycles, and temperatures. The linear viscoelastic Burgers model is helpful in this regard, because interprets the thermoplastic features of the road pavement material, such as the asphalt mixtures, well.
5. The numerical analysis using the finite element method allows us to identify the contribution of viscous deformations relative to the total, and show the significant variation of these deformations for two tests, according to the temperature.
6. The rheological parameters also depend on the composition of the bituminous mixture. For the optimal asphalt content (4.5%), the highest values of rheological parameters were obtained, demonstrating the best mechanical features and resistance to permanent deformations. For the increased asphalt content, viscosity coefficients clearly decrease, which corresponds to the increase in the value of phase angle ϕ and material damping values.
7. The obtained rheological parameters from both methods proved to be sensitive to a change in asphalt content, which means that the methods can be used to design the optimal asphalt mixtures composition—e.g., due to permanent deformation of road surfaces.

In further publications, the calculations using the finite element method for both tests, taking into account the Burgers model, will be verified. In addition, the Burgers parameters will be analysed in the dynamic fatigue test. These parameters change due to the dissipation processes and structural variation in the material.

Author Contributions: Conceptualization, P.M. and A.S.; methodology, P.M.; software, P.M.; validation, P.M., and A.S.; formal analysis, A.S.; investigation, P.M.; resources, P.M.; data curation, P.M.; writing—original draft preparation, P.M.; writing—review and editing, P.M., and A.S.; visualization, P.M.; supervision, P.M., and A.S.; project administration, P.M.; funding acquisition, A.S.

References

1. Judycki, J. Bending Test of Asphaltic Concrete Mixtures Under Statical Loading. In Proceedings of the IV International RILEM Symposium Mechanical Tests for Bituminous Mixes—Characterization, Design and Quality Control, Budapest, Hungry, 20 September 1990; pp. 207–233.
2. Bonaquist, R. *Refining the Simple Performance Tester for Use in Routine Practice*; Report 614; NCHRP, Transportation Research Board: Washington, WA, USA, 2008.
3. Reiner, M.; Ward, A.G. Building Materials Their Elasticity and Inelasticity. North Holland Publishing Company: Amsterdam, The Netherlands, 1954.
4. Brodnyan, J.G.; Gaskin, F.H.; Philppoff, W.; Lendart, E.G. The rheology of asphalt I. *Trans. Soc. Rheol.* **1958**, 2, 285–306. [CrossRef]
5. Van der Poel, C. Road Asphalt of the chapter IX. In *Building Materials Their Elasticity and Inelasticity*; Reiner, M., Ward, A.G., Eds.; North Holland Publishing Company: Amsterdam, The Netherlands, 1954.

6. Monismith, C.L.; Alexander, R.L.; Secor, K.E. Rheologic behavior of asphalt concrete. *Assoc. Asph. Paving Technol. Proc.* **1966**, *35*, 400–450.

7. Vakili, J. Creep behavior of asphalt-concrete under tension. *J. Rheol.* **1984**, *25*, 573–580. [CrossRef]

8. Goodrich, J.L. Asphalt binder rheology, asphalt concrete rheology and asphalt concrete mix properties. *J. AAPT.* **1991**, *60*, 80–119.

9. Kisiel, I.; Lysik, B. *An Outline of Soil Rheology, the Effect of Static Load on the Ground*; Wydawnictwo Arkady: Warszawa, Poland, 1966. (In Polish)

10. Nowacki, W. *Creep Theory*; Wydawnictwo Arkady: Warszawa, Poland, 1963. (In Polish)

11. Jakowluk, A. *Creep and Fatigue Processes in Materials*; WKŁ: Warszawa, Poland, 1993. (In Polish)

12. Hopman, P.C.; Nilsson, R.N.; Pronk, A.C. Theory, Validation and Application of the Visco-Elastic Multilayer Program VEROAD. In Proceedings of the eight International Conference on Asphalt Pavements, Seattle, WA, USA, 10–14 August 1997.

13. Collop, A.C.; Cebon, D.; Hardy, M.S. Viscoelastic approach to rutting in flexible pavements. *J. Transp. Eng.* **1995**, *121*, 82–93. [CrossRef]

14. Rowe, G.M.; Brown, S.F.; Sharrock, M.J.; Bouldin, M.G. Visco-elastic analysis of hot mix asphalt pavement structures. *Transp. Res. Rec.* **1995**, *1482*, 44–51.

15. Sousa, J.B.; Weissman, S.L.; Sackman, J.L.; Monismith, C.L. A non-linear elastic viscous with damage model to predict permanent deformation of asphalt concrete mixes. In Proceedings of the 72nd Annual Meeting of the Transportation Research Board, Washington, WA, USA, 10–14 January 1993.

16. Mackiewicz, P.; Szydło, A. Effect of load repeatability on deformation resistance of bituminous mixtures in creep and rutting tests. *Arch. Civ. Eng.* **2003**, *49*, 35–51.

17. Szydło, A.; Mackiewicz, P. Verification of bituminous mixtures' rheological parameters through rutting test. *Road Mater. Pavement Des.* **2003**, *4*, 423–428. [CrossRef]

18. Szydło, A.; Mackiewicz, P. Asphalt mixes deformation sensitivity to change in rheological parameters. *J. Mat. Civ. Eng.* **2005**, *17*, 1–9. [CrossRef]

19. Jahangiri, B.; Karimi, M.; Tabatabaee, N. Relaxation of hardening in asphalt concrete under cyclic compression loading. *J. Mat. Civ. Eng.* **2017**, *29*, 1–8. [CrossRef]

20. Blab, R.; Harvey, J.T. Viscoelastic rutting model with improved loading assumtions. In Proceedings of the Ninth International Conference on Asphalt Pavements, Copenhagen, Danmark, 17–22 August 2002.

21. You, L.; Yan, K.; Hu, Y.; Ma, W. Impact of interlayer on the anisotropic multi-layered medium overlaying viscoelastic layer under axisymmetric loading. *Appl. Math. Model.* **2018**, *61*, 726–743. [CrossRef]

22. Chen, E.Y.; Pan, E.; Green, R. Surface Loading of a Multilayered Viscoelastic Pavement: Semianalytical Solution. *J. Eng. Mech.* **2009**, *135*, 517–528. [CrossRef]

23. Karimi, M.; Tabatabaee, N.; Jahangiri, B.; Darabi, M.K. Constitutive modeling of hardening-relaxation response of asphalt concrete in cyclic compressive loading. *Constr. Build. Mat.* **2017**, *137*, 169–184. [CrossRef]

24. Yin, H.M.; Buttlar, W.G.; Paulino, G.H.; Di Benedetto, H. Assessment of Existing Micro-mechanical Models for Asphalt Mastics Considering Viscoelastic Effects. *Road Mater. Pavement Des.* **2008**, *9*, 31–57. [CrossRef]

25. You, L.; You, Z.; Yan, K. Effect of anisotropic characteristics on the mechanical behavior of asphalt concrete overlay. *Front. Struct. Civ. Eng.* **2019**, *13*, 110–122. [CrossRef]

26. Karimi, M.; Tabatabaee, N.; Jahanbakhsh, H.; Jahangiri, B. Development of a stress-mode sensitive viscoelastic constitutive relationship for asphalt concrete: Experimental and numerical modeling. *Mech. Time-Depend. Mat.* **2017**, *21*, 383–417. [CrossRef]

27. Kim, Y.R.; Little, D.N. One-dimensional constitutive modeling of asphalt concrete. *J. Eng. Mech.* **1990**, *116*, 751–772. [CrossRef]

28. Lee, H.J.; Kim, Y.R. A viscoelastic constitutive model for asphalt concrete under cyclic loading. *J. Eng. Mech.* **1998**, *124*, 32–40. [CrossRef]

29. Lee, H.J.; Kim, Y.R. Viscoelastic continuum damage model of asphalt concrete with healing. *J. Eng. Mech.* **1998**, *124*, 1224–1232. [CrossRef]

30. Souza, F.V.; Soares, J.B.; Allen, D.H.; Evangelista, F., Jr. Model for predicting damage evolution in heterogeneous viscoelastic asphaltic mixtures. *Transp. Res. Rec.* **2004**, *1891*, 131–139. [CrossRef]

31. Tashman, L.; Masad, E.; Zbib, H.; Little, D.; Kaloush, K. Microstructural viscoplastic continuum model for permanent deformation in asphalt pavements. *J. Eng. Mech.* **2005**, *131*, 47–57. [CrossRef]

32. Zafir, Z.; Siddharthan, R.; Sebaaly, P.E. Dynamic pavement strain histories from moving traffic load. *J. Transp. Eng.* **1994**, *120*, 821–842. [CrossRef]
33. Siddharthan, R.; Yao, J.; Sebaaly, P.E. Pavement strain from moving dynamic 3D load distribution. *J. Transp. Eng.* **1998**, *124*, 557–566. [CrossRef]
34. Collop, C.; Scarpas, A.T.; Kasbergen, C.; de Bondt, A. Development and finite element implementation of a stress dependent elasto-visco-plastic constitutive model with damage for asphalt. *Transp. Res. Rec.* **2003**, *1832*, 96–104. [CrossRef]
35. Judycki, J. Comparison of fatigue criteria for design of flexible and semi-rigid road pavements. In Proceedings of the Eight International Conference on Asphalt Pavements, Seattle, WA, USA, 10–14 August 1997; pp. 919–937.
36. Jaczewski, M.; Judycki, J. Effects of deviations from thermo-rheologically simple behaviour of asphalt mixes in creep on developing of master curves of their stiffness modulus. In Proceedings of the 9th International Conference Environmental Engineering, Vilnius, Lithuania, 22–23 May 2014.
37. *European Standard EN 12697–24:2012. Bituminous Mixtures—Test Methods for Hot Mix Asphalt—Part 24: Resistance to Fatigue*; PKN: Warsaw, Poland, 2012.
38. Manual of Nottingham Asphalt Tester. Available online: https://www.cooper.co.uk (accessed on 28 June 2019).
39. ARA, Inc. ERES Consultants Division. Guide for Mechanistic-Empirical Design of New and Rehabilitated Pavement Structures, Final Report, Part 3—Design and Analysis. Transportation Research Board (TRB), National Cooperative Highway Research Program (NCHRP): Washington, WA, USA, 2004.
40. Mario, P. *Structural Dynamics*, 1st ed.; Springer: Boston, MA, USA, 1991. [CrossRef]
41. Saouma, V.; Miura, F.; Lebon, G.; Yagome, Y. A simplified 3D model for soil-structure interaction with radiation damping and free field input. *Bull. Earthq. Eng.* **2011**, *9*, 1387–1402. [CrossRef]

11

The use of the Acoustic Emission Method to Identify Crack Growth in 40CrMo Steel

Aleksandra Krampikowska [1], **Robert Pała** [2]🆔, **Ihor Dzioba** [2]🆔 **and Grzegorz Świt** [1,*]

[1] Department of Strength of Materials, Concrete and Bridge Structures, Kielce University of Technology,
     Al. 1000-lecia PP 7, 25-314 Kielce, Poland
[2] Faculty of Mechatronics and Mechanical Engineering, Department of Machine Design,
     Kielce University of Technology, Al. 1000-lecia PP 7, 25-314 Kielce, Poland
*   Correspondence: gswit@tu.kielce.pl

Abstract: The article presents the application of the acoustic emission (AE) technique for detecting crack initiation and examining the crack growth process in steel used in engineering structures. The tests were carried out on 40CrMo steel specimens with a single edge notch in bending (SENB). In the tests crack opening displacement, force parameter, and potential drop signal were measured. The fracture mechanism under loading was classified as brittle. Accurate AE investigations of the cracking process and SEM observations of the fracture surfaces helped to determine that the cracking process is a more complex phenomenon than the commonly understood brittle fracture. The AE signals showed that the frequency range in the initial stage of crack development and in the further crack growth stages vary. Based on the analysis of parameters and frequencies of AE signals, it was found that the process of apparently brittle fracture begins and ends according to the mechanisms characteristic of ductile crack growth. The work focuses on the comparison of selected parameters of AE signals recorded in the pre-initiation phase and during the growth of brittle fracture cracking.

Keywords: pattern recognition; acoustic emission; Structural Health Monitoring; brittle fracture; diagnostics

1. Introduction

Diagnosing and monitoring the condition of structures is an important and very topical area in construction. Aging infrastructure and increasing service load of engineering structures are the main drivers of fast-progressing research in the new interdisciplinary field of knowledge called structural health monitoring (SHM), which is closely related to the durability and safe operation over the life of the structural elements.

The load is a typical random load that is difficult to model with currently known fatigue calculation procedures, and can be modelled only for steady loads with low-level amplitudes. As a result, the fatigue life calculation models for steel structures fail to provide accurate information about the risk of fatigue failure. Where calculation methods provide only limited information, the existing structures are subjected to tests. Fatigue damage testing is carried out during inspection. Cracks, which occur for a number of reasons, are detected and, in the case of fatigue cracks, the rate of their growth is assessed.

Depending on the operating conditions, the loading and materials used, the failure takes place according to a ductile or brittle fracture mechanism.

Ductile cracking usually occurs in steels and is preceded by significant plastic deformation that manifests itself long before the failure of the element, thus allowing preventive measures to be undertaken. The brittle fracture mechanism is much more dangerous.

Brittle cracking proceeds without visible deformation of the element and occurs almost immediately. The occurrence of brittle fracture can be expected mainly in high-strength structural steels or in low-temperature operation and overloading of the element with simultaneous increase in the load speed. Welded joints are also exposed to brittle fracture due to their inhomogeneous microstructure, welding defects and residual stresses. The presence of these factors together with the impact of fatigue loads and corrosive environment leads to the degradation of the material and development of micro-cracks, resulting in the brittle failure of steel.

Leading research centers worldwide have devoted a great deal of attention to the development of methods for assessing the condition of structural elements and preventing their failure. As a result, many methods for evaluation of component durability [1–5] are currently in wide use. These methods are based on aversion to change assuming prior knowledge of the element microstructure, load history, and in-service conditions. However, more accurate results will be obtained when as much information as possible can be collected on in-service parameters and material.

Hence, the modelling of bearing capacity and durability of a structure under real operating conditions requires that:

- real operational loads are defined;
- the material model is defined, in particular welded joints material that changes over time under the influence of operating conditions;
- the load around the defect after its initiation and during development (redistribution of stresses) is determined; and
- the interaction of various damage types is identified.

As determining the factors above is very difficult, if not impossible, most analyses are characterized a high degree of conservatism, which significantly reduces assessment accuracy. Appropriate solution should include developing an NDT-based monitoring system able to signal the structural safety risk.

Finding and determining the "destructive characterization" of all hot spots in large-sized structures using the NDT methods is nearly impossible. In steel structures, the volume of "hot spot" is of the order of cubic centimeters, with dimensions of the structure reaching tens and more meters. In our opinion, the solution to the problems of diagnosing engineering structures is the use of continuous, long-term monitoring using the passive NDT methods. One of them is the acoustic emission (AE) method [4–14].

The paper presents a proposal for the use of acoustic emission for the diagnosis of steel structures vulnerable to brittle fracture.

The choice of acoustic emission as the research method was determined by its advantages in relation to other non-destructive methods. These are:

- locating the faults that are undetectable with conventional methods;
- recording only active damage, i.e., the defect growth as it occurs;
- continuous monitoring of structures while in service or during load tests, with continuous data recording;
- detecting all types of damage, whereas most other methods focus on particular defects;
- characterising the rate of damage development during the operation of the structure; and
- enabling characterization of AE signal sources.

The main task of the measurement system consisting of an AE processor is to detect, record, filter and analyze the signals generated by AE sources.

Each destructive process is a source of acoustic emission. The source is described by the parameters of the recorded AE signal. The values of the parameters are used to classify the signals (and, thus, damage processes). The similarity of signals is used to attribute particular signals to the defects caused by specific damage processes, and then by applying statistical grouping methods to identify the existing defects.

For the statistical methods used in identification, an important issue is the optimal selection of recorded 13 AE parameters (counts, counts to peak, amplitude, RMS—root means square voltage, ASL—average signal level, energy, absolute energy, signal strength, rise time, duration, initiation frequency, average frequency, reverberation frequency). The parameters must be characterized by low mutual correlation. A set of diagnostic variables must describe the most important aspects of the studied phenomenon [15–18]. Techniques such as hierarchical and non-hierarchical clustering methods and Kohonen's neural networks are used to build the reference signals data base for identification of destructive processes in steel structures (IPDKS) [4,5].

This paper uses unsupervised pattern recognition methods to characterize different AE activities corresponding to different fracture mechanisms. A sequential feature selection method based on a k-means clustering algorithm is used to achieve high classification accuracy. Fatigue damage propagation represents the main failure factor. To study the contributions of different types of damage at progressive fatigue stages, a tool with the ability to detect the damage initiation and to monitor failure progress online is needed. Acoustic emission (AE) testing has become a recognized suitable and effective non-destructive technique to investigate and evaluate failure processes in different structural components. The main advantage of AE over other condition monitoring techniques is that detected AE signals can be used to characterize the different damage mechanisms. The approach of using parameter distribution has the advantage of real-time damage detection, but can also lead to false conclusions due to noise effects in the AE signals [18]. This is especially true for materials that are working under the fatigue conditions which are usually present in damage mechanism interactions. Therefore, AE techniques are needed to account for more intricate wave propagation features caused by the anisotropic nature of materials and to enable the identification of a large variety of failure modes. It is now possible to detect and capture very large numbers of AE signals, driving a trend for seeking computationally complex algorithms, such as pattern recognition, to determine the onset of significant AE. For a real structure it is not possible to provide a set of training patterns belonging to multiple damage mechanisms, which thus makes the use of unsupervised pattern recognition techniques more appropriate for these studies [18].

The first step of developing an unsupervised pattern recognition process is to classify signals into groups based on similarities. This process involves statistical effects, and the key point of successful feature selection to construct fine classification accuracy. A paper by Doan et al. [16] presented a feature selection method that introduced a sequential method based on the Gustafson–Kessel clustering algorithm. In these the method, the subset of features is selected by minimizing the Davies-Bouldin (DB) index, which is a metric for the evaluation of classification algorithms. The signals are classified into four groups by comparing their features and deciding upon their similarity.

The assignment of the clustering results to the fracture mechanisms is achieved by a detailed analysis of the physical meaning of the data [19–22]. This is the first time that the pattern recognition technique has been applied to a database acquired from such a complex structure in a fatigue testing environment [16]. The applied feature selection algorithm proves to be a powerful tool providing relevant clustering when used together with a k-means algorithm. When a crack occurs in the material, it results in a rapid release of energy, transmitting in the form of an elastic wave, namely acoustic emission (AE). The AE-based detection method has been intensively used also in non-destructive assessments of cracks in papers [23–25]. Additionally, Rabiei and Modarres revealed a log-linear relationship between the AE features and crack growth rate, and presented an end-to-end approach for structural health management [10]. Qu et al. presented a comparative study of the damage level diagnostics of gearbox tooth using AE and vibration measurements; the results indicated that vibration signals were easily affected by mechanical resonance, while the AE signals showed a more stable performance [26]. Zhang et al. have studied defect detection of rails using AE and wavelet transform at a high speed [27]. Li and al. created a template library of cracking sounds and designed a detection device using voiceprint recognition with an accuracy of 77% [28]. Hase et al. combined the Hurst exponent and the neural network to develop a crack detection algorithm of carbide anvils [27]. In the above methods, the AE sensors are usually attached to the surface of the monitoring object, what makes measurement difficult.

A more practical crack identification and detection method is still lacking. Aiming to improve recognition accuracy and generalization, in papers [29–31] a novel crack identification method based on acoustic emission and pattern recognition is proposed. In these methods, the sound pulses from cracks are firstly separated from the original signal by pre-processing. The high-dimensional features are reduced adaptively by using principal component analysis (PCA). The algorithm combines a k-nearest neighbor (kNN) classifier with a support vector machine (SVM) to refine the classification outcome. While debris monitoring does not require any electronics, it is simple to interpret and has excellent sensitivity to wear-related failure; this method is insufficient to non-benign cracks as no debris is produced. Acoustic emission (AE) and vibration signals have more quantitative results to detect the earliest stage of damage in rotating machinery. AE and vibration methods are based on recording transient signals in two different frequency spectrums. While vibration method is based on features that are extracted from time and frequency domain signals recorded by low frequency accelerometers in order to assess the changes in vibrational properties as related to the damage [29–31], the AE method is based on detecting propagating elastic waves released from active flaws. Once transient signals are collected, signal processing methods, such as wavelet decomposition [11], empirical mode decomposition [11], and multivariate pattern recognition [20], are applied.

Typical parameters extracted from the transient signals are root mean square value, frequency domain characteristics, energy, spectral kurtosis, and peak-to-peak vibration level. Due to the difference in the frequency bandwidth, the AE method is more sensitive to microcracks as compared to the vibration method. Typical AE data acquisition approach is based on threshold: an AE signal is detected when the signal level is above threshold. As crack growth is a stochastic process, it is considered that while some data will be lost due to the idle time of data acquisition system between waveform recording intervals, crack information will be stochastically detected. However, it is important to identify how to analyze long-duration signals in order to reduce the influence of background noise from the extracted features. In studies, the AE signals, which accompanying of the fatigue crack growth is important obtained from the scaled laboratory experiments. Acoustic emission (AE) is a health monitoring approach which acts as a passive receiver to record internal activities in structures.

This method is capable of continuous monitoring, which is not the case with most traditional methods. In addition, AE sensors are very sensitive and can capture signals due to micro-scale defect formations coming from the internal regions of structures rather than only those at the surface [18,23]. An unsupervised and supervised pattern recognition algorithm was employed to classify the AE signals. Different damage mechanisms for specimens during cracking were identified using reference database AE signals created from signals registered in the tests described below.

2. Materials and Methods

The tests discussed below were carried out using Zwick-100 testing machine (ZwickRoel Ulm, Germany) on SENB (Single Edge Notch in Bending) specimens (Figure 1) made of 40CrMo steel. The specimens with dimensions 12.5 mm × 25 mm × 110 mm were subjected to heating at 850 °C (15 min), quenching in oil, tempering at 250 °C (3 h) and cooling in oil. As a result, a material with a tempered martensite microstructure was obtained (Figure 2a). The signals of loading, specimen deflection and acoustic emission (AE) were recorded during testing.

Figure 1. The view of the SENB specimen with AE sensors (Zwick-100, ZwickRoel Ulm, Germany) during test: 1—a sample, 2—support, 3—load cylinder, 4—COD extensometer, 5—AE sensors—100–1200 kHz, 6—AE sensors—30–80 kHz.

SENB type specimens with a total fracture length, which includes a notch + previously fatigue crack, equal to about $0.5 \cdot a/W$ are recommended by the ASTM and PN-EN standards [32–35] and commonly used to determine fracture toughness characteristics—critical values of the stress intensity factor (SIF) K_{IC} or the critical J-integral, J_{IC}. If during the entire process of loading the SENB specimen in the net-section area before the crack tip there is a definite predominance of the linear-elastic nature of the stress and strain fields before the crack tip are described by SIF and the fracture toughness is characterized by the critical value of K_{IC}. The condition allowing the use of SIF is to limit the size of the yielding before the crack tip: $r_p \leq 0.01 \cdot a$, where for a plane strain $r_p = (1/6\pi) \cdot (K_I/\sigma_y)^2$. If the plastic zone is larger, it means that the material in front of the fracture tip is elastic-plastic and for the description of mechanical fields an integral J must be used, and fracture toughness is represented by the critical integral $J - J_{IC}$ [36,37].

For determining material strength characteristics, cylindrical specimens 10 mm in diameter were prepared with the measuring section of 50 mm. For determining the critical value of fracture toughness, K_{IC}, the specimens with dimensions 12.5 mm × 25 mm × 110 mm with one-sided notch and crack fatigue (SENB) with a total length of about 12.5 mm were made. Figure 2b shows an example plot of loading a cylindrical specimen. On the basis of the graphs, strength characteristics of 40CrMo steel were determined: $R_e = \sigma_y = 1475$ MPa; $R_m = \sigma_{uts} = 1800$ MPa; $E = 205$ GPa. During the loading of the SENB specimen, it was determined that the crack growth process takes place according to the brittle fracture mechanism. All standard conditions were met. The obtained critical value of fracture toughness is a material characteristic and is equal to $K_{IC} = 39$ MPa·m$^{1/2}$.

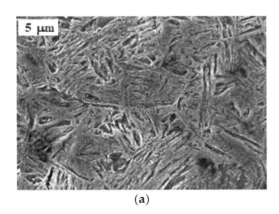
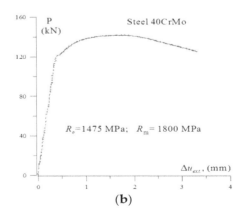

(a)   (b)

Figure 2. (**a**) Microstructure of tempered martensite of 40CrMo steel; and (**b**) loading plot of a cylindrical specimen in uniaxial tensile test.

The reference signal database in the IPDKS method for steel structures was developed using the Fuzzy k-means algorithm. The k-means algorithm belongs to a group of non-hierarchical clustering approaches. Its essence lies in the random choice of initial centres. In each iteration further approximations of the patterns are searched for and enumerated using the given methods. Depending on the assumptions made, the reference element may be one of the elements of the X population or belong to a certain universe $U \supseteq X$.

In metric spaces, a reference element can be calculated as an arithmetic mean to represent the centre of gravity of the cluster.

In general, the *k-clustering* algorithm input is a set of X objects and the expected number of clusters k, and the output is the division into subsets $\{C_1, C_2, \ldots, C_k\}$. Frequently, **k-clustering** algorithms belong to the category of optimization algorithms. Optimization algorithms assume that there is a loss function $k: \{x \mid X \subseteq S\} \to R +$ specified for each S subset. The aim is to find a branch due to minimizing the sum of losses described by Equation (1):

$$E_q = \sum_{i=1}^{k} k(C_i) \tag{1}$$

In order to apply the iterative algorithm, the distance measure used should be determined in the grouping process. Instead of the median, you can use a point that is the resultant point for a given cluster (representation element) calculated, e.g., as a geometric or arithmetic mean. Then we deal with k-means (k-means) algorithms described by Equation (2):

$$k(C_i) = \sum_{r=1}^{|C_i|} d\left(\bar{x}^i, x_r^i\right) \tag{2}$$

where: $\bar{x}^i$—the arithmetic mean (or geometric) of the cluster.

Doing so means that the centroids search for their correct positions using Equation (3) [4,5]:

$$\underline{\mu}_j = \frac{\sum_{j=1}^{n} P\left(\omega_i \big| \underline{x}_j\right)^b \underline{x}_j}{\sum_{j=1}^{n} P\left(\omega_i \big| \underline{x}_j\right)^b}, \tag{3}$$

where $P\left(\omega_i \big| \underline{x}_j\right)$ is the conditional probability of belonging j-th element to the i-th group, b is the parameter that has to take values other than 1, x_{-j} is the j-th element.

The probability function is normalized according to Equation (4):

$$\sum P\left(\omega_i \big| x_{-j}\right) = 1, \; where \; j = 1, \ldots, n, \tag{4}$$

The probability of the membership of the element in each of the clusters $P\left(\omega_i \big| \underline{x}_j\right)$ is calculated from Equation (5):

$$P(\omega_i \big| \underline{x}_j) = \frac{\left(\frac{1}{d_{ij}}\right)^{\frac{1}{b-1}}}{\sum_{r=1}^{c} \left(\frac{1}{d_{rj}}\right)^{\frac{1}{b-1}}}, \tag{5}$$

where $d_{ij}^2 = \|\underline{x}_j - \underline{\mu}_i\|^2$ is the distance of a data point $\underline{x}_j$ from the center of the group $\underline{\mu}_i$.

The *k-means* algorithm consists of the following steps:

1. Randomly select initial centroids.
2. Compute the distances between the data points and the cluster centroids.

3. Compute the membership function value of all elements $P\left(\omega_i|\underline{x}_j\right)$.

4. Compute cluster centroids $\underline{\mu}_i$.

5. If:

- there are no changes in $\underline{\mu}_i$ and $P\left(\omega_i|\underline{x}_j\right)$- return $\mu_{-1} \dots \mu_{-c}$,

- otherwise go back to Step 2.

When this algorithm is used, the number of clusters is pre-determined. However, the speed of computation compensates for this inconvenience. The reference signals database is created with 13 correlated AE parameters (rise time, durability, counts, counts to peak, energy, RMS, ASL, amplitude, average frequency, initiation frequency, reverberation frequency, absolute energy, signal strength). Reference signals obtained in this way make it possible to identify individual destructive processes in all tested structures. The database can be supplemented as per the diagram above whenever new measurement data are obtained.

The use of the grouping method is particularly important for the identification of fracture mechanisms as it allows for timely response to the possible averse effects.

The study used a 24-channel "μSamos" acoustic emission processor, 40 dB preamplifiers and two resonance sensors with flat characteristics in the 30–80 kHz range and two broadband sensor in the 100–1200 kHz range. The use of four sensors with flat characteristics was aimed at determining the lower and upper discrimination thresholds of the AE signal, while the broadband sensor had the task of controlling whether during the measurement there are no signals generated in other frequency bands.

3. Results and Discussion

During the loading of the SENB specimen, the crack growth process takes place according to the brittle fracture mechanism and critical value of fracture toughness is a material characteristic equal to $K_{IC} = 39$ MPa·m$^{1/2}$. At the critical moment, the length of the plastic zone according to the Irwin model for plane strain equals: $2r_p = 2(1/6\pi)\cdot(K_{IC}/\sigma_y)^2 = 0.074$ mm. According to Williams' formulas in front of the plastic zone in the crack plane direction, the level of stress tensor components is $\sigma_{xx} = \sigma_{yy} = 1808$ MPa, $\sigma_{zz} = \nu\cdot(\sigma_{xx} + \sigma_{yy}) = 1193$ MPa. Directly in the plastic zone the stress components are significantly higher and the stress opening component, σ_{yy}, can obtain values higher than $3\sigma_y$, i.e., over 4500 MPa [38]. The σ_{zz} and σ_{xx} components are lower, but also exceed the level of $2\sigma_y$.

High stress levels in front of the crack tip cause material damage processes, cracking development through various mechanisms: brittle fracture on the cleavage planes, brittle intergranular cracking, ductile fracture through nucleation, growth and coalescence of voids. The predominance of any fracture mechanism depends on the stress-strain state in the tested element, on the microstructure type, loading method, the temperature of the environment, and on other effects. Here, the situation was seemingly simple—SENB specimens crack according to the brittle mechanism. The AE method was used to monitor the processes during loading of steel specimens.

Five specimens were tested. The load plots and the character of AE signals distribution were similar in all tested samples. Figure 3b shows the characteristic loading curve and numerous points of the absolute energy of AE signal and the graph illustrating the cumulative absolute energy vs. load on time from 3.5 to 6.2 s. The graphs show AE parameters without using the clustering method (identification) of destructive processes.

Figure 3a shows the change in the absolute energy parameter (aJ), and the specimen loading as a function of time and COD (crack opening displacement). Acoustic emission signals generated during the study showed an increase in their numbers and increase in the value of absolute energy parameter at COD of 0.045 (mm) and 4.2 kN of load. It can be noticed that the increase in the value of the absolute energy parameter occurs when the cracking process begins (initiation, propagation).

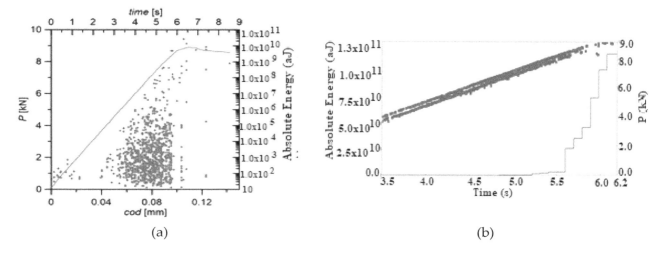

(a)   (b)

Figure 3. The graph of specimen loading (**a**) with absolute energy of AE signals (aJ) and (**b**) graph cumulative absolute energy vs. load on time from 3.5 to 6.2 s. The graphs show AE parameters without using the clustering method (identification) of destructive processes.

During the observations of the fracture surface of the specimens examined using the scanning microscope, three basic mechanisms in the crack development process were identified. The images shown in Figure 4 are presents the same fragment the fracture surface, but at different magnitude. In Figure 4a, at 500×, we can see the general character of brittle fracture surface. In Figure 4b, at 1500×, the details of fracture mechanisms are shown. At the fracture surface, two types of brittle fracture were present: *intercrystallite* cracking and *transcrystallite cleavage*. (Figure 4a,b). There were also areas where cracking developed according to the ductile mechanism, that is, where voids nucleated, grew, and coalesced around the small particles of precipitations (Figure 4b).

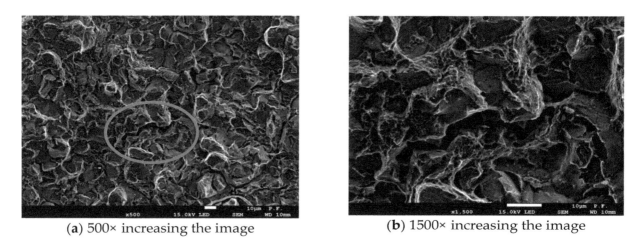

(**a**) 500× increasing the image   (**b**) 1500× increasing the image

Figure 4. Morphology of the fracture surface the SENB specimen breakthrough: (**a**) general character of fracture surface; and (**b**) details of fracture mechanisms.

Observations of the process of cracking over time shows that at first cracking proceeds according to brittle mechanisms, while cracking due to sudden immediate coalescence of voids is the last stage of crack growth. These observations were confirmed by the analysis of AE signals generated in the tested SENB specimen. However, in order to be able to detect the differences in the processes accompanying the cracking of steel specimens, the signal grouping method must be used. Figure 4a, b shows the selected AE parameters (absolute energy) not analyzed using the clustering methods. It can be noticed that it is difficult to interpret the changes in the work of the tested element. It is impossible to assess the moment of initiation of the crack or its subsequent propagation. It can be said that individual parameters not analyzed by clustering methods do not help in the interpretation of changes in the tested samples. That is why it is so important to use Big Data analysis to create and identify destructive processes on the basis of the pattern recognition methods.

Using one of the grouping methods, namely the fuzzy k-means algorithm, we developed the reference signal data base describing the processes of brittle fracture during destructive tests of steel structures. The AE signals generated and recorded in the crack initiation and development area were subjected to a grouping analysis using the NOESIS 5.8 program, which allowed to separate four basic classes. Figure 5a–c present the effects of using method of grouping signals—a non-hierarchical k-means method to identify mechanisms accompanying the brittle fracture.

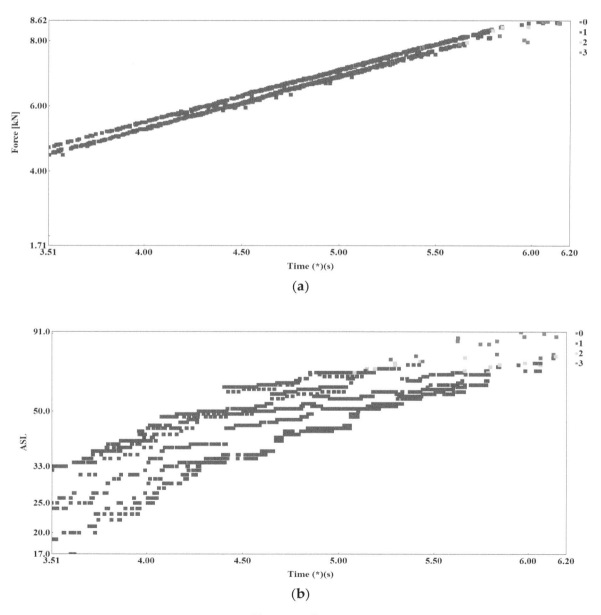

Figure 5. Cont.

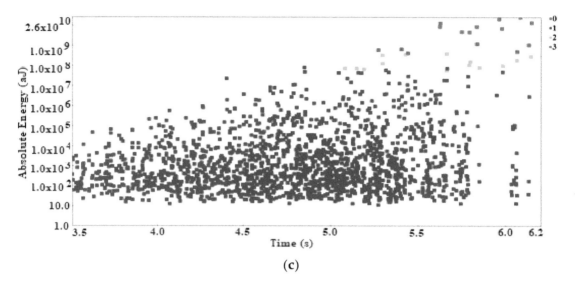

(c)

Figure 5. Scatterplots of acoustic emission signals: (**a**) Force (kN) vs. time (s); (**b**) ASL (dB) vs. time (s); and (**c**) absolute energy (a) vs. time (s).

Analyzing the division of AE signals into the classes marked in Figure 5 with colors and figures and using photographs (Figure 4) from the scanning microscope, it can be noticed that cracking due to sudden immediate combination of the voids, breaking bridges between them are pink signals marked with the number 3. The same type of signals recorded in the last stage of cracking the sample tensile steel S355JR, when there was no cracking at the brittle mechanisms (Figure 6). Signals marked in blue and number 1 come mainly from elastic and plastic deformations, as well as from individual noise generated by the strength machine and the loading system. These signals, therefore, probably can be equated with the complex, rapid dislocation motion process in ferrite and the act of fracture.

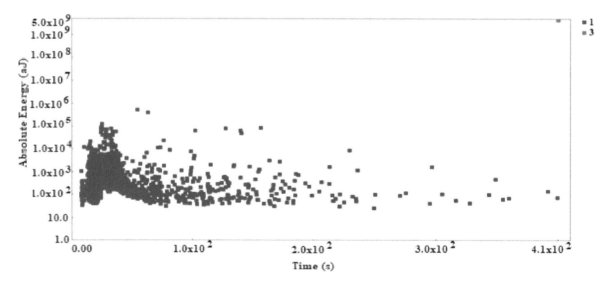

Figure 6. Scatterplot: Absolute energy (aJ) vs. time (s) for uniaxial tested specimen.

Figure 7a,b presents the parameters of the acoustic emission signal: initiation frequency (IF) (kHz) and reverberation frequency (RF) (kHz) vs. time and signal classes. Analysing these graphs, it can be seen that green signals have a similar frequency range in the initial stage (IF) parameter as well as in the later stage (RF) parameter during signal duration. This range falls within the frequency range of pink signals.

We can, therefore, assume that the signals of green color characterize the process of dislocation motion, and cracking on the cleavage surfaces in ferrite, i.e., brittle cleavage fracture. The red signals show different frequencies at different stages of the load frequency band. Parameter (IF) signals occurring in the earlier stage of the load have much higher than others, which allows to assume that these signals are generated in a different environment than ferrite. These may be signals from the cracking of precipitates segregated at the ferrite grain boundaries or connected with the process of de-bonding from the ferrite matrix. These signals can be imitated, in ferrite, and in precipitations. That is why there is a different frequency in parameter (IF).

The signal imitated in the precipitations should be short. However, the signal imitated in ferrite is longer and therefore parameter (RF) is similar. In the next stage of loading, red signals characterize the development of intercrystallite fracture growth and are imitated in ferrite.

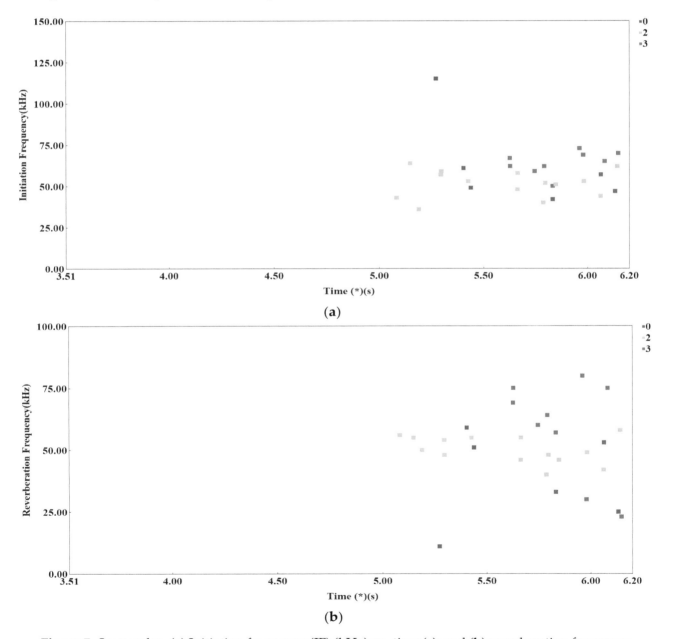

Figure 7. Scatterplot: (**a**) Initiation frequency (IF) (kHz) vs. time (s); and (**b**) reverberation frequency (RF) (kHz) vs. time (s).

4. Conclusions

4.1. General Conclusions

To raise AE technology to the next level, we must improve it from several fronts. Only then can AE become a serious player in structural integrity assessment. We need to better characterize AE sources; to extract more from AE signals reaching sensors; to devise effective AE parameters; to integrate new analysis methods, such as wave transformation analysis(WT), moment tensor analysis (MTA), and pattern recognition analysis (PRA); to include simulation tools in analyses; to accumulate basic data on structures with standardized procedures; and to devise combinatorial approach between localized damage evaluation and long-range detection and to develop regional or global database under international cooperation. Using systematic approach along with NDT, fracture mechanics, etc., improved AE can offer more value to industrial users and to contribute substantially to structural integrity assessment. Development of the proposed AE monitoring technique reported in this paper facilitates for the prognostics and life predictions of the structure. The developed methodology can be utilized for continuous in-service monitoring of structures and has proven to be promising for use in practice.

4.2. Specific Applications

1. Application of the k-means grouping method based on the analysis in the 13 parameter space of AE signals provides positive results in the classification of destructive processes in fracture toughness K_{Ic} tests.

2. Application of the k-means grouping method allows differentiating between brittle fractures mechanisms.

3. The process of brittle fracture was shown to be a sequential process composed of inter-grain cracking, brittle-cleavage and ductile fracture.

4. Acoustic emission technology was found to be able to successfully monitor and detect real-time crack initiation and also separate and identify AE signals that correspond to different fracture mechanisms

5. The grouping analysis shows that some AE parameters noticeably change after the initiation of the crack (absolute energy, initial frequency, reverberation frequency), which indicates their high independence and suitability for monitoring the cracking process.

6. The developed base of reference signals forms the theoretical basis for using pattern-based AE technology to detect and monitor crack propagation in structures that do not have a "load history" to assess the safety of steel structures using the principles of fracture mechanics.

Author Contributions: Conceptualization: G.Ś. and I.D.; methodology: G.Ś., A.K.; software: R.P.; validation: G.Ś., I.D.; formal analysis: A.K.; investigation: A.K., R.P.; resources: G.S, R.P.; data curation: A.K., R.P.; writing—original draft preparation: G.Ś., I.D.; writing—review and editing: G.Ś., I.D.; visualization: R.P., A.K.; supervision: G.Ś.; project administration: G.Ś.; funding acquisition: G.Ś., R.P.

References

1. FITNET: Fitness-for-Service. *Fracture-Fatigue-Creep-Corrosion*; Koçak, M., Webster, S., Janosch, J.J., Ainsworth, R.A., Koerc, R., Eds.; GKSS Research Centre Geesthacht GmbH: Stuttgart, Germany, 2008.
2. Ono, K. Application of Acoustic Emission for Structure Diagnosis. *Diagn.-Diagn. Struct. Health Monit.* **2011**, 2, 3–18.
3. Adamczak, A.; Świt, G.; Krampikowska, A. Application of the Acoustic Emission Method in the Assessment of the Technical Condition of Steel Structures. *IOP Conf. Ser. Mat. Sci. Eng.* **2019**, 471, 032041. [CrossRef]

4. Świt, G.; Krampikowska, A. Influence of the number of acoustic emission descriptors on the accuracy of destructive process identification in concrete structures. In Proceedings of the 7th IEEE Prognostics and System Health Management Conference (PHM-Chengdu), Chengdu, China, 19–21 October 2016; Article no. 7819756.

5. Świt, G. Acoustic emission method for locating and identifying active destructive processes in operating facilities. *Appl. Sci.* **2018**, *8*, 1295. [CrossRef]

6. ASTM E647. *Standard Test Method for Measurement of Fatigue Crack Growth Rates*; ASTM International: West Conshohocken, PA, USA, 2010.

7. Biancolini, M.E.; Brutti, C.; Paparo, G.; Zanini, A. Fatigue cracks nucleation on steel, Acoustic emission and fractal analysis. *Int. J. Fatigue* **2006**, *28*, 1820–1825. [CrossRef]

8. Caesarendra, W.; Kosasih, B.; Tieu, A.K.; Zhu, H.; Moodie, C.A.S.; Zhu, Q. Acoustic emission-based condition monitoring methods: Review and application for low speed slew bearing. *Mech. Syst. Signal Process.* **2016**, *72–73*, 134–159. [CrossRef]

9. Keshtgar, A.; Modarres, M. Detecting Crack Initiation Based on Acoustic Emission. *Chem. Eng. Trans.* **2013**, *33*, 547–552.

10. Rabiei, M.; Modarres, M. Quantitative methods for structural health management using in situ acoustic emission monitoring. *Int. J. Fatigue* **2013**, *49*, 81–89. [CrossRef]

11. Zhang, X.; Feng, N.; Wang, Y.; Shen, Y. Acoustic emission detection of rail defect based on wavelet transform and Shannon entropy. *J. Sound Vib.* **2015**, *339*, 419–432. [CrossRef]

12. Keshtgar, A.; Modarres, M. Acoustic Emission-Based Fatigue Crack Growth Prediction. In Proceedings of the Reliability and Maintainability Symposium (RAMS), Orlando, FL, USA, 28–31 January 2013.

13. Yan, Z.; Chen, B.; Tian, H.; Cheng, X.; Yang, J. Acoustic detection of cracks in the anvil of a large-volume cubic high-pressure apparatus. *Rev. Sci. Instrum.* **2015**, *86*, 124904. [CrossRef]

14. Gao, L.X.; Zai, F.L.; Su, S.B.; Wang, H.Q.; Chen, P.; Liu, L.M. Study and application of acoustic emission testing in fault diagnosis of low-speed heavy-duty gears. *Sensors* **2011**, *11*, 599–611. [CrossRef]

15. Qu, Y.Z.; He, D.; Yoon, J.; Van Hecke, B.; Bechhoefer, E.; Zhu, J.D. Gearbox tooth cut fault diagnostics using acoustic emission and vibration sensors—A comparative study. *Sensors* **2014**, *14*, 1372–1393. [CrossRef] [PubMed]

16. Doan, D.D.; Ramasso, E.; Placet, V.; Zhang, S.; Boubakar, L.; Zerhouni, N. An unsupervised pattern recognition approach for AE data originating from fatigue tests on polymer-composite materials. *Mech. Syst. Signal Process.* **2015**, *64*, 465–478. [CrossRef]

17. Ting-Hua, Y.; Stathis, C.; Stiros, X.-W.Y.; Jun, L. *Structural Health Monitoring—Oriented Data Mining, Feature Extraction, and Condition Assessment*; Hindawi Publishing Corporation: London, UK, 2014.

18. Tang, J.; Soua, S.; Mares, C.; Gan, T.-H. An experimental study of acoustic emission methodology for in service condition monitoring of wind turbine blades. *Renew. Energy* **2016**, *99*, 170–179. [CrossRef]

19. The MathWorks, Inc. *Statistics and Machine Learning Toolbox*; R2017a; The MathWorks, Inc.: Natick, MA, USA, 2017.

20. Li, L.; Lomov, S.V.; Yan, X.; Carvelli, V. Cluster analysis of acoustic emission signals for 2D and 3D woven glass/epoxy composites. *Compos. Struct.* **2014**, *116*, 286–299. [CrossRef]

21. Crivelli, D.; Guagliano, M.; Monici, A. Development of an artificial neural network processing technique for the analysis of damage evolution in pultruded composites with acoustic emission. *Compos. Part B* **2014**, *56*, 948–959. [CrossRef]

22. Desgraupes, B. *Clustering Indices*; University of Paris Ouest-Lab Modal'X: Paris, France, 2013.

23. Eftekharnejad, B.; Mba, D. Monitoring natural pitting progress on helical gear mesh using acoustic emission and vibration. *Strain* **2011**, *47*, 299–310. [CrossRef]

24. Zhu, X.; Zhong, C.; Zhe, J. A high sensitivity wear debris sensor using ferrite cores for online oil condition monitoring. *Meas. Sci. Technol.* **2017**, *28*, 75102. [CrossRef]

25. Li, R.; He, D. Rotational machine health monitoring and fault detection using EMD—Based acoustic emission feature quantification. *IEEE Trans. Instrum. Meas.* **2012**, *61*, 990–1001. [CrossRef]

26. Loutas, T.H.; Roulias, D.; Pauly, E.; Kostopoulos, V. The combined use of vibration, acoustic emission and oil debris on-line monitoring towards a more effective condition monitoring of rotating machinery. *Mech. Syst. Signal Process.* **2011**, *25*, 1339–1352. [CrossRef]

27. Hase, A.; Mishina, H.; Wada, M. Correlation between features of acoustic emission signals and mechanical wear mechanisms. *Wear* **2012**, *292*, 144–150. [CrossRef]

28. Li, R.; Seçkiner, S.U.; He, D.; Bechhoefer, E.; Menon, P. Gear fault location detection for split torque gearbox using AE sensors. *IEEE Trans. Syst. Man Cybern. Part C Appl. Rev.* **2012**, *42*, 1308–1317. [CrossRef]

29. Gu, D.; Kim, J.; An, Y.; Choi, B. Detection of faults in gearboxes using acoustic emission signal. *J. Mech. Sci. Technol.* **2011**, *25*, 1279–1286. [CrossRef]

30. Zhang, L.; Yalcinkaya, H.; Ozevin, D. Numerical Approach to Absolute Calibration of Piezoelectric Acoustic Emission Sensors using Multiphysics Simulations. *Sens. Actuators A Phys.* **2017**, *256*, 12–23. [CrossRef]

31. Zhang, L.; Ozevin, D.; Hardman, W.; Timmons, A. Acoustic emission signatures of fatigue damage in idealized bevel gear spline for localized sensing. *Metals* **2017**, *7*, 242. [CrossRef]

32. ASTM E8. *Standard Test Method for Tension Testing of Metallic Materials*; ASTM International: West Conshohochen, PA, USA, 2003.

33. ASTM E1820-09. *Standard Test. Method for Measurement of Fracture Toughness*; ASTM International: West Conshohochen, PA, USA, 2011.

34. ISO 12135:2002. *Metallic Materials—Unified Method of Test for the Determination of Quasistatic Fracture Toughness*; International Organization for Standardization: Geneva, Switzerland, 2002.

35. Schwalbe, K.H.; Landes, J.D.; Heerens, J. *Classical Fracture Mechanics Methods*; GKSS 2007/14.2007; Elsevier: Amsterdam, the Netherlands, 2007.

36. Anderson, T.L. *Fracture Mechanics: Fundamentals and Applications*; CRC Press: Boca Raton, FL, USA, 2008.

37. Neimitz, A. *Mechanika Pękania*; Powszechne Wydawnictwo Naukowe (PWN): Warsaw, Poland, 1999. (In Polish)

38. Pała, R.; Dzioba, I. Influence of delamination on the parameters of triaxial state of stress before the front of the main crack. *AIP Conf. Proc.* **2018**, *2029*, 020052. [CrossRef]

12

Effect of Freeze–Thaw Cycling on the Failure of Fibre-Cement Boards, Assessed using Acoustic Emission Method and Artificial Neural Network

Tomasz Gorzelańczyk *[ID] and Krzysztof Schabowicz[ID]

Faculty of Civil Engineering, Wrocław University of Science and Technology, Wybrzeże Wyspiańskiego 27, 50-370 Wrocław, Poland
* Correspondence: tomasz.gorzelanczyk@pwr.edu.pl

Abstract: This paper presents the results of investigations into the effect of freeze–thaw cycling on the failure of fibre-cement boards and on the changes taking place in their structure. Fibre-cement board specimens were subjected to one and ten freeze–thaw cycles and then investigated under three-point bending by means of the acoustic emission method. An artificial neural network was employed to analyse the results yielded by the acoustic emission method. The investigations conclusively proved that freeze–thaw cycling had an effect on the failure of fibre-cement boards, as indicated mainly by the fall in the number of acoustic emission (AE) events recognized as accompanying the breaking of fibres during the three-point bending of the specimens. SEM examinations were carried out to gain better insight into the changes taking place in the structure of the tested boards. Interesting results with significance for building practice were obtained.

Keywords: fibre-cement boards; non-destructive testing; acoustic emission; artificial neural networks; SEM

1. Introduction

Fibre-cement boards have been used in construction since the beginning of the last century. Their inventor was the Czech engineer Ludwik Hatschek, who developed and patented the technology of producing fibre cement, then called "Eternit". The material was strong, durable, lightweight, moisture-resistant, freeze–thaw resistant and non-combustible [1]. Fibre cement became one of the most popular roofing materials in the world in the 20th century. This was so until one of its components (i.e., asbestos) was found to be carcinogenic. In the 1990s, asbestos was replaced with environment-friendly fibres, mainly cellulose fibres. The fibre-cement boards produced today are made up of cement, cellulose fibres, synthetic fibres and various additives and admixtures. They are a completely different building product than the original one [2], and still require investigation and improvement. The additional components and fillers of fibre-cement boards are lime powder, mica, perlite, kaolin, microspheres and recycled materials [3,4], whereby fibre-cement boards can be regarded as an innovative product which fits into the sustainable development strategy. At present, such boards are used in construction mainly as ventilated façade cladding [5], as illustrated in Figure 1. In the course of their service life, fibre-cement boards are exposed to various factors, such as chemical (acid rains) and physical aggressiveness (ultraviolet radiation), but mainly to variable environmental impacts, including sub-zero temperatures in the winter season.

Figure 1. Exemplary uses of fibre-cement boards as ventilated façade cladding.

After a few winter seasons, the effect of sub-zero temperatures—especially of temperature (freeze–thaw) cyclicity—needs to be determined in order to establish whether the fibre-cement boards can remain in service as ventilated façade cladding. The knowledge of this effect is essential not only from the scientific point of view, but also for building practice. It is worth noting that research on fibre-cement boards has so far been mostly limited to determining—solely through the bending strength (modulus of rupture, *MOR*) test—their standard physicomechanical parameters and the effect of in-service factors (e.g., soaking–drying cycles, heating–raining cycles and high temperatures) and the various fibres and production processes [6]. Only a few cases of testing fibre-cement boards by non-destructive methods, limited to imperfections arising during the production process, can be found in the literature [7–10]. Besides the effects of high temperature and fire described in [11,12], the impact of sub-zero temperatures is one of the most destructive in-service factors to many building products, particularly composite products containing reinforcement in the form of various fibres, especially fibres of organic origin (to which cellulose fibres belong). In the authors' opinion, freeze–thaw cycles can very adversely affect the durability of such composites. Experiments were carried out in order to prove this thesis. The experiments consisted of subjecting fibre-cement board specimens to 1 and 10 freeze–thaw cycles and then investigating them under three-point bending by means of the acoustic emission (AE) method. Artificial intelligence in the form of an artificial neural network [13] was employed to analyse the experimental results. Previous studies by the authors [11,12,14,15] presented the assessment of the effect of freeze–thaw cycling based solely on bending strength to be inadequate. Whereas in this study, using the acoustic emission technique and analysing the degradation of the specimens, the authors were able to describe the degrading changes in the structure of the tested boards on the basis of not only the mechanical parameters, but also the acoustic phenomena. The registered AE signals provided the basis for developing reference acoustic spectrum characteristics accompanying cement matrix cracking and fibre breaking during bending. Then, an artificial neural network was used to recognize the characteristics in the AE records. In the course of freeze–thaw cycling, the fibres in the boards gradually degraded, which manifested as a fall in the number of events recognized as accompanying the breaking of fibres. This is described in more detail later in this paper. In order to verify the results and gain better insight into the changes taking place in the structure of the fibre-cement boards, they were examined under a scanning electron microscope (SEM).

2. Literature Survey

To-date, research on fibre-cement boards has focused on the effect of in-service factors [16–18] and the effect of high temperatures, determined by testing the physicomechanical parameters of the boards—mainly their bending strength (*MOR*). Only a few cases of testing fibre-cement boards by non-destructive methods, including the acoustic emission method, have been reported in the literature. Ardanuy et al. [6] presented the results of investigations into the effect of high temperatures on fibre-cement boards, but were limited to the bending strength test. Li et al. [19] examined the effect of high temperatures on composites produced using the extrusion method, but solely on the basis of the mechanical properties of the composite. Schabowicz et al. [11,12] used non-destructive methods to assess the effects of high temperature and fire on the degree of degradation of fibre-cement boards on the basis of the physicomechanical parameters. Other reported investigations of fibre-cement boards were devoted to the detection of imperfections arising during the production process. Papers by Drelich et al. [8] and Schabowicz et al. [20] presented the possibility of exploiting Lamb waves in a non-contact ultrasound scanner to detect defects in fibre-cement boards at the production stage. A method of detecting delaminations in composite elements by means of an ultrasonic probe was presented in a study by Stark, Vistap et al. [21]. Ultrasonic devices and a method used to detect delaminations in fibre-cement boards were described by Dębowski et al. [7]. Berkowski et al. [22], Hoła and Schabowicz [23] and Davis et al. [24] proposed the use of the impact-echo method jointly with the impulse response method to recognize delaminations in concrete elements. However, it is not recommended to test fibre-cement boards in this way since the two methods are intended for testing elements which are thicker than 100 mm. The special hammer used in the impulse response method can damage the fibre-cement boards being tested, while in the impact-echo method multiple wave reflections cause disturbances which make it difficult to interpret the obtained image [22]. Therefore, it is inadvisable to use the two methods to test fibre-cement boards, which are about 8 mm thick. There is scant information in the literature on the use of other non-destructive methods to test fibre-cement boards. The preliminary research described in [9,25] showed the terahertz (T-Ray) method to be suitable for testing fibre-cement boards. The character of terahertz signals is very similar to that of ultrasonic signals, but their interpretation is more complicated. Schabowicz et al. [15] and Ranachowski et al. [26] used X-ray microtomography to identify delaminations and low-density regions in fibre-cement boards. This technique was found to precisely reveal differences in the microstructure of such boards. It can be a useful tool for testing the structure of fibre-cement boards in which defects can arise as a result of production errors, but it is applicable only to small boards. As already mentioned, only a few cases of testing fibre-cement boards by means of acoustic emission have been reported so far. Ranachowski et al. [26] carried out pilot tests on fibre-cement boards produced by extrusion, including boards exposed to the temperature of 230 °C for 2 h, using the acoustic emission method to determine the effect of cellulose fibres on the strength of the fibre-cement boards and tried to distinguish the AE events emitted by the fibres from the ones emitted by the cement matrix. The investigations showed this method to be suitable for testing fibre-cement boards. Schabowicz et al. [11,12] and Gorzelańczyk et al. [11] proposed the use of the acoustic emission method to study the effects of fire and high temperatures on fibre-cement boards. The effect of high temperatures on concrete has been studied using the acoustic method (e.g., by Ranachowski [27] and Ranachowski et al. [28–30]), and is described widely in the literature. A large quantity of data are recorded during acoustic emission measurements, and they need to be properly analysed and interpreted. For this purpose, it can be useful to combine the acoustic emission method and artificial intelligence, including artificial neural networks (ANNs). ANNs are used to analyse and recognize signals acquired during the failure of various materials [31]. In [32–34] ANNs were used to analyse the results of testing concrete by means of non-destructive methods. Łazarska et al. [35] and Woźniak et al. [36] in their investigations of steel successfully used the acoustic emission method and artificial neural networks to analyse the obtained results. Rucka and Wilde [37,38], Zielińska and Rucka [39] and Wojtczak and Rucka [40] successfully used the ultrasonic method to investigate damage to concrete structures and masonry pillars. ANNs were also successfully used by Schabowicz et al. [11,12] to analyse the results of tests consisting of exposing fibre-cement boards to fire and high temperature.

Considering the above information, the authors came to the conclusion that the acoustic emission method combined with artificial neural networks would be suitable for assessing the changes taking place in the structure of fibre-cement boards exposed to freeze–thaw cycling.

3. Strength Tests

Two series of fibre-cement boards, denoted respectively A and B, were tested to determine the effect of freeze–thaw cycling. Altogether 60 specimens were tested. The basic specifications of the boards in the two series are given in Table 1.

Table 1. Basic specifications of the tested fibre-cement boards of series A and B.

Series	Board Thickness e (mm)	Board Colour	Application	Board Bulk Density ρ (g/cm^3)	Flexural Strength MOR (MPa)
A	8.0	natural	exterior	1.65	24
B	8.0	full body coloured	exterior	1.58	38

The freeze–thaw cycling of the specimens was conducted as follows. First the specimens were cooled (frozen) at a temperature of $-20 \pm 4\,°C$ in a freezer for 1–2 h and kept at this temperature for the next hour. The specimens were heated (thawed) in a water bath at a temperature of $20 \pm 4\,°C$ for 1–2 h and kept at this temperature for the next hour. One should note that the above temperatures apply to the conducting medium (i.e., air or water). Each freeze–thaw cycle lasted on average about 5–6 h. Air-dry reference specimens (not subjected to freeze–thaw cycling) were denoted as A_R and B_R. The denotations of exemplary series of boards are presented in Table 2. Figure 2 shows exemplary views of the tested (20×100 mm and 8 mm thick) specimens.

Table 2. Series of fibre-cement boards and test cases and their denotations.

Series Name/Test Case	Series A	Series B
Air-dry condition (reference board)	A_R	B_R
1 freeze–thaw cycles	A_1	B_1
10 freeze–thaw cycles	A_{10}	B_{10}

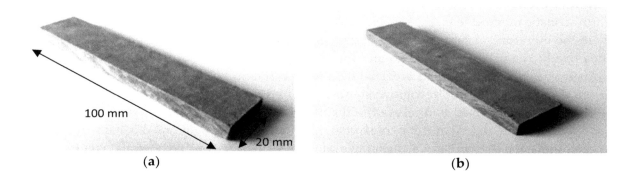

(a) (b)

Figure 2. Tested fibre-cement board specimens: (**a**) board A, (**b**) board B.

In order to determine the effect of freeze–thaw cycling on the fibre-cement boards, the latter were subjected to three-point bending and investigated using the acoustic emission method. Breaking force F, strain ε and AE signals were registered in the course of the three-point bending. Figure 3 shows the three-point bending test stand and the acoustic emission measuring equipment.

Figure 3. (**left**) Three-point bending test stand and acoustic emission (AE) measuring equipment and (**right**) close-up of fibre-cement board specimen during test.

The curve of flexural stress σ_m, the flexural strength (*MOR*), the limit of proportionality (*LOP*) and strain ε were taken into account in the analysis of the experimental results. The *MOR* was calculated from the standard formula [41]:

$$MOR = \frac{3Fl_s}{2b\,e^2},$$

(1)

where:

F is the loading force (N);
l_s is the length of the support span (mm);
b is the specimen width (mm); and
e is the specimen thickness (mm).

Figure 4 shows σ–ε graphs under bending for the specimens of all the tested fibre-cement boards.

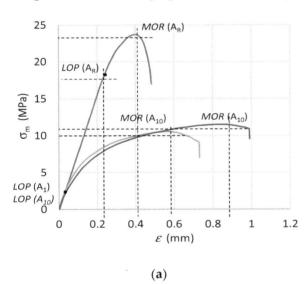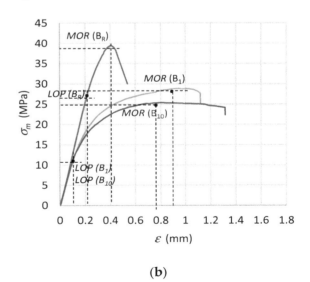

(**a**)   (**b**)

Figure 4. Bending σ–ε relation for specimens of fibre-cement boards: (**a**) series A, (**b**) series B. *LOP*: limit of proportionality; *MOR*: modulus of rupture.

Figure 4 shows that as a result of freeze–thaw cycling, flexural strength (*MOR*) decreased by 35%–50% in comparison with the flexural strength (*MOR*) of the reference fibre-cement boards. No significant difference in the change of bending strength (*MOR*) between the specimens subjected to one or ten freeze–thaw cycles was noticed. Note that the reference specimens were tested in air-dry condition at a mass moisture of 6%–8%. The effect of moisture on the value of flexural strength should be mainly ascribed to the weakening of the bonds between the crystals of the cement matrix structural lattice. The weakening is due to the fact that the bonds partially dissolve at a higher moisture content in the material, whereby the flexural strength (MOR) slightly decreases. As regards the path of the σ–ε, curve, one can see (Figure 4) that it clearly changed with the number of freeze–thaw cycles for both tested series of boards. Therefore it can be concluded that for the tested series of fibre-cement boards it was possible to determine the effect of the number of freeze–thaw cycles, as reflected in not only a reduction in flexural strength (MOR), but also in changes in the path of the σ–ε curve. It was found that when the number of cycles was increased from 1 to 10, the stiffness of the fibre-cement board and its brittleness decreased. In the case of the tested series, as the number of cycles increased, so did the range of the nonlinear increase in flexural stress while the limit of proportionality (LOP) considerably decreased. Thus, one can conclude that destructive changes took place in the structure of the fibre-cement boards. However, in the authors' opinion, knowledge of the mechanical parameters is not enough to determine the damaging effect of freeze–thaw cycles on fibre-cement boards. Therefore, in order to better identify the changes taking place in the structure of the tested fibre-cement boards, the acoustic method and an artificial neural network were used in this research.

4. Investigations Conducted Using Acoustic Emission Method and Artificial Neural Network

The next step in this research on degrading changes in the structure of fibre-cement boards exposed to freeze–thaw cycling was an analysis of the AE signals registered in the course of the three-point bending test. The analysis was based on AE descriptors such as: events rate N_{ev}, events sum $\sum N_{ev}$ and events energy E_{ev}, and on the signal frequency distribution. Figure 5 shows exemplary values of events sum $\sum N_{ev}$ registered for the boards of series A_R–A_{10} and B_R–B_{10}.

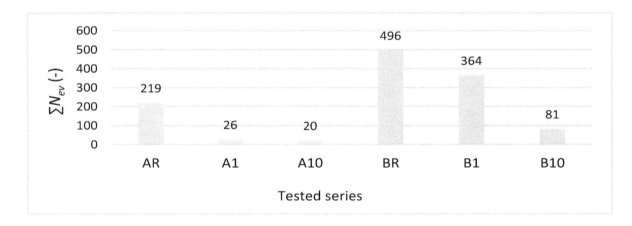

Figure 5. Exemplary events sum $\sum N_{ev}$ values registered for air-dry specimens and specimens subjected to freeze–thaw cycling.

For a more precise analysis of the failure of the boards under bending and the effect of freeze–thaw cycling, events sum $\sum N_{ev}$ and flexural stress σ_m versus time for selected cases are shown in Figure 6

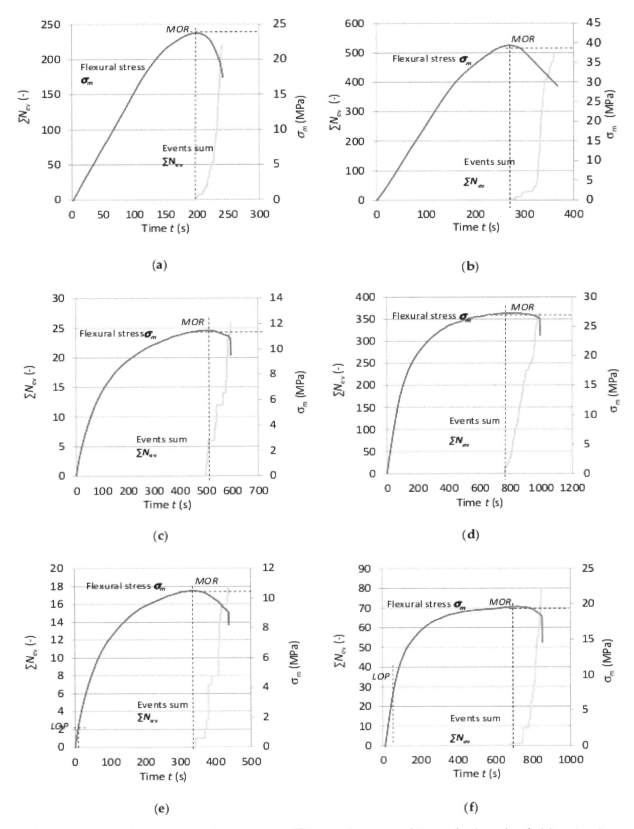

Figure 6. Flexural stress σ_m and events sum $\sum N_{ev}$ as function of time t for boards of: (**a**) series A_R, (**b**) series B_R, (**c**) series A_1, (**d**) series B_1, (**e**) series A_{10}, (**f**) series B_{10}.

A clear fall in registered events and a change in the path of events sum $\sum N_{ev}$ can be seen in Figure 6. The events were registered after the flexural strength (*MOR*) was exceeded. Note that a reduction in the number of events and a decrease in their energy were observed after subjecting the boards to freeze–thaw cycling. This does not mean that no AE events occurred, but one can suppose that the discrimination threshold for the registered events in the case of exposure to freeze–thaw cycling was too high. This can be connected with the measuring capability of the equipment used. Whereas the registered events with much lower energy could indicate a different process of destruction of the fibres—the latter can be pulled out of the wet cement matrix, whereby the event energy declines. These suppositions could be confirmed by SEM image analysis, which is presented further in this paper.

A spectral analysis of the AE event characteristics was carried out to more precisely identify the origins of the registered AE events. Reference acoustic spectra for cement matrix cracking, fibre breaking and the acoustic background were selected on this basis, which is described in detail in [12]. The reference acoustic spectra for cement matrix cracking were selected by analysing the record of acoustic activity in the time-frequency system during the bending of fibre boards previously fired in a laboratory furnace at a temperature of 230 °C for 3 h. It should be mentioned here that the investigations presented in this paper are part of a larger project devoted to the testing of fibre-cement boards. For example, in [11] it was observed that when fibre-cement boards were exposed to a temperature of 230 °C for 3 h, it resulted in the pyrolysis of their cellulose fibres, whereby the obtained fibre-cement board structure was completely devoid of fibres. Owing to this, reference acoustic spectra could be obtained for cement matrix cracking alone. The reference acoustic spectrum characteristic for fibre breaking was selected from the spectra obtained for air-dry reference boards of series A_R to B_R, characterized by a repetitively similar characteristic in the frequency range of 10–24 kHz, clearly distinct from the cement matrix characteristic. The characteristic of the background acoustic spectrum, originating from the testing machine, was determined by averaging the characteristics for all the tested boards of series A and B in the initial phase of bending.

One should note that the selected spectral fibre breaking characteristics are understood as the signal accompanying the cracking of the cement matrix together with the fibres, whereas the spectral matrix characteristic is understood as the signal accompanying the cracking of the cement matrix alone. The selected acoustic spectrum characteristic reference standards were recorded every 0.5 kHz in 80 intervals. Figure 7 shows a record of the reference acoustic spectrum characteristics of the signal accompanying respectively cement matrix cracking and fibre breaking, and of the background.

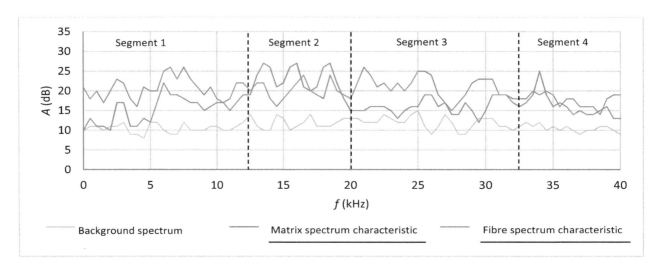

Figure 7. Background, fibre and cement matrix acoustic spectrum characteristics as function of frequency [12].

Figure 7 shows that the background acoustic activity was at the level of 10–15 dB. The cement matrix acoustic spectrum characteristic reached the acoustic activity of 25 dB within frequency ranges 5–10 kHz (segment 1) and 20–32 kHz (segment 3). The acoustic activity of over 25 dB within frequency ranges 12–18 dB (segment 2) and 32–38 kHz (segment 4) was read off for the fibres.

The cement matrix, fibre and background reference standards were implemented in an artificial neural network, and the training and testing of the latter began. A unidirectional multilayer backpropagation structure with momentum was adopted for the ANN. A model of the artificial neuron is shown in Figure 8. The model includes N inputs, one output, a summation block and an activation block.

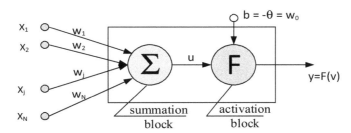

Figure 8. Model of an artificial neuron [13,42].

The following variables and parameters were used to describe the model shown in the Figure 8:

$$x_i = (x_1, x_2, \ldots , x_N) \qquad \text{an input vector,} \qquad (2)$$

$$w_i = (w_1, w_2, \ldots , w_N) \qquad \text{a weight vector,} \qquad (3)$$

$$b = -\theta = w_0 \qquad \text{a bias,} \qquad (4)$$

$$v = u + b = \sum_{j=1}^{N} w_j x_j - \theta = \sum_{j=0}^{N} w_j x_j \text{ a network potential,} \qquad (5)$$

$$F(v) \qquad \text{an activation function.} \qquad (6)$$

A model of the artificial neural network with inputs, information processing neurons and output neurons is shown in Figure 9.

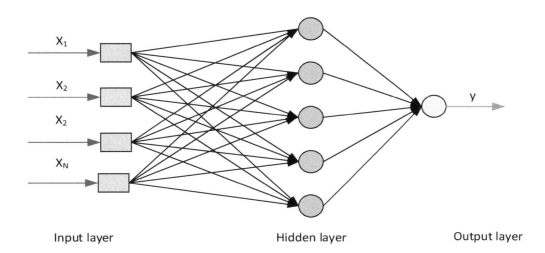

Figure 9. Model of the artificial neural network [13,43].

Eight appropriate learning sequences were adopted iteratively to achieve optimal compatibility of the learned ANN with the training pattern, as presented in Table 3.

Table 3. ANN learning sequences.

No.	Learning Sequences	No.	Learning Sequences
1	$16,000 \times A, 8000 \times B, 4000 \times C, 2000 \times D$	5	$16,000 \times A, 8000 \times B, 4000 \times C$
2	$16,000 \times A, 8000 \times B, 4000 \times C, 2000 \times D$	6	$16,000 \times A, 8000 \times B, 4000 \times C$
3	$16,000 \times A, 8000 \times B, 4000 \times C, 2000 \times D$	7	$16,000 \times A, 8000 \times B, 1000 \times C$
4	$16,000 \times A, 8000 \times B, 4000 \times C$	8	$16,000 \times A, 8000 \times B$

The spectral characteristics of fibre breaking were assigned to input A, the characteristics of cement matrix cracking were assigned to input B and the spectral characteristics of the background were assigned to input C. The spectral characteristics of fibre breaking were reproduced at input D.

After the ANN was trained on the input data, its mapping correctness was verified using the training and testing data. For this purpose, two pairs of input data were fed, that is, the data used for training the ANN, to check its ability to reproduce the reference spectra, and the one used for testing the ANN, to check its ability to identify the reference spectral characteristics originating from the fibres and the cement matrix during the bending test. For the eight performed training sequences, the ANN compliance with the training standard amounted to 0.995. Then, records of the ANN output in the form of recognised acoustic spectra for, respectively, fibre breaking, matrix cracking and the background were obtained. The learning coefficient (accelerating learning) was adopted at the level of 0.01 and momentum (increasing the stability of the obtained network configuration) at the level of 0.1. The following sigmoidal activation function was used: $1/(1 + \exp(-x_i))$.

Figure 10 shows the results of the recognition of the reference acoustic spectrum standards. They are superimposed on the record of events rate N_{ev} and bending stress σ_m versus time. The diagrams are for the reference specimens of series A_R and for the series subjected to 1 and 10 freeze–thaw cycles (respectively A_1 and A_{10}). In order to better illustrate the recognized acoustic spectra, the matrix reference standards are marked green while the fibre reference standards are marked light brown.

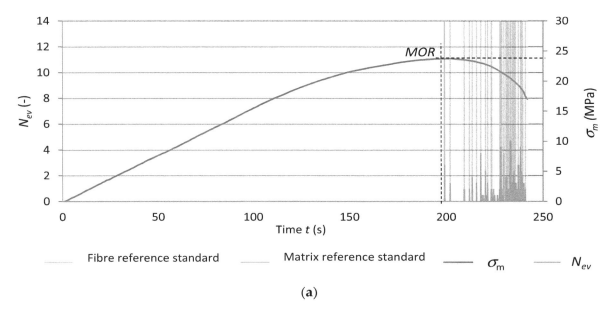

(a)

Figure 10. *Cont.*

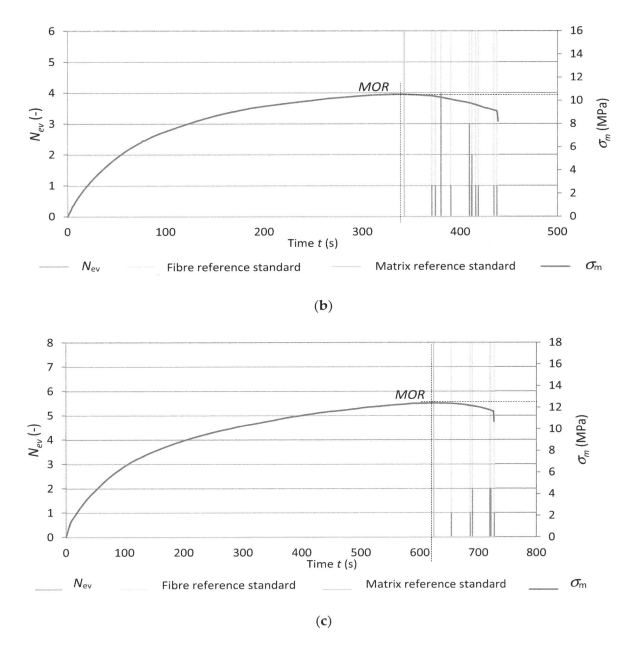

Figure 10. Freeze–thaw cycling diagrams of events rate N_{ev} and bending stress σ_m versus time, with superimposed identified reference spectral characteristics: (**a**) series A_R, (**b**) series A_1, (**c**) series A_{10}.

Figure 10 clearly shows that the number of registered events for the fibre-cement boards subjected to 1 and 10 freeze–thaw cycles was lower in comparison with the reference boards. Slightly fewer events were registered after ten freeze–thaw cycles. These events had very low energy E_{ev} ranging from 10 to 100 nJ. An event originating from a cement matrix fracture initiates next events originating from the breaking of fibres or from their pulling out of the matrix. Under the influence of moisture and freeze–thaw cycles, the cement-fibre board became more plastic, which manifested in the disappearance of the interval in which strains were proportional to stresses. The matrix cracked once the flexural strength (*MOR*) was exceeded. The ten freeze–thaw cycles limited the registered events to solely cement matrix cracking after the exceedance of the flexural strength (*MOR*).

Table 4 shows events recognized as respectively accompanying fibre breaking and cement matrix cracking for the tested fibre-cement boards of series A_R–A_{10}.

Table 4. Events recognized as respectively accompanying fibre breaking and cement matrix cracking for fibre-cement boards of series A_R–A_{10}.

Series	Events Sum $\sum N_{ev}$	Sum of Recognized Events $\sum N_{ev,r}$	Sum of Events Assigned to Fibre Breaking $\sum N_{ev,f}$	Sum of Events Assigned to Matrix Cracking $\sum N_{ev,m}$
A_R	219	201	193	8
A_1	26	24	20	4
A_{10}	20	19	16	3

Graphs of events rate N_{ev} and flexural stress σ_m versus time under freeze–thaw cycling, with superimposed identified reference spectral characteristics for the specimens of series B_R, B_1 and B_{10} are shown in Figure 11.

Figure 11 shows that the course of the AE signals registered for the fibre-cement boards of series B_1 was similar to that for series B_R, whereas the flexural stress σ_m curves differed. The limit of proportionality clearly fell and the interval in which the increment in stress relative to strain was nonlinear was wider. These changes can be ascribed to cement matrix yielding. The numerous events with the high energy of over 1000 nJ, originating from fibre breaking, indicate fibre destruction similar as in the reference boards. Therefore, it can be concluded that the fibres used in the tested boards did not change their properties under the influence of moisture and freeze–thaw cycling. This was confirmed by the high flexural strength (*MOR*) of about 27 MPa. In the case of the fibre-cement boards of series B_{10} subjected to ten freeze–thaw cycles, a slight decrease (about 20%) in flexural strength (*MOR*) and a drop in fibre breaking events in comparison with series B_1 were observed. This means that the long-lasting dampness in conjunction with the freeze–thaw cycling had a degrading effect on the boards of series B_{10}, reducing their strength and increasing their deformability. Moreover, the decrease in the sum of registered events in comparison with the reference boards could also be due to the lower acoustic activity of the events accompanying the breaking of the fibres, emitted below the discrimination threshold.

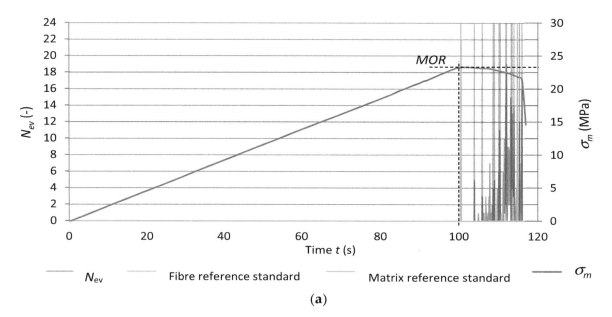

(a)

Figure 11. *Cont.*

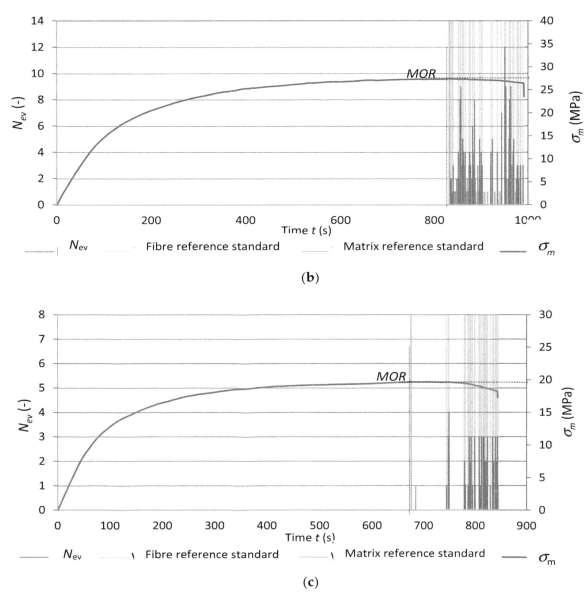

Figure 11. Events rate N_{ev} and flexural stress σ_m versus time, under freeze–thaw cycling, with superimposed identified reference spectral characteristics for: (**a**) series B_R, (**b**) series B_1, (**c**) series B_{10}.

Summing up the results of the investigations, one can conclude that the effect of freeze–thaw cycling manifested itself mainly in a decrease in flexural strength (*MOR*) for all the tested series. The graphs presented in Figures 10 and 11 show that when the flexural strength (*MOR*) was reached, the cement matrix cracked. The cracking of the cement matrix initiated events consisting of the breaking of the fibres. In all the series, the linear increment in stress relative to strain was found to clearly increase. The freeze–thaw cycling reduced the acoustic activity of the events taking place during the bending test. One can suppose that the different ways in which the fibres are destroyed as a result of freeze–thaw cycling can contribute to the decrease in the energy of fibre breaking events. Fibres can be damp and swollen and when undergoing destruction emit events characterized by very low energy. Whereas dampness-resistant PVA fibres (polyvinyl alcohol) instead of breaking can be pulled out of the damp cement matrix, which also will not generate high energy events. One can suppose that by optically examining the image of the fractured surface one can gain information about the mode of failure of the fibres.

Table 5 shows the events recognized as accompanying fibre breaking and cement matrix cracking for the boards of series B_R–B_{10}.

Table 5. Events recognized as accompanying fibre breaking and cement-matrix cracking for boards of series B_R–B_{10}.

Series	Events Sum $\sum N_{ev}$	Sum of Recognized Events $\sum N_{ev,r}$	Sum of Events Assigned to Fibre Breaking $\sum N_{ev,f}$	Sum of Events Assigned to Matrix Cracking $\sum N_{ev,m}$
B_R	496	487	439	48
B_1	364	350	332	18
B_{10}	81	76	68	8

5. Investigations Conducted Using Scanning Electron Microscope (SEM)

A high-resolution environmental scanning electron microscope (SEM) Quanta 250 FEG, FEI with an EDS analyser was used for the investigations. Figures 12 and 13 show exemplary images obtained by means of the SEM for respectively series A_R–A_{10} and B_R–B_{10}. The examined fibre-cement boards had previously been subjected to the three-point bending test.

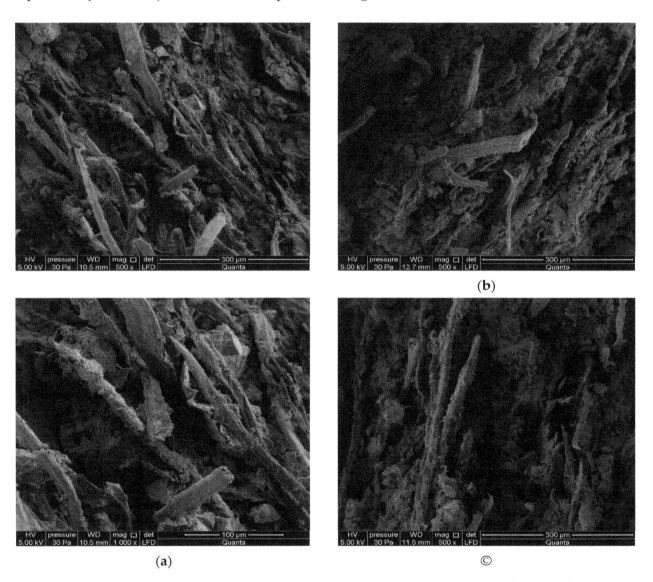

Figure 12. SEM images of boards: (**a**) series A_R (magnification 1000× at top, 500× at bottom), (**b**) series A_1, (**c**) series A_{10}.

Figure 13. SEM images of boards: (**a**) series B_R (magnification 1000× at top, 500× at bottom), (**b**) series B_1, (**c**) series B_{10}.

On the basis of an analysis of the SEM and EDS images, the macrostructure of the fibre-cement boards of series A_R and B_R can be described as compact. The microscopic examinations revealed the structure to be fine-pore, with pore size of up to 50 µm. Cavities up to 500 µm wide and grooves were visible in the places in the fractured surface where fibres had been pulled out. In the images one can see cellulose and PVA fibres. Various forms of hydrated calcium silicates of the C-S-H type occurred, with an "amorphous" phase and a phase comprised of strongly adhering particles predominating. An analysis of the composition of the fibres showed elements native to them and elements native to cement. The surface of the fibres was coated with a thin layer made up of cement matrix and hydration products. The fact that there were few places with a space between the fibres and the matrix indicates that they were strongly bonded. An examination of the fibre-cement boards subjected to one or ten freeze–thaw cycles reveals that most of the fibres were pulled out of the cement matrix. Numerous cavities, most left by the pulled-out cellulose fibres, and grooves were visible. The matrix structure was more granular and included numerous delaminations. Strongly compact structures having an irregular shape were found to be present.

Summing up the results of the investigations, the authors conclude that the freeze–thaw cycling of fibre-cement boards affected their structure. Fibres wholly pulled out of the matrix and the voids left by them in the cement matrix were visible in the fibre-cement boards subjected to freeze–thaw cycling.

This was especially noticeable in the fibre-cement boards subjected to 10 freeze–thaw cycles, mainly in the boards of series A. In the case of the reference boards, much more fibres (whose ends were firmly "anchored" in the cement matrix) were ruptured. This conclusively proves that freeze–thaw cycling significantly weakened the structure of fibre-cement boards.

6. Conclusions

The impact of low temperatures, in the form of freeze–thaw cycling, is by nature destructive to most building products. The degree of resistance to this impact is measured by the number of freeze–thaw cycles after which such products retain the properties specified by the standards. The investigations of the fibre-cement boards of series A and B carried out as part of this study showed that the boards differed in their degree of resistance to the impact of freeze–thaw cycling. Thanks to the use of the acoustic emission method during the three-point bending of the boards, the course of their failure for one or ten freeze–thaw cycles could be observed and compared with that of the reference boards. An artificial neural network was employed to analyse the results yielded by the acoustic emission method. The investigations conclusively proved that freeze–thaw cycling affected the way fibre-cement boards failed, as reflected mainly in the decrease in the number of AE events recognized as accompanying the breaking of fibres during the three-point bending of the specimens. SEM examinations were carried out in order to obtain better insight into the changes taking place in the structure of the tested boards. They confirmed that significant changes took place in the structure of the boards, especially after 10 freeze–thaw cycles. The structure became less compact (more granular). Most of the fibres were not ruptured during the bending test, but pulled out of the cement matrix, as confirmed by the low energy of the registered AE events and their small number. This was particularly evident in the case of the tested fibre-cement boards of series A. In the authors' opinion, the above findings are important for building practice since they indicate that it is inadvisable to use fibre-cement boards whose flexural strength (*MOR*) considerably decreases under the influence of freeze–thaw cycling for the cladding of ventilated façades, especially in high buildings located in zones of high wind load.

Author Contributions: T.G. prepared the specimens, completed the experiments, analysed the test results and performed paper editing; K.S. conceived and designed the experimental work.

References

1.   Schabowicz, K.; Gorzelańczyk, T. Fabrication of fibre cement boards. In *The Fabrication, Testing and Application of Fibre Cement Boards*, 1st ed.; Ranachowski, Z., Schabowicz, K., Eds.; Cambridge Scholars Publishing: Newcastle upon Tyne, UK, 2018; pp. 7–39. ISBN 978-1-5276-6.

2.   Bentchikou, M.; Guidoum, A.; Scrivener, K.; Silhadi, K.; Hanini, S. Effect of recycled cellulose fibres on the properties of lightweight cement composite matrix. *Constr. Build. Mater.* **2012**, *34*, 451–456. [CrossRef]

3.   Savastano, H.; Warden, P.G.; Coutts, R.S.P. Microstructure and mechanical properties of waste fibre–cement composites. *Cem. Concr. Compos.* **2005**, *27*, 583–592. [CrossRef]

4.   Coutts, R.S.P. A Review of Australian Research into Natural Fibre Cement Composites. *Cem. Concr. Compos.* **2005**, *27*, 518–526. [CrossRef]

5.   Schabowicz, K.; Szymków, M. Ventilated facades made of fibre-cement boards (in Polish). *Mater. Bud.* **2016**, *4*, 112–114. [CrossRef]

6.   Ardanuy, M.; Claramunt, J.; Toledo Filho, R.D. Cellulosic Fibre Reinforced Cement-Based Composites: A Review of Recent Research. *Constr. Build. Mater.* **2015**, *79*, 115–128. [CrossRef]

7.   Dębowski, T.; Lewandowski, M.; Mackiewicz, S.; Ranachowski, Z.; Schabowicz, K. Ultrasonic tests of fibre-cement boards (in Polish). *Przegląd Spaw.* **2016**, *10*, 69–71. [CrossRef]

8.   Drelich, R.; Gorzelanczyk, T.; Pakuła, M.; Schabowicz, K. Automated control of cellulose fibre cement boards with a non-contact ultrasound scanner. *Autom. Constr.* **2015**, *57*, 55–63. [CrossRef]

9.   Chady, T.; Schabowicz, K.; Szymków, M. Automated multisource electromagnetic inspection of fibre-cement boards. *Autom. Constr.* **2018**, *94*, 383–394. [CrossRef]
10. Schabowicz, K.; Jóźwiak-Niedźwiedzka, D.; Ranachowski, Z.; Kudela, S.; Dvorak, T. Microstructural characterization of cellulose fibres in reinforced cement boards. *Arch. Civ. Mech. Eng.* **2018**, *4*, 1068–1078. [CrossRef]
11. Schabowicz, K.; Gorzelańczyk, T.; Szymków, M. Identification of the degree of fibre-cement boards degradation under the influence of high temperature. *Autom. Constr.* **2019**, *101*, 190–198. [CrossRef]
12. Schabowicz, K.; Gorzelańczyk, T.; Szymków, M. Identification of the degree of degradation of fibre-cement boards exposed to fire by means of the acoustic emission method and artificial neural networks. *Materials* **2019**, *12*, 656. [CrossRef] [PubMed]
13. Leflik, M. Some aspects of application of artificial neural network for numerical modelling in civil engineering. *Bull. Pol. Acad. Sci. Tech. Sci.* **2013**, *61*, 39–50. [CrossRef]
14. Gorzelańczyk, T.; Schabowicz, K.; Szymków, M. Non-destructive testing of fibre-cement boards, using acoustic emission (in Polish). *Przegląd Spaw.* **2016**, *88*, 35–38. [CrossRef]
15. Adamczak-Bugno, A.; Gorzelańczyk, T.; Krampikowska, A.; Szymków, M. Non-destructive testing of the structure of fibre-cement materials by means of a scanning electron microscope (in Polish). *Bad. Nieniszcz. I Diagn.* **2017**, *3*, 20–23. [CrossRef]
16. Claramunt, J.; Ardanuy, M.; García-Hortal, J.A. Effect of drying and rewetting cycles on the structure and physicochemical characteristics of softwood fibres for reinforcement of cementitious composites. *Carbohydr. Polym.* **2010**, *79*, 200–205. [CrossRef]
17. Mohr, B.J.; Nanko, H.; Kurtis, K.E. Durability of kraft pulp fibre-cement composites to wet/dry cycling. *Cem. Concr. Compos.* **2005**, *27*, 435–448. [CrossRef]
18. Pizzol, V.D.; Mendes, L.M.; Savastano, H.; Frías, M.; Davila, F.J.; Cincotto, M.A.; John, V.M.; Tonoli, G.H.D. Mineralogical and microstructural changes promoted by accelerated carbonation and ageing cycles of hybrid fibre–cement composites. *Constr. Build. Mater.* **2014**, *68*, 750–756. [CrossRef]
19. Li, Z.; Zhou, X.; Bin, S. Fibre-Cement extrudates with perlite subjected to high temperatures. *J. Mater. Civ. Eng.* **2004**, *3*, 221–229. [CrossRef]
20. Schabowicz, K.; Gorzelańczyk, T. A non-destructive methodology for the testing of fibre cement boards by means of a non-contact ultrasound scanner. *Constr. Build. Mater.* **2016**, *102*, 200–207. [CrossRef]
21. Stark, W. Non-destructive evaluation (NDE) of composites: Using ultrasound to monitor the curing of composites. In *Non-Destructive Evaluation (NDE) of Polymer Matrix Composites. Techniques and Applications*, 1st ed.; Karbhari, V.M., Ed.; Woodhead Publishing: Limited, UK, 2013; pp. 136–181. ISBN 978-0-85709-344-8.
22. Berkowski, P.; Dmochowski, G.; Grosel, J.; Schabowicz, K.; Wójcicki, Z. Analysis of failure conditions for a dynamically loaded composite floor system of an industrial building. *J. Civ. Eng. Manag.* **2013**, *19*, 529–541. [CrossRef]
23. Hoła, J.; Schabowicz, K. State-of-the-art non-destructive methods for diagnostic testing of building structures–anticipated development trends. *Arch. Civ. Mech. Eng.* **2010**, *10*, 5–18. [CrossRef]
24. Davis, A.; Hertlein, B.; Lim, K.; Michols, K. Impact-echo and impulse response stress wave methods: Advantages and limitations for the evaluation of highway pavement concrete overlays. In Proceedings of the Conference on Nondestructive Evaluation of Bridges and Highways, Scottsdale, AZ, USA, 4 December 1996; pp. 88–96.
25. Chady, T.; Schabowicz, K. Non-destructive testing of fibre-cement boards, using terahertz spectroscopy in time domain (in Polish). *Bad. Nieniszcz. I Diagn.* **2016**, *1*, 62–66.
26. Schabowicz, K.; Ranachowski, Z.; Jóźwiak-Niedźwiedzka, D.; Radzik, Ł.; Kudela, S.; Dvorak, T. Application of X-ray microtomography to quality assessment of fibre cement boards. *Constr. Build. Mater.* **2016**, *110*, 182–188. [CrossRef]
27. Ranachowski, Z.; Ranachowski, P.; Dębowski, T.; Gorzelańczyk, T.; Schabowicz, K. Investigation of structural degradation of fibre cement boards due to thermal impact. *Materials* **2019**, *12*, 944. [CrossRef] [PubMed]
28. Ranachowski, Z.; Schabowicz, K. The contribution of fibre reinforcement system to the overall toughness of cellulose fibre concrete panels. *Constr. Build. Mater.* **2017**, *156*, 1028–1034. [CrossRef]
29. Ranachowski, Z. The application of neural networks to classify the acoustic emission waveforms emitted by the concrete under thermal stress. *Arch. Acoust.* **1996**, *21*, 89–98.

30. Ranachowski, Z.; Jóźwiak-Niedźwiedzka, D.; Brandt, A.M.; Dębowski, T. Application of acoustic emission method to determine critical stress in fibre reinforced mortar beams. *Arch. Acoust.* **2012**, *37*, 261–268. [CrossRef]

31. Yuki, H.; Homma, K. Estimation of acoustic emission source waveform of fracture using a neural network. *NDT E Int.* **1996**, *29*, 21–25. [CrossRef]

32. Schabowicz, K. Neural networks in the NDT identification of the strength of concrete. *Arch. Civ. Eng.* **2005**, *51*, 371–382.

33. Asteris, P.G.; Kolovos, K.G. Self-compacting concrete strength prediction using surrogate models. *Neural Comput. Appl.* **2019**, *31*, 409–424. [CrossRef]

34. Lee, S.C. Prediction of concrete strength using artificial neural networks. *Eng. Struct.* **2003**, *25*, 849–857. [CrossRef]

35. Łazarska, M.; Woźniak, T.; Ranachowski, Z.; Trafarski, A.; Domek, G. Analysis of acoustic emission signals at austempering of steels using neural networks. *Met. Mater. Int.* **2017**, *23*, 426–433. [CrossRef]

36. Woźniak, T.Z.; Ranachowski, Z.; Ranachowski, P.; Ozgowicz, W.; Trafarski, A. The application of neural networks for studying phase transformation by the method of acoustic emission in bearing steel. *Arch. Metall. Mater.* **2014**, *59*, 1705–1712. [CrossRef]

37. Rucka, M.; Wilde, K. Experimental study on ultrasonic monitoring of splitting failure in reinforced concrete. *J. Nondestruct. Eval.* **2013**, *32*, 372–383. [CrossRef]

38. Rucka, M.; Wilde, K. Ultrasound monitoring for evaluation of damage in reinforced concrete. *Bull. Pol. Acad. Sci. Tech. Sci.* **2015**, *63*, 65–75. [CrossRef]

39. Zielińska, M.; Rucka, M. Non-Destructive Assessment of Masonry Pillars using Ultrasonic Tomography. *Materials* **2018**, *11*, 2543. [CrossRef]

40. Wojtczak, E.; Rucka, M. Wave Frequency Effects on Damage Imaging in Adhesive Joints Using Lamb Waves and RMS. *Materials* **2019**, *12*, 1842. [CrossRef]

41. EN 12467–Cellulose Fibre Cement Flat Sheets. Product Specification and Test Methods. 2018. Available online: https://standards.globalspec.com/std/10401496/din-en-12467 (accessed on 9 June 2019).

42. Osowski, S. *Neural Networks for Information Processing (in Polish)*; OWPW: Warsaw, Poland, 2000.

43. Estêvão, J.M.C. Feasibility of using neural networks to obtain simplified capacity curves for seismic assessment. *Buildings* **2018**, *8*, 151. [CrossRef]

13

Investigation of Structural Degradation of Fiber Cement Boards Due to Thermal Impact

Zbigniew Ranachowski [1],*⑩, Przemysław Ranachowski [1], Tomasz Dębowski [1],
Tomasz Gorzelańczyk [2]⑩ and Krzysztof Schabowicz [2]⑩

[1] Experimental Mechanics Division, Institute of Fundamental Technological Research, Polish Academy of
    Sciences, Pawińskiego 5B, 02-106 Warszawa, Poland; pranach@ippt.pan.pl (P.R.); tdebow@ippt.pan.pl (T.D.)
[2] Faculty of Civil Engineering, Wrocław University of Science and Technology, Wybrzeże Wyspiańskiego 27,
    50-370 Wrocław, Poland; Tomasz.Gorzelanczyk@pwr.edu.pl (T.G.); krzysztof.schabowicz@pwr.edu.pl (K.S.)
*   Correspondence: zranach@ippt.pan.pl

Abstract: The aim of the present study was to investigate the degradation of the microstructure and mechanical properties of fiber cement board (FCB), which was exposed to environmental hazards, resulting in thermal impact on the microstructure of the board. The process of structural degradation was conducted under laboratory conditions by storing the FCB specimens in a dry, electric oven for 3 h at a temperature of 230 °C. Five sets of specimens, that differed in cement and fiber content, were tested. Due to the applied heating procedure, the process of carbonization and resulting embrittlement of the fibers was observed. The fiber reinforcement morphology and the mechanical properties of the investigated compositions were identified both before, and after, their carbonization. Visual light and scanning electron microscopy, X-ray micro tomography, flexural strength, and work of flexural test W_f measurements were used. A dedicated instrumentation set was prepared to determine the ultrasound testing (UT) longitudinal wave velocity c_L in all tested sets of specimens. The UT wave velocity c_L loss was observed in all cases of thermal treatment; however, that loss varied from 2% to 20%, depending on the FCB composition. The results obtained suggest a possible application of the UT method for an on-site assessment of the degradation processes occurring in fiber cement boards.

Keywords: cement-based composites; fiber cement boards; durability; ultrasound measurements

1. Introduction

Fiber cement board (FCB) is a versatile, green, and widely-applied building material. It acts as a substitute for natural wood and wood-based products, such as plywood or oriented strand boards (OSB). The properties of FCB, as a construction material, make it preferable for use as a ventilated, façade cladding for newly-built and renovated buildings, interior wall coverings, balcony balustrade panels, base course and chimney cladding, and enclosure soft-fit lining [1]. FCB can be applied to unfinished, painted, or simply-impregnated surfaces. Fiber cement components have been used in construction for over 100 years, mainly as roofing covers, in the form of corrugated plates or non-pressurized tubes. FCB façades are exposed to a variety of different environmental hazards. Adverse factors can include visual an ultraviolet light radiation, wind and ice-clod impacts, and thermal stresses evoked by temperature changes, etc. [2,3]. These hazards may result in board embrittlement, shrinkage, or bending. An example of a FCB façade, showing considerable damage after ten years of exposure to climate hazards, is presented in Figure 1. The fracture process that can develop in building materials is complex because the strains are not uniformly distributed during the fracture, particularly in regions where there are cracks. The facade boards are usually fixed to the wall-construction on their edgings, which exposes them to flexural stresses. The currently-applied fiber cement boards are designed to carry the mechanical load by the cellulose and polyvinyl alcohol (PVA) fiber reinforcements. The fibers reinforce the FCB component only when they are added in a specific quantity (5–10% wt.) and when they are uniformly dispersed throughout the cementitious matrix. A highly complex procedure is required to achieve this goal, as well as to avoid faults under efficient industrial conditions. Hatschek solved the problem by inventing a machine with a rotating sieve and a vat containing a diluted fiber slurry, Portland cement, and mineral components [4,5]. A thin film of FCB is formed on a moving belt, partially wrapped around the sieve, similar to the procedure used in paper sheet-making [6].

Figure 1. An example of a FCB façade showing considerable damage after ten years of exposure to climate hazards.

As the service performance of fiber cement boards may be affected by the improper function of reinforcements (i.e., damaged fibers, inhomogeneous concentrations, or poor quality fibers), several methods were proposed for testing the performance of the boards. These include:

- emitting and receiving the ultrasonic Lamb waves [7,8],
- the impact-echo method combined with the impulse-response method [9],
- the ultrasound (UT) longitudinal wave velocity c_L method [10,11],
- the acoustic emission method [12],
- X-ray micro tomography [13,14].

The UT wave velocity c_L method is one of the few methods that can be applied in situ to control the degradation processes in building façades. Some authors [15,16] have performed experiments that resulted in the degradation of fiber reinforcements by pyrolyzing the fibers; i.e., by exposing them to an elevated temperature for several hours. These authors reported that a loss of material elasticity can be observed when thermal treatment is conducted using temperatures exceeding 100 °C.

It is also possible to analyze the morphology and distribution of fibers in the cementitious matrix using microscopy, by applying different kinds of visible light during the testing process. A detailed description of this procedure can be found in [17]. Cellulose fibers are usually thicker than PVA fibers. They are beige-colored, like wood, while PVA fibers are pale and transparent. After thermal treatment, both types of fibers become brown, which is evidence of their structural dehydrogenation, so that what remains in the fibers is mostly dark-colored structural carbon.

The quality of the boards was evaluated using an exact measure to determine their mechanical toughness, understood here as the integrated product of applied stress and strain per unit of cross section of the investigated board, i.e., the work of flexural test W_f. The latter parameter can be determined as the work made over the deflection curve during the bending test [12]. In this study, the authors began testing with an initial force F_0 of 2 N and continued to break the fiber reinforcements so that the final decrease was registered at 40% of the maximum load—$F_{0.4\ MAX}$. Under these conditions, the flexural test W_f can be calculated by applying the formula:

$$W_f \frac{1}{S} \int_{F_0}^{F_{0.4\ MAX}} F\, da \tag{1}$$

where S = specimen cross section; a = specimen deflection under the loading pin.

The loss of FCB elasticity can also be determined by applying the ultrasound testing (UT) method. In large objects with small thicknesses, like the flat boards, the following dependence combines UT longitudinal wave velocity c_L and the modulus of elasticity E:

$$c_L = \sqrt{\frac{E}{\rho(1-v^2)}} \tag{2}$$

where ρ = bulk material density; v = Poisson ratio.

2. Materials and Methods

In this study, five sets of specimens, made of five different types of FCB, were prepared for examination. These were labeled A, B, C, D and E. All of the specimens were fabricated by applying the Hatschek, or flow-on, forming method. Different matrix fillers and concentrations of fibers were determined for each set and resulted in different flexural strengths, which was measured by applying the standard EN 12467 three-point bending test [18]. The specimens of all FCB types, listed above, were stored in a dry, electric oven for 3 h at a temperature of 230 °C. The parameters of the heating procedure were chosen experimentally, after performing some preliminary tests to evoke considerable changes in the microstructure of the investigated materials. That treatment resulted in the decomposition (i.e., numerous broken chemical C–O bonds) of fiber reinforcements due to the process of carbonization. The carbonization process mostly concerns the linear chains of dehydrated glucose molecules, which are responsible for building up the cellulose fiber system.

The specimens, which underwent the high temperature treatment, were labeled A_T, B_T, C_T, D_T and E_T. Small pieces of each FCB type were prepared for microscopic observation, both before and after the elevated temperature treatment, and are presented in Figure 2. The mechanical properties of the specimens are shown in Table 1.

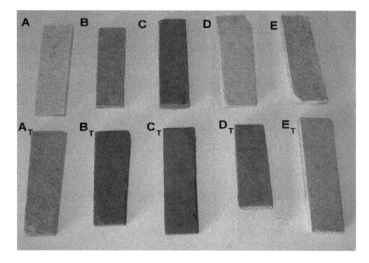

Figure 2. View of the small FCB (fiber cement board) pieces used for microscopic observation. Upper row: specimens of FCB types A, B, C, D and E before the elevated temperature treatment. Lower row: specimens of FCB types A, B, C, D and E after the elevated temperature treatment.

Table 1. Mechanical properties of the tested FCB compositions.

Board Symbol	Remarks	Board Thickness [mm]	Apparent Density [kg/m³]	Flexural Strength [MPa]
A	low cement content in the board matrix, approved for internal use	8.4	1000	14
B	low cement content in the board matrix, approved for internal use	7.4	1600	21
C	colored brown and approved for internal use	7.8	1700	23
D	approved for external use	8	1600	36
E	approved for internal use	9.3	1700	23

For detailed insight into each specimen's microstructure, the authors applied an X-ray microtomography (micro-CT) technique, which is described in more detail in [13]. A Nanotom 30, made by General Electric (Baker Hughes GE, Houston, TX, USA) was used in the investigation. The system included a micro-focal source of X-ray radiation, a movable table on which to place a specimen, and a flat FCB panel with a radiation detector, having a resolution of 2000×2000 pixels. The microstructure of each FCB sample was observed on the cross sections (tomograms) of the investigated specimens, using a grey scale convention that was directly related to the amount of the local radiation absorption of the materials. The grey scale covers a wide range of grey levels and is ordered from pure white, related to maximum absorption, to pure black, related to minimum absorption, respectively. Un-hydrated cement particles and aggregate grains are objects in the cement matrix that demonstrate the highest degree of absorption ability. The hydration products that make up the major components of the cement matrix present a slightly lower absorption ability. Next in line are the hydrated calcinates, which demonstrate an even lower absorption ability and, at the end of the scale, are the organic fibers (if present) and the regions of high porosity. To obtain the optimal X-ray penetration and absorption of the investigated specimens, the following parameters of the scanning procedure were set: X-ray lamp voltage—115 kV, lamp current—95 microamperes, and shot exposition time—750 ms. Scanning, performed by the authors in this study, resulted in a large set of tomograms (specimen cross sections), performed for every 5 µm of the specimen height.

The authors prepared a dedicated instrumentation set to determine the UT longitudinal wave velocity c_L in the boards made of fibrous materials. The Wave velocity c_L was determined using the UT material tester, which was capable of measuring the time of flight T of the elastic wave front, across a board of known thickness d, with the application of the formula $c_L = d/T$. The investigation was done using the ultrasonic material tester, UTC110, produced by Eurosonic (Vitrolles, France) [19]. A report in the related literature indicates that low-frequency ultrasound (50–200 kHz) was routinely used to characterize defects (a few centimeters in size) in the concrete structures on site. However, ultrasound at low-frequency ranges cannot be used to test fiber cement boards. Some experiments [11] have revealed that the sensitivity of the ultrasound parameters required to determine the structural properties of FCB is achieved when the ultrasound wavelength becomes comparable to the dimensions of the local delaminations and the lengths of the fiber inclusions. This wavelength λ remains in the following relation to the frequency f of the emitting source and the propagation velocity of the traveling ultrasonic longitudinal waveform:

$$\lambda = \frac{c_L}{f} \tag{3}$$

Thus, taking into account the propagation velocity of 1000–2000 m/s registered in the FCB, the authors recommend the application of a frequency of 1 MHz to achieve the propagation of wavelengths in the range of 1–2 mm. The instrumentation included an Olympus Videoscan [20] transmitting and receiving transducer, which emitted an ultrasonic beam measuring 19 mm in diameter, at a frequency of 1 MHz. The parameters were designed for coupling with low-density (i.e., 1000–2000 kg/m³) materials and, thus, exhibited a low-acoustic impedance of 10 MegaRayl. The contact between the rough surface of the FCB and the face of the transducer was achieved by using a 0.6 mm thick layer of polymer jelly interfacing foil (PM-4-12) produced by Olympus (Waltham, MA, USA) [20]. The custom-designed holder, with articulated joints and a compression spring, was prepared to ensure the correct coupling of the ultrasonic transducers to both surfaces of the investigated boards. A detailed view of the holder is presented in Figure 3.

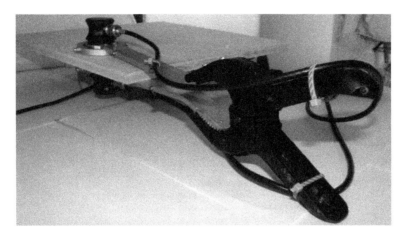

Figure 3. Detailed view of the custom-designed holder for correct coupling between the ultrasonic transducers and both sides of the rough surface board.

3. Results

3.1. Optical Light Microscopy

The morphology and surface views of all of the investigated compositions were analysed by visual light microscopy (AM4113ZTL 1.3 Megapixel Dino-Lite Digital Microscope with integral LED lighting, AnMo Electronics Corp. (Hsinchu, Taiwan). The magnified surface views of the investigated specimens are presented in Figure 4. Highly-diverse fiber distributions, due to the different fiber compositions, are noticeable in the micrographs.

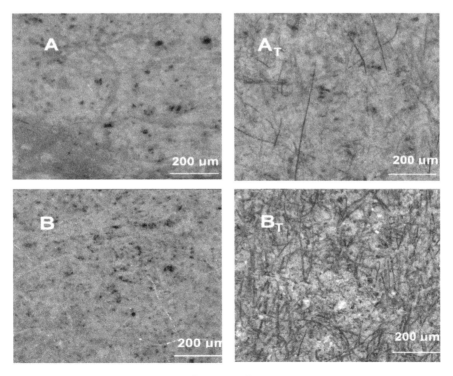

Figure 4. *Cont.*

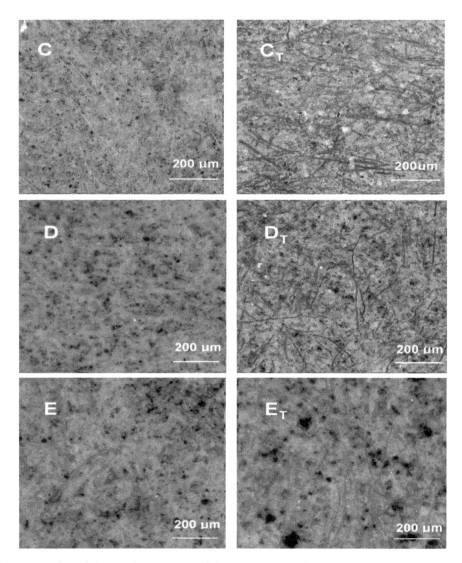

Figure 4. Micrographs of the surface views of the investigated specimens. **Left**: before the elevated temperature treatment; **Right**: after the elevated temperature treatment.

3.2. Scanning Electron Microscopy

A high-resolution environmental scanning electron microscope, Quanta 250 FEG, FEI (Hillsboro, OR, USA), was used in the investigations, along with an energy dispersive X-ray spectroscopy (EDS) analyzer. Figure 5 shows the exemplary SEM images for the tested fiber cement boards. Samples that were not exposed to a high temperature are shown on the left, while those that were exposed to a temperature of 230 °C are shown on the right. Exemplary elemental composition results for board C, which were obtained using the EDS analyzer, are shown in Figure 6.

An analysis of the images obtained from the scanning electron microscope and the EDS analyzer shows that the fiber cement boards, in the A–E series, have a compact macrostructure. Microscopic examinations revealed a fine-pore structure, with pores of up to 50 μm in size. Cavities and grooves, up to 500 μm wide, were visible in the fracture areas where the fibers had been pulled out. Cellulose fibers and, in some boards (B and D), PVA fibers, are clearly visible in the images. Various forms of hydrated calcium silicates of the C-S-H type occur. Both an "amorphous" phase and a phase built of strongly-adhering particles predominate. An analysis of the fiber composition showed that fiber elements and some cement elements were present. An analysis of the chemical composition of the matrix showed elements that are typical of cement. The surface of the fibers was covered with a thin layer of cement paste and hydration products. The fact that there are very few areas with a space between the fibers and the cement paste, indicates that the fiber-cement bond is strong.

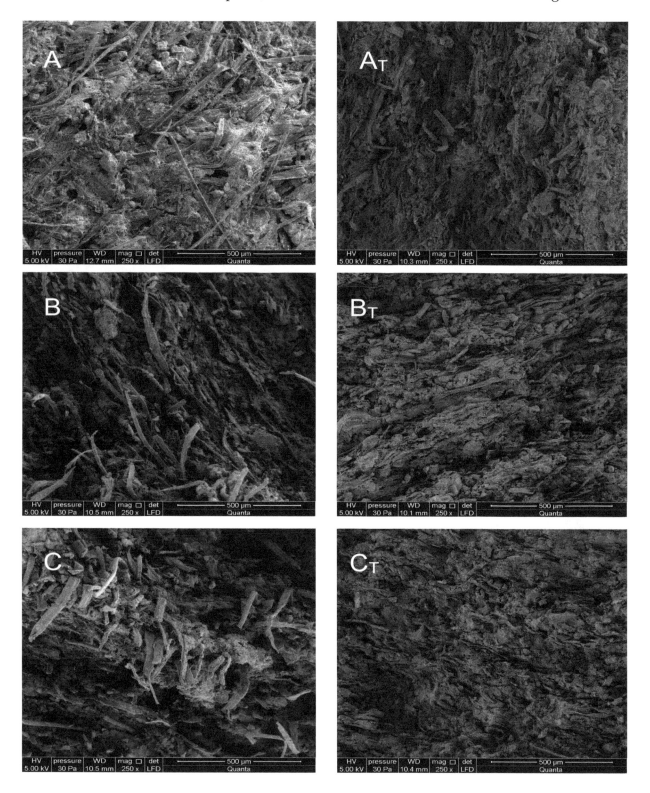

Figure 5. *Cont.*

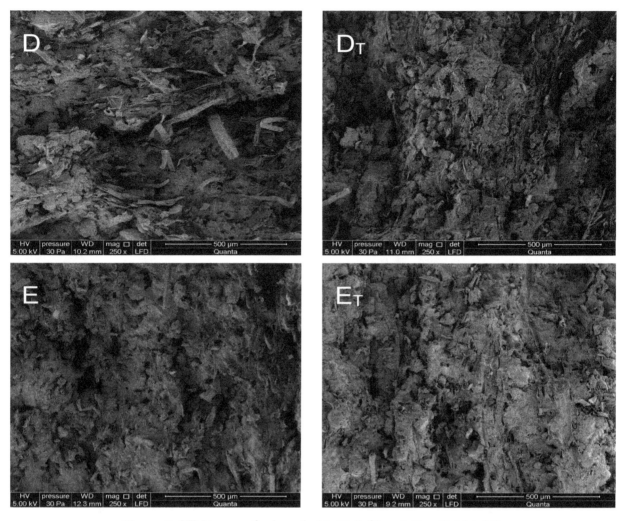

Figure 5. SEM images for boards (**A–E**) (left) and (**A$_T$–E$_T$**) (right).

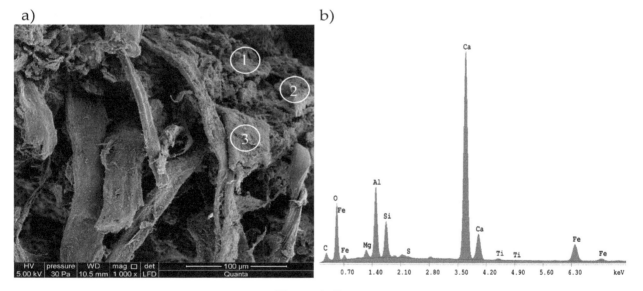

Figure 6. *Cont.*

c)   d)

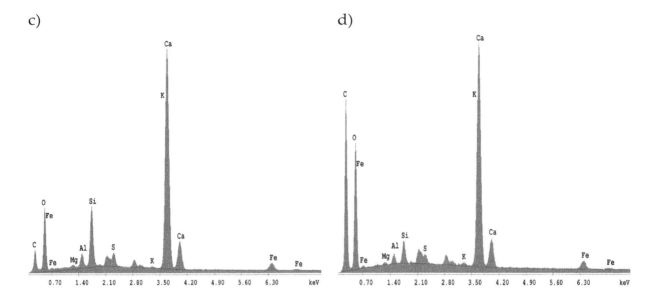

Figure 6. Results obtained using the energy dispersive X-ray spectroscopy (EDS) analyzer for board C: (**a**) areas of elemental composition analysis, (**b**) results of EDS in point 1, (**c**) results of EDS in point 2, (**d**) results of EDS in point 3.

A macroscopic analysis of the fiber cement boards in the A_T–E_T series, which were exposed to a temperature of 230 °C for 3 h, shows a clear change in the color of the samples. Examinations of the A_T–E_T fiber cement boards yielded consistent results. Most of the fibers in the boards were found to be burnt-out, or melted into the matrix, leaving cavities and grooves which were visible in all of the tested boards. The structure of the few remaining fibers was strongly degraded. Examination of the cement particles in further fractures revealed that burning-out also degrades their structure. The structure of the matrix was found to be more granular, showing many delaminations. Numerous caverns and grooves left by the pulled-out fibers, as well as the pulled-out cement particles, were observed.

3.3. X-ray Microtomography

Examples of virtually-cut, three-dimensional projections of $4 \times 4 \times 4$ mm^3 cubes, from specimens D, D_T and B_T, are presented in Figure 7. Specimen D, i.e., before the elevated temperature treatment, shows the regular microstructure without faults such as delaminations. The results of the elevated temperature impact is visible within the volume of specimen D_T, in the form of significant delaminations. The size and number of these delaminations are even more visible in the image of specimen B_T, which was made of a material with lower mechanical performance.

Another way to present information about the specimens, derived from the X-ray scanning procedure, is to determine the brightness distribution of all of the examined voxels (i.e., volumetric pixels). The magnitudes of brightness of the different microstructural elements were included in the following ranges: the area of voids and fibers: 0–50 arbitrary units, a.u., fillers: 50–140 a.u., and dense phases; i.e., un-hydrated cement and fine aggregate grains: 140–170 a.u. The greyscale brightness distribution (GBD) of all voxels belonging to the three specimens presented in Figure 7, are shown in Figure 8. It is worth noting that the occurrence of the delamination processes caused a shift of the affected GBD curves to the left, i.e., in the direction of a region of voids.

Figure 7. Examples of virtually-cut, three-dimensional projections of $4 \times 4 \times 4$ mm^3 cubes from specimens "D" (**left**), "D$_T$" (**center**) and "B$_T$" (**right**). Specimen "D" shows the regular microstructure without faults such as delaminations. The results of the elevated temperature impact is visible within the volume of specimen "D$_T$" and "B$_T$" in the form of significant delaminations.

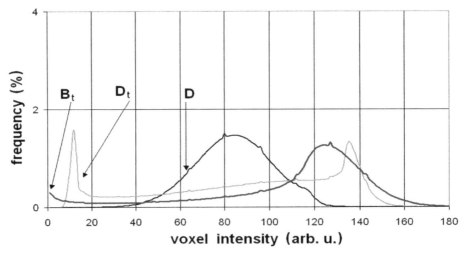

Figure 8. The greyscale brightness distribution (GBD) of all voxels belonging to the three examined specimens: "D", "D$_T$" and "B$_T$". The occurrence of the delamination processes caused a shift of the affected GBD curves to the left, i.e., in the direction of a region of voids.

3.4. Assessment of Ultrasound Wave Velocity c_L Loss and Presentation of the Results of Mechanical Tests

The results of the UT longitudinal wave velocity c_L tests are presented in Figure 9. For each 30×30 cm^2 FCB board (A, B, C, D, E, A$_T$, B$_T$, C$_T$, D$_T$ and E$_T$), two series, of ten measurements each, were performed in order to determine the dispersion of the results. The measurements were done, randomly, at different locations, over the entire surface of the boards. The standard deviation of a single series of measurements was included in the range of 2–3% of the average value of the readings. The UT longitudinal wave velocity c_L loss was observed in all cases of thermal treatment; however, that loss varied from 2% to 20% depending on the FCB composition. It is worth mentioning that the time to perform ten UT measurements took approximately 5 min, suggesting that this may be a good method for in-situ application.

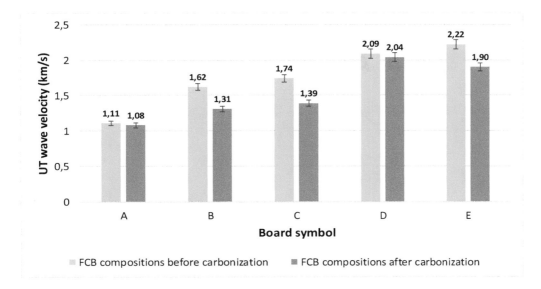

Figure 9. Results of the UT longitudinal wave velocity c_L tests, measured in all FCB compositions, before and after carbonization. The black marks depict standard deviations (2–3%), determined in populations of the measurements.

To determine the changes in the mechanical toughness of the specimens, the authors performed bending tests using three 30×30 cm^2 samples of each kind of board, following the requirements of the ISO 8336 [21]. The average results of the mechanical tests performed on the three samples are presented in Table 2.

Table 2. The results of the mechanical tests performed on the investigated FCB compositions.

Board Symbol	F_{MAX} in the State As-Delivered [N]	F_{MAX} after Thermal Treatment [N]	Work of Flexural Test W_f in the State As-Delivered [J/m^2]	Work of Flexural Test W_f after Thermal Treatment [J/m^2]
A	216	78	1330	29
B	250	82	681	32
C	475	115	848	190
D	330	384	1104	37
E	424	446	790	331

Figure 10 and Table 2 present the results of the mechanical tests. The blue curve shows the behavior of the specimens in an as-delivered state, and the red curve shows the behavior of the specimens with the pyrolyzed reinforcing fibers. Based on an analysis of the curves, the board specimens in an as-delivered state are capable of withstanding a load close to F_{MAX} for the time required to destroy the fiber reinforcement system. The energy required for that damage can be estimated as W_f, by applying Formula (1). The specimens that underwent the cellulose fiber carbonizing process demonstrated the brittle characteristics of the rupture, i.e., the break of the material cross section appeared immediately after reaching the critical F_{MAX} level. Work of flexural test, calculated for the carbonized specimens, equaled approximately 2–5% of the value estimated for the as-delivered state of the FCB specimens. It is also worth mentioning that, in the composition of boards with a low cement content, the brittle matrix broke under a relatively low loading force, while its internal fiber reinforcement system could withstand more stages of the damage process.

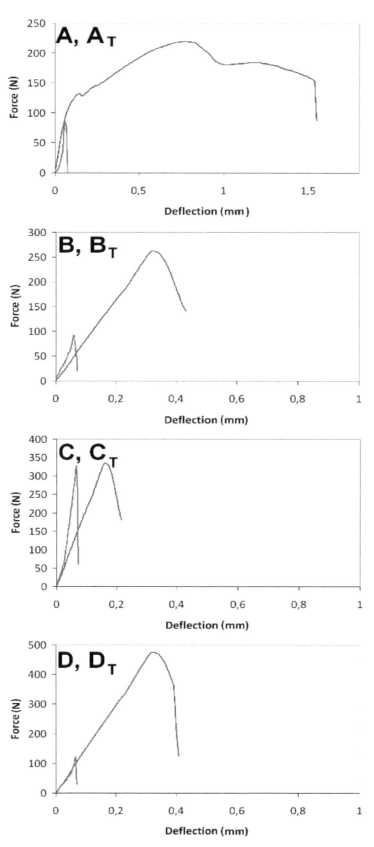

Figure 10. *Cont.*

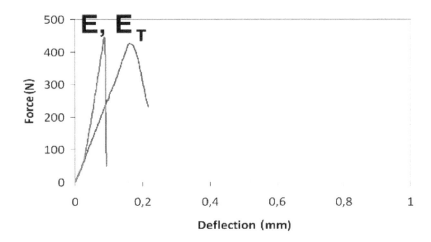

Figure 10. Typical load-deflection curves of the tested FCB compositions. The blue curves show the behavior of the specimen in an as-delivered state and the red curves show the behavior of the specimens with the carbonized reinforcing fibers.

4. Conclusions

The authors investigated five different compositions of fiber cement boards. Two of these compositions, A and B, contained a low amount of cement, which resulted in low flexural strength (14–21 MPa). The other three compositions contained more cement and their flexural strength was determined at the higher range of 23–36 MPa. The fibers applied in compositions A and D, having the best quality and proper length (approximately 3 mm), resulted in the highest value of the work of flexural test W_f, before carbonization. The carbonization process, designed in the laboratory to simulate the long exposure of FCBs to environmental hazards, significantly influenced the mechanical properties of all of the investigated compositions. The micrograph images of the carbonized specimens show the transition of the fibers from their original color into brown. The SEM examinations confirmed the marked changes in the structure that took place as a result of the exposure to a temperature of 230 °C for 3 h. In all of the tested fiber boards, most of the fibers were found to be burnt out, or melted into the matrix, leaving cavities and grooves. The structure of the few remaining fibers was highly degraded. The decrease of the W_f parameter was considerable for all of the tested compositions, as a result of the embrittlement of the fiber reinforcements. The delaminations within the microstructure of the specimens, due to the thermal treatment, was clearly visible in the three-dimensional projections obtained by applying the micro-CT technique. The delaminations also caused a shift in the affected GBD curves to the left, i.e., into the region signalling the presence of loose phases.

In the opinion of the authors, the ultrasound method has proven its applicability for testing the quality of fiber cement boards. The dedicated UT transducers, with low acoustic impedance and polymer jelly interface, were capable of achieving the required propagation of UT waves in order to determine their velocity in the investigated materials. The UT wave velocity c_L in compositions with low levels of flexural strength (A, B) was in the range of 1.1–1.6 km/s (1100–1600 m/s), whereas the wave velocity c_L in compositions with higher flexural strength (B, C, D, E) was in the range of 1.7–2.22 km/s (1700–2220 m/s). The decrease of wave velocity c_L after carbonization occurred in all tested compositions; however, its magnitude was diverse and was included in the range of 2–20%, in relative units. The lowest decrease of c_L occurred in the board made with the best quality components, i.e., the board intended for external use. All of these characteristics lead the authors to recommend the UT method as a useful tool for the on-site assessment of the degradation processes occurring in fiber cement boards.

Author Contributions: Z.R. analyzed the UT and micro-CT test results and performed paper editing; P.R. performed the microscopic observations; T.D. prepared the specimens; T.G. completed the experiments; K.S. conceived and designed the experimental work.

References

1. Schabowicz, K.; Gorzelańczyk, T. Applications of fibre cement boards. In *The Fabrication, Testing and Application of Fibre Cement Boards*, 1st ed.; Ranachowski, Z., Schabowicz, K., Eds.; Cambridge Scholars Publishing: Newcastle upon Tyne, UK, 2018; pp. 107–121. ISBN 978-1-5276-6.
2. Mohr, B.J.; Nanko, H.; Kurtis, K.E. Durability of kraft pulp fibre-cement composites to wet/dry cycling. *Cement Concrete Compos.* **2005**, *27*, 435–448. [CrossRef]
3. Coutts, R.S.P. A Review of Australian Research into Natural Fibre Cement Composites. *Cement Concrete Compos.* **2005**, *27*, 518–526. [CrossRef]
4. Schabowicz, K.; Gorzelańczyk, T. Fabrication of fibre cement boards. In *The Fabrication, Testing and Application of Fibre Cement Boards*, 1st ed.; Ranachowski, Z., Schabowicz, K., Eds.; Cambridge Scholars Publishing: Newcastle upon Tyne, UK, 2018; pp. 7–39. ISBN 978-1-5276-6.
5. Bledzki, A.K.; Gassan, J. Composites reinforced with cellulose based fibres. *Prog. Polym. Sci.* **1999**, *24*, 221–274. [CrossRef]
6. Cooke, T. Formation of Films on Hatcheck Machines. Bonded Wood and Fibre Composites. Available online: www.fibrecementconsulting.com/publications/publications.htm (accessed on 16 November 2018).
7. Drelich, R.; Gorzelańczyk, T.; Pakuła, M.; Schabowicz, K. Automated control of cellulose fibre cement boards with a non-contact ultrasound scanner. *Autom. Constr.* **2015**, *57*, 55–63. [CrossRef]
8. Kaczmarek, M.; Piwakowski, B.; Drelich, R. Noncontact Ultrasonic Nondestructive Techniques: State of the Art and Their Use in Civil Engineering. *J. Infrastruc. Syst.* **2017**, *23*, 45–56. [CrossRef]
9. Mori, K.; Spagnoli, A.; Murakami, Y.; Kondo, G.; Torigoe, I. A new non-destructive testing method for defect detection in concrete. *NDT&E Int.* **2002**, *35*, 309–406.
10. McCann, D.M.; Forde, M.C. Review of NDT methods in the assessment of concrete and masonry structures. *NDT&E Int.* **2001**, *34*, 71–84.
11. Ranachowski, Z.; Schabowicz, K.; Gorzelańczyk, T.; Lewandowski, M.; Cacko, D.; Katz, T.; Dębowski, T. Investigation of Acoustic Properties of Fibre-Cement Boards. In Proceedings of the IEEE Joint Conference—ACOUSTICS, Ustka, Poland, 11–14 September 2018; pp. 275–279. [CrossRef]
12. Ranachowski, Z.; Schabowicz, K. The contribution of fiber reinforcement system to the overall toughness of cellulose fiber concrete panels. *Constr. Build. Mater.* **2017**, *156*, 1028–1034. [CrossRef]
13. Schabowicz, K.; Jóźwiak-Niedźwiedzka, D.; Ranachowski, Z.; Kudela, S., Jr.; Dvorak, T. Microstructural characterization of cellulose fibres in reinforced cement boards. *Arch. Civil Mech. Eng.* **2018**, *18*, 1068–1078. [CrossRef]
14. Ranachowski, Z.; Schabowicz, K.; Gorzelańczyk, T.; Kudela, S., Jr.; Dvorak, T. Visualization of Fibers and Voids Inside Industrial Fiber Concrete Boards. *Mater. Sci. Eng. Int. J.* **2018**, *1*, 00022. [CrossRef]
15. Ardanuy, M.; Claramunt, J.; Filho, R.D.T. Cellulosic fiber reinforced cement-based composites: A review of recent research. *Constr. Build. Mater.* **2015**, *79*, 115–128. [CrossRef]
16. Li, Z.; Zhou, X.; Shen, B. Fiber-Cement Extrudates with Perlite Subjected to High Temperatures. *J. Mater. Civil Eng.* **2004**, *16*, 221–229. [CrossRef]
17. Jarząbek, D. Application of advanced optical microscopy. In *The Fabrication, Testing and Application of Fibre Cement Boards*, 1st ed.; Ranachowski, Z., Schabowicz, K., Eds.; Cambridge Scholars Publishing: Newcastle upon Tyne, UK, 2018; pp. 89–106. ISBN 978-1-5276-6.
18. *BS EN 12467:2012 Fibre-Cement Flat Sheets—Product Specification and Test Methods*; British Standards Institution: London, UK, 31 July 2016.
19. MISTRAS Products & Systems. Available online: http://www.eurosonic.com/en/products/ut-solutions/utc-110.html (accessed on 19 November 2018).
20. Olympus Ultrasonic Transducers. Available online: http://www.olympus-ims.com/en/ultrasonic-transducers/contact-transducers/#! (accessed on 19 November 2018).
21. *ISO 8336:2009. Fibre-Cement Flat Sheets—Product Specification and Test Methods*; ISO Standards: Geneva, Switzerland, May 2009; Available online: https://www.iso.org/standard/45791.html (accessed on 20 March 2019).

14

Accuracy of Eddy-Current and Radar Methods used in Reinforcement Detection

Łukasz Drobiec *⑩, Radosław Jasiński⑩ and Wojciech Mazur⑩

Department of Building Structures, Silesian University of Technology; ul. Akademicka 5, 44-100 Gliwice, Poland; radoslaw.jasinski@polsl.pl (R.J.); wojciech.mazur@polsl.pl (W.M.)
* Correspondence: lukasz.drobiec@polsl.pl

Abstract: This article presents results from non-destructive testing (NDT) that referred to the location and diameter or rebars in beam and slab members. The aim of paper was to demonstrate that the accuracy and deviations of the NDT methods could be higher than the allowable execution or standard deviations. Tests were conducted on autoclaved aerated concrete beam and nine specimens that were specially prepared from lightweight concrete. The most advanced instruments that were available on the market were used to perform tests. They included two electromagnetic scanners and one ground penetrating radar (GPR). The testing equipment was used to analyse how the rebar (cover) location affected the detection of their diameters and how their mutual spacing influenced the detected quantity of rebars. The considerations included the impact of rebar depth on cover measurements and the spread of obtained results. Tests indicated that the measurement error was clearly greater when the rebars were located at very low or high depths. It could lead to the improper interpretation of test results, and consequently to the incorrect estimation of the structure safety based on the design resistance analysis. Electromagnetic and radar devices were unreliable while detecting the reinforcement of small (8 and 10 mm) diameters at close spacing (up to 20 mm) and of large (20 mm) diameters at a close spacing and greater depths. Recommendations for practical applications were developed to facilitate the evaluation of a structure.

Keywords: NDT methods; rebar location; eddy-current method; GPR method

1. Introduction

Non-destructive testing that refers to the location and geometry of reinforcement in the structure is currently quite common while using different types of measuring equipment. The obtained test results are often used to make verifying calculations to determine the resistance of elements and decide whether they should not be further used or whether they should be reinforced. Rebar diameters and the precise location of reinforcement in the structure is required to calculate its resistance. For elements subjected to bending (slabs, beams) and compression and bending (columns), the reinforcement location affects internal forces, and thus the element strength. Even minor errors in detecting its location significantly change the result of resistance calculations and affect the structure safety. Moreover, non-destructive testing is employed at the acceptance of the performed facilities. It is particularly important that the accuracy of measurements is higher than the allowable execution deviations specified in standards. Therefore, tests on the structure reinforcement require information on the precision and limitations of the applied method and test equipment. All the above are useful in developing the programme for estimating the reliable building conditions.

Manufacturers are improving their products, and the measurement accuracy is increasing. It is very important to specify the measurement accuracy and to do this before the beginning of evaluating works. Non-destructive testing (NDT) methods are particularly useful for testing a wide surface or many elements. Electromagnetic and radar methods are currently the most often used NDT methods in detecting reinforcement in the structure [1–4]. Both of the methods have their advantages and disadvantages. It can be assumed that advantages of the electromagnetic method are: the accuracy of measurements and the possibility for determining the diameter of reinforcement. Its disadvantages include a short range of measurements and some errors that are triggered by resolution, which are significant when the rebars are closely arranged (lap splices, bundles of rebars) [5,6]. The advantage of the radar method is the possibility of localising the reinforcement at great depths; and its disadvantages cover difficulties in measuring diameters measurements and some measurement errors of damp structures [7,8].

Despite the continuous development, electromagnetic and radar methods still have present limitations. Many papers in the literature [3–5,8] present advantages and possibilities for using electromagnetic and radar methods. Information regarding their limitations and measurement accuracy is rarely published. According to the definition [9], the accuracy is a compatibility level between the obtained result of a single measurement and the expected value that is related to the systematic and accidental errors. However, design standards and standards specifying structures reinforced with rebars permit some execution deviations. Thus, it is interesting whether the real accuracy of methods is within standard limits of execution deviations. When the measurement accuracy is higher than the allowable execution deviations, then the non-destructive testing methods should not be used.

The accuracy of non-destructive tests on the reinforcement geometry can be analysed when considering the measurements of rebar diameter and their position in perpendicular and parallel direction to the plane the tested element [3]. The measurement accuracy mainly depends on the dimension of concrete cover and the mutual spacing of adjacent rebars. This article describes results from testing the measurement accuracy of reinforcement diameter and geometry while using the most advanced instruments that are available on the market. The main aim of paper was to demonstrate whether the measurement accuracy and related measurement uncertainties could be higher than the allowable execution deviation. The tests were conducted on reinforced precast beam made of autoclaved aerated concrete beam and nine specimens specially prepared from lightweight concrete. The analysis involved the effect of the cover thickness and horizontal spacing of rebars on determining the number of rebars and measurement results for their diameters. Additionally, the aspect of measurement accuracy of concrete cover thickness and the effect of cover dimensions on the measurement accuracy were analysed. The obtained results were compared to deviations accepted by standards. The measurement accuracy of diameters and covers was analysed in accordance with the uncertainty of results when assuming suitable estimators of uncertainty. The comparison of tests results for the same elements that were obtained by different methods and the direct analysis of obtained results had a significant contribution to the development of methods of NDT.

2. Accuracy and Possibilities of Non-Destructive Testing

2.1. Electromagnetic Method

The accuracy of electromagnetic tests mainly depends on the depth, spacing, and arrangement of rebars toward the direction of scanning, the type of reinforcement, and the quality of concrete surface [10,11]. A type of spectral analysis and compensation procedures of systematic measuring errors also affect the accuracy of the tests.

The depth of rebars location has a significant impact on the accuracy of diameter measurements, and even on the accuracy of rebars location. The maximum depth, at which reinforcement can be detected, depends on the diameter of rebars. For typical rebar diameters of 6–25 mm, depending on the employed device, the biggest depth at which the reinforcement can be detected is 100–200 mm. The closer location of rebars to the scanned surface evidently ensures a greater accuracy of the measured and real diameter. The acceptable measurement accuracy (5%) of the reinforcement diameter is obtained when the rebars are located at a depth of ca. 60 mm. Unfortunately, electromagnetic methods are not effective in detecting the non-metallic reinforcement, which is becoming increasingly common [12,13].

Drobiec et al. [11] describe tests to detect plain and ribbed rebars of 12 diameter. The tests were done with four different devices. Variations of the measured thickness of the concrete cover were within 4 mm, and the measured reinforcement diameter differed by not more than one gradation. Sivasubramanian et al. [14] presented the analysis of measurement errors of the electromagnetic device in slabs having dimensions of $400 \times 400 \times 250$ mm, which were made of high compressive strength concrete (average 71.3 N/mm^2). One rebar was placed in each slab. Rebars of 12, 16, 20, 25, and 32 mm diameter and 70 and 85 mm covers were used (in total 10 of test models). Figure 1 illustrates the graph of the relative measuring error of the reinforcement diameter as a function of the concrete cover size. The acceptable error (<10%) was obtained for the cover thickness of up to ca. 40 mm. The measurement error of diameter of the order of 100% was obtained in tests regarding reinforcement at the depth of over 70 mm. It should be emphasized that, although the paper [15] is relatively recent, the used device is not a modern one. As no tests on models of concrete with lower strength were conducted, it is difficult to estimate the impact of the high compressive strength concrete. In typical reinforced concrete structures (beams, slabs, columns) with standard covers ($c \sim$ 20–45 mm), electromagnetic devices that are available in practice can be used to perform tests with an accuracy of the measurement that is less than 10%. The gradation of rebar diameters causes difficulties in diameter differentiation.

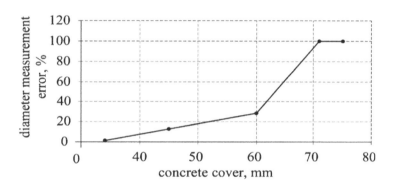

Figure 1. The error of the reinforcement diameter measured with an electromagnetic device, depending on the cover size according to [14].

2.2. Radar Method

In the radar method, the measuring range of the structure depth depends on concrete structure, a type of an antenna, and the frequencies of the excited impulse [15–17]. In typical devices, this range is up to 750 mm. The image contrast that was obtained from radar tests depends on the relative difference between dielectric constant values at the contact area between materials. There are no difficulties in interpreting the obtained image due to considerable differences in constant values for concrete and steel. Tests on reinforcement location that were conducted with radar method generate radargrams, that is, the record of all reflected signals registered during the passage of a measuring probe on the element surface. The reinforcement image on radargram is a distortion of course contour lines in the shape of hyperbola arms directed the radargram down.

The modern measurement systems perform automatic analyses of radargrams, convert radargrams taken close to each other in one image, and visualize reinforcement in the construction in a legible way for users. The additional software can be used to take a spatial image of the structure with the reinforcement. The radar devices do not provide direct information regarding reinforcement diameters. Perhaps measurements in a three-dimensional (3D) model can be a solution for this problem. However, some published articles present some mathematical correlations between the shape of a hyperbola that is shown on the radargram and the reinforcement diameter. Devices with the option of determining the reinforcement diameter are soon likely to appear on the market. However, the accuracy of the measurement method might be still a problem.

Currently, the accuracy of the cover measurements with the radar method performed with the best devices is defined as ±5–10 mm.

The papers [18–24] describe that the shape of hyperbola illustrating the reinforcement on the radargram depends on the wave propagation velocity v and passage time t_0 of measuring the probe above the tested reinforcement. The time of the measuring probe passage is the time passing from the moment when the device records the reinforcement for the first time until the moment when it stops to record it. The velocity of the wave propagation affected the hyperbola curve, whereas the time of the probe passage affected the range of its arms. Knowing the parameter values a and b of the hyperbola b (Figure 2), their relation with the reinforcement diameter ϕ, wave propagation velocity v, and time of probe passage t_0, on the tested reinforcement can be expressed, as following:

$$a = t_0 + \frac{\phi}{v}, \tag{1}$$

$$b = \frac{v}{2}\left(t_0 + \frac{\phi}{v}\right) \tag{2}$$

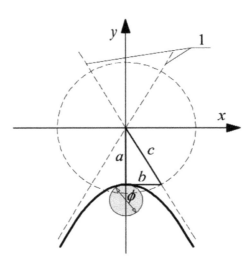

Figure 2. Assumptions to derive Equations (5) and (6), 1—hyperbola asymptotes.

The wave propagation velocity v and passage time t_0 of the measuring probe were recorded by a measuring device; and, parameters a and b were read from the radargram. Therefore, there were no obstacles to determine the reinforcement diameter ϕ from Equations (5) and (6). The paper [25] confirmed the possible use of that method for defining the reinforcement, as under laboratory conditions, the obtained measuring error of diameters of steel pipes and cables was in the order of 1.7–5.3%.

Another way of defining the diameter was suggested in the paper [26]. It was assumed that the impulse distribution angle from the transmitting antenna was 90° (Figure 3). Such an assumption was useful in determining the rebar diameter without taking into account the wave propagation velocity v and passage time t_0 of the measuring probe above the tested reinforcement. The reinforcement diameter ϕ can be determined according to:

$$\phi = 2\sqrt{2}L - 2X - 2Y \tag{3}$$

where: L, X, Y—geometric data according to Figure 3.

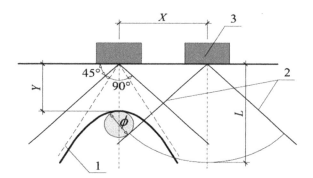

Figure 3. Assumptions to derive the Equation (7), 1—hyperbola asymptotes, 2—the angle of impulse distribution from the antenna, 3—a measuring probe.

2.3. Combination of Electromagnetic and Radar Methods

Wiwatrojanagul et al. [27] and Wiwatrojanagul et al. [28] suggested connecting the advantages of the electromagnetic and radar methods. Tests were conducted on specimens that were placed in a pen and in concrete. Rebar diameters were tested in rebar laboratories. The electromagnetic method was found to not give correct test results, because the reinforcement was placed too close. On the basis of conducted tests, the empirical equations were developed to estimate the diameter of contact rebars in the rebar laboratory. The accuracy of the determined diameters was obtained at the level of 2.35%. Empirical equations were found to have some application limitations regarding the tested reinforcement and concrete.

To sum up, the advantage of devices working according to the electromagnetic method was their high accuracy at small depths of measurements. A small range of diameter measurements, which was limited to about 6 cm, was their disadvantage Radar devices had much bigger range (even up to 75 cm). However, devices that are available on market do not offer the possibility of diameter measurements. In the world, some methods for diameter determining on the basis of radargram measurements obtained from the radar tests are being developed. Radar devices with the option of diameter measuring are likely to appear soon on the market.

3. Tested Specimens

3.1. Autoclaved Areated Concrete (AAC) Precast Lintels

To verify the accuracy of NDT methods, some tests were performed on widely used lintels that were made of AAC with the width of $b = 180$ mm, height $h = 240$ mm and the total length of $L = 2000$ mm. The strength of AAC elements was $f = 4$ N/mm². Detailed results from the material tests on lintels can be found in the papers [29–32]. The lintel reinforcement was made of steel with a yield strength of 500 N/mm² (B class according to EN 1992-1-1:2008 [33]). Longitudinal rebars had a diameter of 8 mm (three rebars down and two rebars up). Longitudinal reinforcement in the form of open stirrups that were made of rebars having a diameter of 4.5 mm. Stirrups were placed along the whole element at a constant spacing of 150 mm (Figure 4). The longitudinal reinforcement and stirrups were welded and then covered with corrosion-resistant protective coating made of resin.

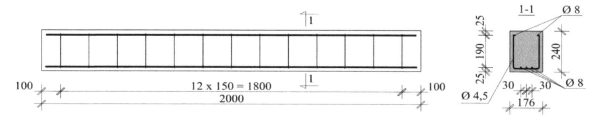

Figure 4. Dimensions and reinforcement of the tested lintel.

Two beams were selected for tests. To ensure the precise location of the reinforcement, both ends of tested beams were broken down and the reinforcement was measured, as in Figure 5. Tiny openings (diameter of 2 mm) were made in 11 places on lateral surfaces and at eight bottom points (between stirrups) to measure the real reinforcement cover—c_{obs} with the accuracy of $\Delta c_{obs} = \pm 0,1$ mm. As the location of both beams was the same, tests were only conducted on one of them.

Figure 5. Tested elements with the reinforcement broken down at its ends.

3.2. Lightweight Concrete Specimens

The impact of bar diameter and spacing on measurement accuracy was tested on nine models (Figure 6) that were prepared from lightweight concrete with density of 0.9 kN/m^3 and strength of 10.2 N/mm^2 after seven days, and of 18.1 N/mm^2 after 28 days. Each model had three rebars that were arranged to ensure the distance between rebars equal to 20, 30, and 40 mm (dimension "a" in Figure 6). The diameter of used rebars was ø = 10; 16; and, 20 mm. The dimensions of models in a plain view were 240 × 440 mm and thickness of 40, 60, and 80 mm. Different thickness was to ensure the correct (two-sided) cover of rebars with concrete assuming that rebar distance from one of the surface was 20 mm. It is the minimum thickness that is accepted by the standard EN 1996-1-1 [33] for the exposure class XC1 (indoor) and the structural class S3. The wooden elements were precisely cut to provide the cover thickness of 20 mm. Reinforcement was laid using wooden spacers of relevant thicknesses (Figure 7a), and then stabilized in washers with screws (Figure 7b). Concrete was laid, mechanically compacted, and properly cured. Figure 8 shows models before and after concreting. Specimens symbols contained a letter S, diameter d, and space between rebars a. For example, S-15-30 identified a specimen that was reinforced with three rebars having a diameter of 16 mm and space between rebars that is equal to 30 mm.

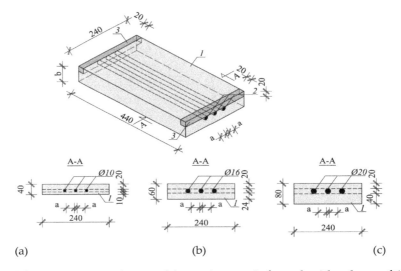

(a)   (b)   (c)

Figure 6. Lightweight concrete specimens: (**a**) specimen reinforced with rebars of 10 mm diameter, (**b**) specimen reinforced with rebars of 16 mm diameter, (**c**) specimen reinforced with rebars of 20 mm diameter, 1—lightweight concrete, 2—rebars, 3—wooden washer.

(a)   (b)

Figure 7. Stabilization of reinforcement in formwork: (**a**) distance resulting from wooden spacers, (**b**) rebar fastening to wooden washers with screws.

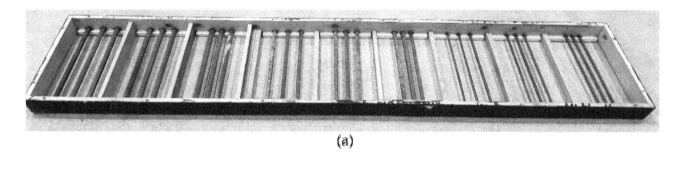

(a)

(b)

Figure 8. Specimen reinforcement in formwork (**a**) and specimens after concreting (**b**).

4. Applied Test Equipment

The tests were conducted using two electromagnetic scanners: PS 200 (manufacturer Hilti Corp., Schaan, Lichtenstein), Profometer 630 AI (manufacturer Proceq AG, Schwerzenbach, Switzerland) and one GPR device –GPR Live (manufacturer Proceq AG, Schwerzenbach, Switzerland). Figure 9 shows the testing apparatus. A scanning transducer of the electromagnetic device 1 was equipped with one circumferential transmitting coil and seven pairs of receiving coils. The receiving coils induce current with microammeters, and the received signal is then processed and analysed. The electromagnetic device no. 2 is similar in principle. The radar equipment 3 was equipped with antennas to perform tests, while using a signal with a variable frequency within a range of 0.2–4.0 GHz. Frequency was changed progressively in an automatic way during tests, and the max. acquisition time was 20 ns.

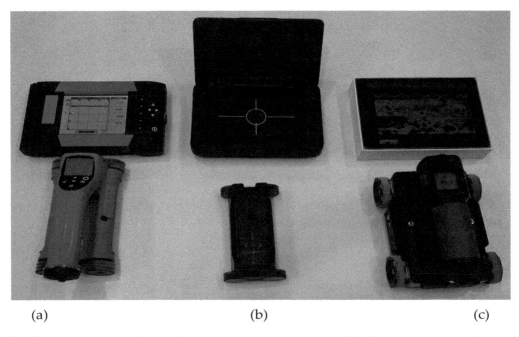

(a)   (b)   (c)

Figure 9. Devices used in tests, (**a**) electromagnetic device (1), (**b**) electromagnetic device (2), and (**c**) radar device (3).

The measuring accuracy of each device was the same, $\Delta c_{obs} = \pm 1.0$ mm (according to information provided by device manufacturers).

5. Testing Locations

5.1. AAC Precast Lintels

The tests were performed with each device by scanning the same points on the lateral and bottom surface of beams. Stirrup rebars were detected on the lateral surface, whereas the main reinforcement was tested from the bottom. During tests, line and area scans were conducted. Thus, the test consisted in moving the transducer along lines or areas that were routed on the surface of the tested element (Figure 10). Figure 11a illustrates scanned areas for stirrup testing and Figure 11b presents scanned areas for testing the main reinforcement. Figure 11c shows the location of a line scan and the point of measuring stirrups, whereas Figure 11d shows the location of a line scan for the main reinforcement.

Figure 10. Transducer on the surface of the tested element.

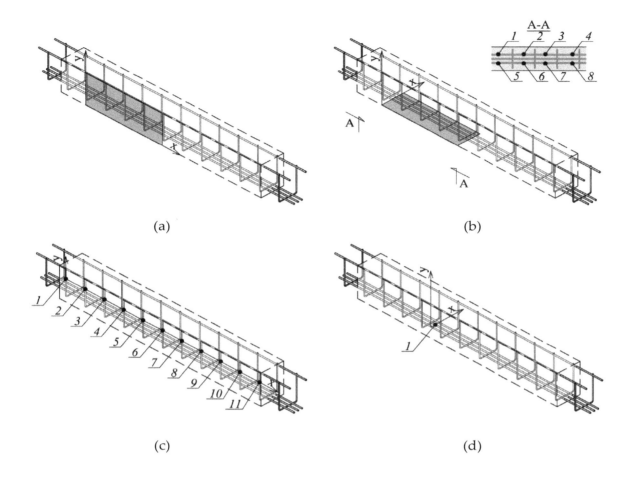

Figure 11. Places of lintel tests: (**a**) Stirrup tests—area scan (measurement in the grey-coloured field), (**b**) main reinforcement test—area scan (measurement in the grey-coloured field, measurement at points 1–8), (**c**) Stirrup tests—line scan (measurement at points 1–11), and (**d**) main reinforcement test—line scan (measurement at point 1).

5.2. Lightweight Concrete Specimens

Tests on lightweight concrete specimens, just as tests on lintels, consisted in scanning along lines or on surfaces that were routed on a given element using the transducer. The lightweight concrete specimens were tested using the electromagnetic Equipment (1) over the area of 150 × 300 mm illustrated in Figure 12a. The electromagnetic Equipment (2) was used to perform line scans at the mid-length of the specimen (Figure 12b), while area scans were conducted with the radar equipment over the area of 200 × 300 mm, as shown in Figure 12c. At the first stage, scans were conducted on the element surface where the concrete thickness was 20 mm, and then the element was turned to conduct further scans. Concrete cover was 10, 24, and 40 mm for rebars having a diameter of 10, 16, and 20 mm. Figure 13 shows the types of used equipment.

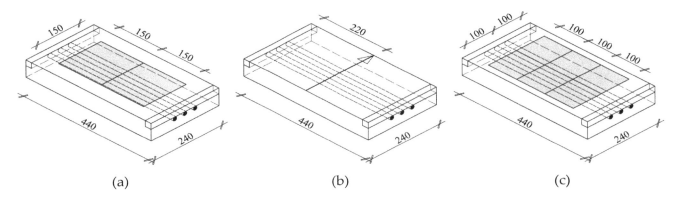

Figure 12. Location of tests performed with: (**a**) electromagnetic device (1), (**b**) electromagnetic device (2), (**c**) radar device (3).

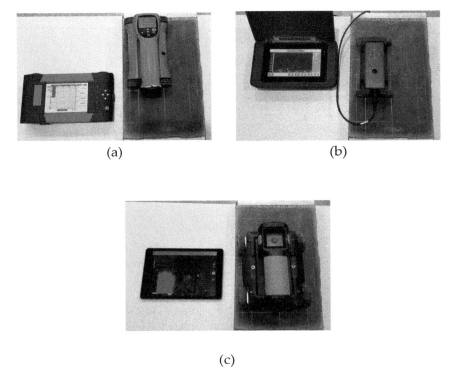

Figure 13. Testing specimens using: (**a**) electromagnetic device (1), (**b**) electromagnetic device (2), and (**c**) radar device (3).

6. Test Results

6.1. AAC Precast Lintels (AAC)

6.1.1. Stirrups Tests

The stirrups were tested at the beam side as line and area scans. Line scans could be only used to determine the reinforcement location (spacing and concrete cover). Figure 14 illustrates the comparison of exemplary results from line scans that were taken with electromagnetic devices and the radar device (radargram). A great conformity was found between the test results and real measurements of rebar location. Figure 15 presents examples of area scans taken with the electromagnetic device 1 and the radar device 3. The area scans conducted with two methods very clearly showed both stirrup reinforcement and longitudinal reinforcement.

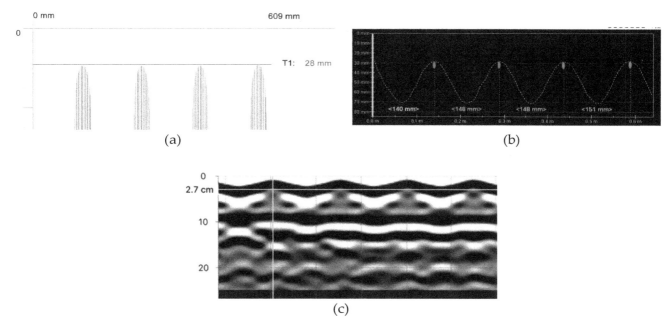

Figure 14. Line scans (measurement at points shown in Figure 10c) with devices: (**a**) electromagnetic 1, average cover of visible results—27.8 mm, (**b**) electromagnetic 2, average cover—28.6 mm, (**c**) radar 3, average cover 27 mm.

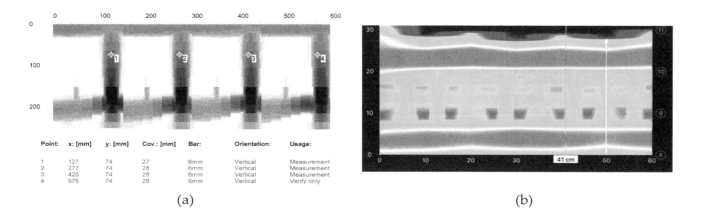

Figure 15. Area scans conducted with electromagnetic devices (measurement on the surface shown in Figure 10a), (**a**) electromagnetic 1, average cover of visible results– 27.8 mm, (**b**) radar 3,– 27 mm.

6.1.2. Tests on Longitudinal Rebars

The aim of testing the main reinforcement was to determine the cover size, the number, and diameters of rebars in the main reinforcement (diameters were determined only in case of electromagnetic scans). Figure 16 shows the comparison of results from electromagnetic tests. The area scan (Figure 16a) was taken with the electromagnetic device 1 and the line scan (Figure 10b) was taken with the electromagnetic device 2. Both of the devices did not detect the correct number of rebars. Device 1 detected two rebars instead of three, but it measured the diameter quite correctly. However, device 2 detected only one rebar and the measured diameter was equal to 18 mm. Figure 16 shows the results from tests performed with the GPR device. Area and line scans were taken. The line scan is presented as a radargram (Figure 17a) and the area scan as a map (Figure 17b). Only one longitudinal rebar was detected during both tests while using the radar method.

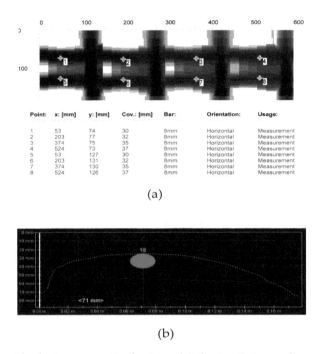

(a)

(b)

Figure 16. Scans taken with electromagnetic devices, (**a**) device 1, two rebars detected, average cover of visible results—33.6 mm, diameter 6-8 mm (measurement on the surface shown in Figure 10b), (**b**) device 2, one rebar detected, average cover of visible results—28 mm, diameter 18 mm (measurement at the point shown in Figure 11d).

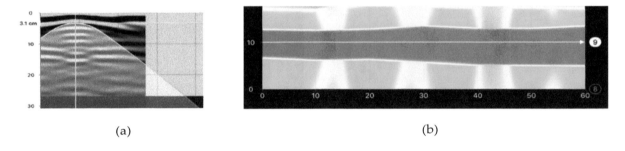

(a)   (b)

Figure 17. Scans taken with the radar (3), one rebar detected, (**a**) line scan (radargram—measurement at the point shown in Figure 10d), (**b**) area scan (measurement on the surface shown in Figure 10b).

To sum it up, the test results for the longitudinal reinforcement, in which the rebars were closer to each other (an axial spacing of rebars was 30 mm, as shown Figure 4) were poorer than in the case of tests that were performed on stirrup reinforcement as described in point 6.1.1. None of devices used during various testing methods, correctly detected the number of longitudinal rebars. The most satisfactory result was observed for tests that were performed with the electromagnetic instrument 1.

6.2. Lightweight Concrete Specimens (Tests with Concret Cover 20 mm)

6.2.1. Tests Using a 20 mm Cover

Tests on lintel beams indicated significant errors in measuring the quantity of reinforcement and the number of rebars in the main reinforcement. Thus, additional tests were performed on lightweight models that were prepared for this purpose. The aim of those tests was to capture the relationship between rebar diameters and spacing and the accuracy of measurements. Figure 18 shows the results from tests conducted with the electromagnetic equipment (1). They contain the measured number and diameter $\varnothing_{obs}$ of rebars and average cover c_{obs}.

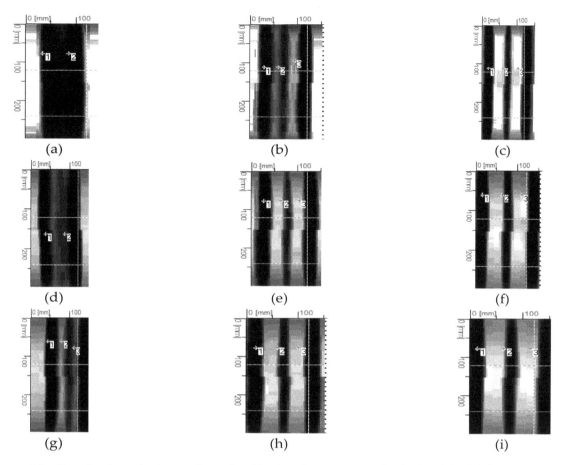

Figure 18. Results from tests conducted with the electromagnetic device (1): (**a**) S-10-20 (result: two rebars, $\varnothing_{obs}$ = 8 mm, c_{obs} = 25 mm), (**b**) S-10-30 (three rebars, $\varnothing_{obs}$ = 8 mm, c_{obs} = 20 mm), (**c**) S-10-40 (three rebars, $\varnothing_{obs}$ = 10 mm, c_{obs} = 21 mm), (**d**) S-16-20 (two rebars, $\varnothing_{obs}$ = 30 mm, c_{obs} = 25 mm), (**e**) S-16-30 (three rebars, $\varnothing_{obs}$ = 16 mm, c_{obs} = 18 mm), (**f**) S-16-40 (three rebars, $\varnothing_{obs}$ = 20 mm, c_{obs} = 20 mm), (**g**) S-20-20 (three rebars, $\varnothing_{obs}$ = 20 mm, c_{obs} = 25 mm), (**h**) S-20-30 (three rebars, $\varnothing_{obs}$ = 20 mm, c_{obs} = 18 mm), and (**i**) S-20-40 (three rebars, $\varnothing_{obs}$ = 20 mm, c_{obs} = 22 mm).

As in case of lintel beams, the tests failed to detect the correct number of rebars spaced at a_{nom} = 20 mm for specimens that were reinforced with rebars having a diameter $\varnothing_{nom}$ = 10 mm. In the specimen with rebars at the similar spacing, but reinforced with rebars $\varnothing_{nom}$ = 16 mm, three rebars could be noticed, but only two of them were detected with the equipment, which also overestimated the diameter of 30 mm. When rebars with a 20 mm diameter were used, the device identified as many as three rebars, even at the minimum spacing; however, the indicated diameter was twice smaller (10 mm) than the real value. At a larger spacing a_{nom} = 30 and 40 mm, the electromagnetic equipment (1) detected the correct diameter of rebars, but some inaccuracies were found while measuring the covers.

Figure 19 shows the results from tests conducted with the electromagnetic equipment (2). As above, the results include the number of rebars, their measured diameters d, and the average size of tested concrete cover c. For the smallest diameter and spacing, the equipment detected only one rebar, just as in case of tests on precast beam. The equipment found three rebars in specimens that were reinforced with rebars of 16 and 20 mm diameter, spaced at more than 20 mm. Some deviations were observed in measured diameters and covers. Contrary to the electromagnetic equipment (1), the equipment (2) provided more accurate diameters with less precisely measured thickness of concrete cover. Differences in cover measurements were even in the order of up to 5 mm.

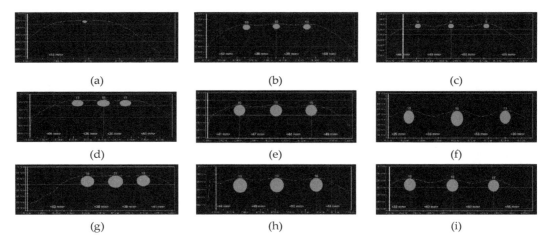

Figure 19. Results from tests performed with the electromagnetic device (2): (**a**) S-10-20 (result: one bar, $\varnothing_{obs}$ = 8 mm, c_{obs} = 20 mm), (**b**) S-10-30 (three rebars, $\varnothing_{obs}$ = 10 mm, c_{obs} = 19.3, mm), (**c**) S-10-40 (three rebars, $\varnothing_{obs}$ = 9 mm, c_{obs} = 19.3 mm), (**d**) S-16-20 (three rebars, $\varnothing_{obs}$ = 17 mm, c_{obs} = 15.5, mm), (**e**) S-16-30 (three rebars, $\varnothing_{obs}$ = 15 mm, c_{obs} = 16.4, mm), (**f**) S-16-40 (three rebars, $\varnothing_{obs}$ = 14 mm, c_{obs} = 16.5, mm), (**g**) S-20-20 (three rebars, $\varnothing_{obs}$ = 19 mm, c_{obs} = 15 mm), (**h**) S-20-30 (three rebars, $\varnothing_{obs}$ = 19 mm, c_{obs} = 15.4 mm), and (**i**) S-20-40 (three rebars, $\varnothing_{obs}$ = 18 mm, c_{obs} = 18.3 mm).

Figure 20 shows results from tests that were conducted with the electromagnetic equipment (3). The results include the number and average size of tested concrete cover c_{obs}. The correct number of rebars was detected in all of the tested specimens. The measured value of covers also showed high compliance with the true value (20 mm) The radar equipment could not directly measure diameters, which was its drawback. Therefore, there are no rebar diameters in Figure 20.

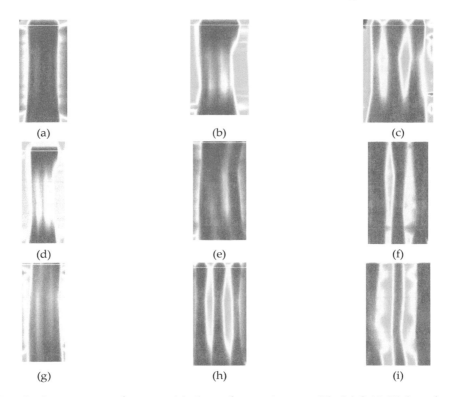

Figure 20. Results from tests performer with the radar equipment (3): (**a**) S-10-20 (result: three rebars, c_{obs} = 22 mm), (**b**) S-10-30 (three rebars, c_{obs} = 22 mm), (**c**) S-10-40 (three rebars, c_{obs} = 21 mm), (**d**) S-16-20 (three rebars, c_{obs} = 22 mm), (**e**) S-16-30 (three rebars, c_{obs} = 21 mm), (**f**) S-16-40 (three rebars, c_{obs} = 20 mm), (**g**) S-20-20 (three rebars, c_{obs} = 22 mm), (**h**) S-20-30 (three rebars, c_{obs} = 20 mm), (**i**) S-20-40 (three rebars, c_{obs} = 20 mm).

6.2.2. Tests Using Covers of Variable Thickness

Figure 21 shows the results from tests that were conducted with the electromagnetic equipment (1). The obtained results were comparable to tests using the cover with constant thickness (Figure 18). However, some differences were noticed. All of the rebars were detected in specimens with rebars of 10 mm diameter and concrete thickness reduced by 50%. However, at a spacing of 20 mm (Figure 21a) and 30 mm (Figure 21b), the rebar diameter was identified with a serious error of the order of 40% and 20%, respectively. Correct results were obtained for rebars at a clear spacing of 40 mm, just as for twice as big cover.

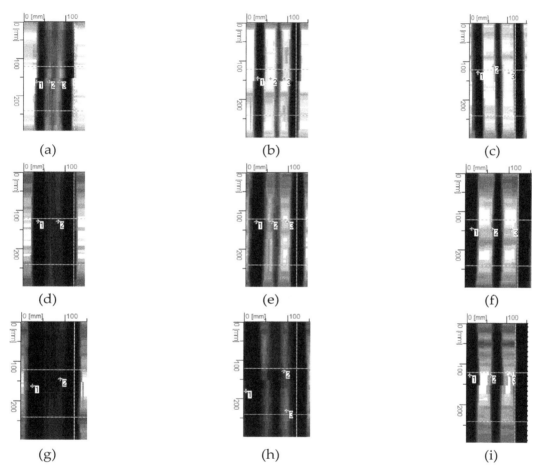

Figure 21. Results from tests conducted with the electromagnetic device (1): (**a**) S-10-20 (result: three rebars, $\varnothing_{obs}$ = 6 mm, c_{obs} = 11 mm), (**b**) S-10-30 (three rebars, $\varnothing_{obs}$ = 12 mm, c_{obs} = 12 mm), (**c**) S-10-40 (three rebars, $\varnothing_{obs}$ = 10 mm, c_{obs} = 11 mm), (**d**) S-16-20 (two rebars, $\varnothing_{obs}$ = 30 mm, c_{obs} = 32 mm), (**e**) S-16-30 (three rebars, $\varnothing_{obs}$ = 16 mm, c_{obs} = 27 mm), (**f**) S-16-40 (three rebars, $\varnothing_{obs}$ = 16 mm, c_{obs} = 24 mm), (**g**) S-20-20 (two rebars, $\varnothing_{obs}$ = 20 mm, c_{obs} = 48 mm), (**h**) S-20-30 (three rebars, $\varnothing_{obs}$ = 20 mm, c_{obs} = 47 mm), (**i**) S-20-40 (three rebars, $\varnothing_{obs}$ = 20 mm, c_{obs} = 35 mm).

Tests on specimens with rebar of 16 mm diameter and the cover of 24 mm produced results that were comparable to those for the same specimens, but on the other side (at cover thickness of 20 mm). For rebars at a closer spacing, the device wrongly identified two rebars having a diameter of 30 mm (Figure 21d). It overmeasured the thickness of concrete cover by more than 30%. For rebars at a moderate spacing, the measurement errors were comparable to those from testing the concrete cover of 20 mm thickness. In the specimens with rebars at the largest spacing, the measured diameters of rebars were the same.

A doubled increase in cover thickness in specimens with rebars of 20 mm caused a more serious errors event at the closest spacing of rebars (Figure 21g) because their number was incorrectly identified. Higher measurement deviations were found in two other specimens when compared to tests using the cover with twice as small thickness.

Figure 22 shows the results from tests that were conducted with the electromagnetic equipment (2). Similarly as for the electromagnetic device (1), some differences were noticed in test results when compared to tests using the concrete cover of 20 mm thickness. For rebars with a diameter of 10 mm at the closest spacing (Figure 22a), their correct number was not identified. At a greater spacing, the equipment identified the correct number but overmeasured the rebar diameters. The results for specimens with rebars having a diameter of 16 mm were similar to those from testing the other side of the element at the concrete cover of 20 mm. Tests on specimens with rebars having a diameter of 20 mm indicated more significant inaccuracies in measurements in comparison to tests on rebars having a diameter of 10 mm. The device failed to identify the correct number of rebars at spacing of 20 mm (Figure 22g), whereas their correct number was detected when twice the concrete thickness was used. The test results were more satisfactory at larger spacing of rebars. The diameters were slightly undermeasured and the thickness of concrete cover was correct.

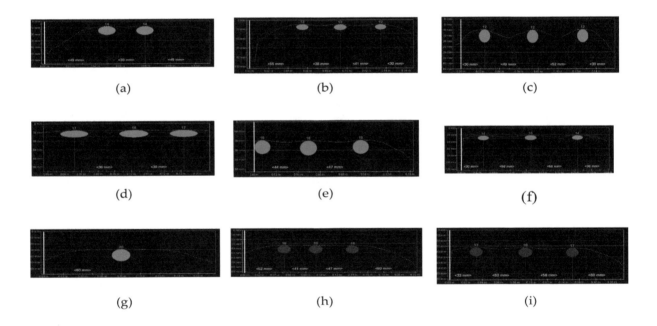

Figure 22. Results from tests performed with the electromagnetic device (2): (**a**) S-10-20 (result: two rebars, $\varnothing_{obs}$ = 14 mm, c_{obs} = 7.8 mm), (**b**) S-10-30 (three rebars, $\varnothing_{obs}$ = 13 mm, c_{obs} = 9.7 mm), (**c**) S-10-40 (three rebars, $\varnothing_{obs}$ = 12 mm, c_{obs} = 9.7 mm), (**d**) S-16-20 (three rebars, $\varnothing_{obs}$ = 17 mm, c_{obs} = 19.5, mm), (**e**) S-16-30 (three rebars, $\varnothing_{obs}$ = 16 mm, c_{obs} = 19.8 mm), (**f**) S-16-40 (three rebars, $\varnothing_{obs}$ = 14 mm, c_{obs} = 21.9 mm), (**g**) S-20-20 (one rebar, $\varnothing_{obs}$ = 20 mm, c_{obs} = 29 mm), (**h**) S-20-30 (three rebars, $\varnothing_{ob}$ = 18 mm, c_{obs} = 33.5 mm), (**i**) S-20-40 (three rebars, $\varnothing_{obs}$ = 17 mm, c_{obs} = 33.1 mm).

Figure 23 shows results from tests that were conducted with the radar equipment (3). The number of rebars was wrongly identified only in the case of specimens with rebars having a diameter of 10 mm and 20 mm, and at the closest spacing. However, measurements of concrete cover were quite accurate.

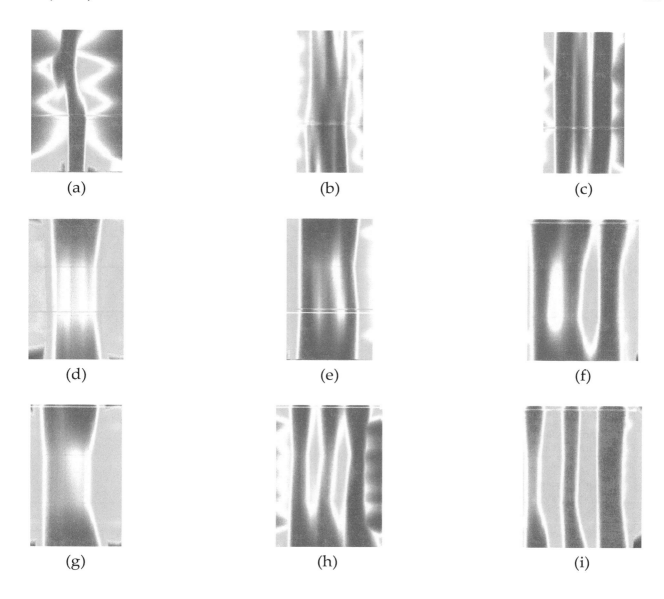

Figure 23. Results from tests performed with the radar equipment (3): (**a**) S-10-20 (result: one rebar, $c_{\mathrm{obs}} = 11$ mm), (**b**) S-10-30 (three rebars, $c_{\mathrm{obs}} = 12$ mm), (**c**) S-10-40 (three rebars, $c_{\mathrm{obs}} = 11$ mm), (**d**) S-16-20 (three rebars, $c_{\mathrm{obs}} = 23$ mm), (**e**) S-16-30 (three rebars, $c_{\mathrm{obs}} = 24$ mm), (**f**) S-16-40 (three rebars, $c_{\mathrm{obs}} = 23$ mm), (**g**) S-20-20 (two rebars, $c_{\mathrm{obs}} = 38$ mm), (**h**) S-20-30 (three rebars, $c_{\mathrm{obs}} = 40$ mm), (**i**) S-20-40 (three rebars, $c_{\mathrm{obs}} = 41$ mm).

7. Discussion of Results

7.1. AAC Precast Lintels

7.1.1. Stirrups Tests

Direct measurements were taken to analyse the accuracy of conducted tests. The thickness of reinforcement cover in holes made with a 2 mm drill was subjected to direct measurements at the tested locations (Figure 7). Results from these measurements are presented in column 2 of Table 1.

Table 1. Measurement results for stirrup reinforcement using each method.

Measuring Point no.	Cover Measured with the Method, mm			
	Direct Method, c_{obs}	Electromagnetic Device (1)	Electromagnetic Device (2)	Radar Device (3)
1	30.1	27	28.9	27
2	28.1	28	28.6	27
3	28.3	28	29.1	27
4	28.2	28	28.1	27
5	29.1	27	28.6	28
6	27.3	27	28.2	27
7	27.9	27	28.7	27
8	26.9	28	28.3	27
9	28.3	27	28.4	27
10	28.2	27	28.6	27
11	29.0	28	28.7	28
Average cover, c_{obs}:	28.3	27	28.6	27
Nominal cover, c_{nom}:	15 mm $\leq c_{nom} \leq$ 25 mm			
Calibration uncertainties Δc_{obs}	0.1	1.0	0.1	1.0
Standard deviation, s:	0.87	0.52	0.30	0.40
Standard deviation of the mean value $\bar{s} = \frac{s}{\sqrt{n}}$	0.3	0.2	0.1	0.1
Standard deviation of the calibration $S_d = \frac{\Delta c_{obs}}{\sqrt{3}}$	0.1	0.6	0.1	0.6
Standard uncertainty $S_c = \sqrt{S_d^2 + \bar{s}^2}$	0.3	1	0.1	1
Result with standard uncertainty $c_{mv} \pm S_c$	28.3 ± 0.3	27 ± 1	28.6 ± 0.1	27 ± 1
Maximum uncertainty $\Delta c = \Delta c_{obs} + 3\bar{s}$	0.9	1	0.4	1
Result with maximum uncertainty $c_{mv} \pm \Delta c$	28.3 ± 0.9	27 ± 1	28.6 ± 0.4	27 ± 1
Minimum cover c_{min}	27.4	26	28.2	26

Results from non-destructive testing of 11 inner stirrups of the beam are presented in Table 1 (three last columns of the Table), along with estimators of uncertainty according to JCGM 100:2008 [34]. The uncertainty of calibration was assumed to be equal to the accuracy of particular devices, and the uncertainty of an experimenter and the random uncertainty were neglected. The real average cover of stirrups was c_{obs} = 28.3 mm, with a standard deviation of 0.87 mm and variation coefficient of 3%. The estimated average reinforcement cover with the standard uncertainty was 3 ± 0.3 mm (the confidence level 68.3%), and, while taking into account the maximum uncertainty at the confidence level of 99.7%, the cover was equal to 28.3 ± 0.9 mm. The reinforcement nominal cover was c_{nom}

= 25 mm, and the execution deviation allowed by the standard EN 13670 [35] was Δ_{minus} = 10 mm. The result obtained from direct measurements was within the determined limits. When considering other methods, the obtained average cover was 27 mm, 28.6 mm, and 27 mm, while the standard and maximum deviations did not exceed ±0.3 mm and ±1.0 mm. The obtained minimum covers measured with every NDT methods were bigger than the determined minimum cover while taking into account the dimension deviation. The reinforcement diameter was only determined with the electromagnetic scanner 1. The obtained result was equal to 6 mm, with the real diameter of 4.5 mm. However, the device manufacturer declared the detection of rebars with a diameter from 6 mm.

The diameters of stirrups were calculated on the basis of the obtained radargram and while using the Equations (1)–(3). Moving time t_o was determined with the device, and other quantities that are required for equations were read from the radargram. For Equations (1) and (2), the calculated stirrup diameter was 5.6 mm, and in case of the Equation (3), the calculated diameter was 5.9 mm. Thus, the determined diameter was greater by 33% and 41%, respectively, than the real diameter.

7.1.2. Tests on Longitudinal Rebars

Similarly, as in case of testing stirrups, rebar cover was directly measured and column 2 of Table 2 shows the results. Measurement results are presented in Table 2 (three last columns of the Table) along with estimators of uncertainty according to JCGM 100:2008 [34]. The same assumptions were made as for the uncertainty analysis of stirrup location. There were some problems with the comparison of measurement results with the real results, because each device detected fewer rebars than the real number. Finally, it was decided to compare the test results of the electromagnetic device 1 (Figure 10a) to real measurements made on extreme rebars in holes that were drilled in places where electromagnetic measurements were taken (eight measuring points). In case of devices 2 and 3 devices that detected only one rebar, values from holes were compared to the value measured for one rebar. Thus, the same values are presented for measuring points 1 and 5, 2 and 6, 3 and 7, 4, and 8 (Figure 10a) in Table 2. The real average cover of longitudinal rebars was c_{obs} = 29.2 mm, with the standard deviation of 0.7 mm and the variation coefficient of 2.4 mm. The estimated average reinforcement cover with the standard uncertainty was 29.2 ± 0.3 mm (the confidence level 68.3%), and taking into account the maximum uncertainty at the confidence level of 99.7%, the cover was equal to 29.2 ± 0.8 mm.

The nominal cover of reinforcement was c_{nom} = 25 mm, and the execution deviation allowed by the standard EN 13670:2011 [35] was Δ_{minus} = 10 mm. Thus, the obtained result from the direct tests was within the determined limits. For other methods, medium-sized covers equal to 34 mm, 27.5 mm, and 31 mm were obtained. Standard deviations mostly did not exceed ± 1 mm and at a maximum ± 4 mm. The obtained minimum covers were bigger than minimum values while taking into account executive deviations.

In case of main reinforcement testing (beams at the bottom), the obtained deviations were considerably greater than in the case of stirrup tests (beams on the lateral surface). It should be emphasized that every device that was used in the tests did not detect the real number of rebars in the main reinforcement. Consequently, the result was burdened with a serious error resulting from the wrong reading of rebars number by research devices.

Table 2. Measurement results for the main reinforcement using each method.

Measuring Point no.	Cover Measured with the Method, mm			
	Direct Method, c_{obs}	Electromagnetic Device (1)	Electromagnetic Device	Radar Device (3)
1	29.1	30	26.2	31
2	28.8	32	27.3	30
3	28.7	35	28.1	31
4	29.2	37	28.4	32
5	29.1	30	26.2	31
6	28.9	32	27.3	30
7	29.2	35	28.1	31
8	30.9	37	28.4	32
Average cover, c_{obs}:	**29.2**	**34**	**27.5**	**31**
Nominal cover, c_{nom}:	$15\ mm \leq c_{nom} \leq 25\ mm$			
Calibration uncertainties Δc_{obs}	0.1	1.0	0.1	1.0
Standard deviation, s:	0.70	2.9	0.9	0.8
Standard deviation of the mean value, $\bar{s} = \frac{s}{\sqrt{n}}$	0.2	1.0	0.3	0.3
Standard deviation of the calibration $S_d = \frac{\Delta c_{obs}}{\sqrt{3}}$	0.1	0.6	0.1	0.6
Standard uncertainty $S_c = \sqrt{S_d^2 + \bar{s}^2}$	0.3	1	0.3	0.6
Result with standard uncertainty $c_{mv} \pm S_c$	29.2 ± 0.3	34 ± 1	27.5 ± 0.3	31 ± 0.6
Maximum uncertainty $\Delta c = \Delta c_{obs} + 3\bar{s}$	0.8	4	1.1	2
Result with maximum uncertainty $c_{mv} \pm \Delta c$	29.2 ± 0.8	34 ± 4	27.5 ± 1.1	31 ± 2
Minimum cover c_{min}	28.4	30	26.4	29

7.2. Lightweight Concrete Specimens

Table 3 presents the results for lightweight concrete models that were tested on the side where the cover thickness was constant and equal to 20 mm, whereas Table 4 shows the results for the variable thickness of the cover. A shaded field shows measured values that were the same as the nominal values in models. The analysis of values in the table indicates that the number of rebars having a diameter of 10, 16, and 20 mm could be identified at a spacing greater than 20 mm and cover thickness greater than 10 mm while using the electromagnetic equipment (1) and (2). Measured values closest to nominal values of diameters were observed at a rebar spacing of 30 mm. The equipment showed similar diameters, but the false number of rebars at a closer spacing. The average ratio $\varnothing_{obs}/\varnothing_{nom}$ in accordance with electromagnetic methods (1) and (2), was 1.1 and 0.93, respectively, for the cover of 20 mm, and 1.08 and 1.02 for the cover of variable thickness. If the accuracy of indications of the electromagnetic equipment of the order of ca. ±1.0 mm was taken into account during the analysis of measurements, then the results from measuring diameters with these methods could be considered as reliable at a spacing larger than 20 mm and for diameters that are greater than 10 mm. Rebar diameters could not be read in the case of using the radar equipment (3). Therefore, the diameters were calculated in a similar way to precast lintel beam, while using Equations (1)–(3). More satisfactory compliance with nominal values was achieved using Equations (1) and (2). Table 3 shows those values. The measured diameters were greater by ca. 15% than nominal values.

Table 3. Test results for lightweight concrete specimens (at constant cover thickness of 20 mm).

Geometry of Models				Results from Tests Conducted with the Equipment (1), (2) or (3)								
Reinforcement Diameter (mm)	Number of Rebars n_{nom}	Rebar Spacing a_{nom}, mm	Reinforcement Cover c_{nom}, mm	$\varnothing_{obs}$, mm			n_{obs}, mm			c_{obs}, mm		
				(1)	(2)	(3) *	(1)	(2)	(3)	(1)	(2)	(3)
		20		8	8	11.9	2	1	3	25	20	22
10	3	30	20	8	10	11.8	3	3	3	20	19.3	22
		40		10	9	11.5	3	3	3	21	19.3	21
		20		30	17	18.8	2	3	3	25	15.5	22
16	3	30	20	16	15	18.2	3	3	3	18	16.4	21
		40		20	14	17.8	3	3	3	20	16.5	20
		20		20	19	23.3	3	3	3	25	15	22
20	3	30	20	20	19	23.1	3	3	3	18	15.4	20
		40		20	18	22.8	3	3	3	22	18.3	20

(1)– electromagnetic device (Figure 13a), (2) – electromagnetic device (Figure 13b), (3) – radar device (Figure 13c), * – values obtained from Equations (1) and (2).

Table 4. Test results for lightweight concrete specimens (at variable cover).

Geometry of Models				Results from Tests Conducted with the Equipment (1), (2) or (3)								
Reinforcement Diameter (mm)	Number of Rebars n_{nom}	Rebar Spacing a_{nom}, mm	Reinforcement Cover c_{nom}, mm	$\varnothing_{obs}$, mm			n_{obs}, mm			c_{obs}, mm		
				(1)	(2)	(3) *	(1)	(2)	(3)	(1)	(2)	(3)
		20		6	14	12.0	3	2	1	11	7.8	11
10	3	30	10	12	13	11.9	3	3	3	12	9.7	12
		40		10	12	11.6	3	3	3	11	9.7	11
		20		30	17	18.9	2	3	3	32	19.5	23
16	3	30	24	16	16	18.8	3	3	3	27	19.8	24
		40		16	14	18.2	3	3	3	24	21.9	23
		20		20	20	23.4	2	1	2	48	29	38
20	3	30	40	20	18	23.2	3	3	3	47	33.5	40
		40		20	17	22.7	3	3	3	35	33.1	41

(1) – electromagnetic device (Figure 13a), (2) – electromagnetic device (Figure 13b), (3) – radar device (Figure 13c), * – values obtained from Equations (1) and (2).

The rebar covers could be measured while using all of those methods. At the nominal cover with thickness of 10 mm, average values of 11.3, 9.07, and 11.3 could be read from the equipment (1), (2), and (3). For specimens with a 20 mm cover, the average measured values of cover thickness were 21.6, 17.3, and 21.1 mm, depending on the used device. for slightly greater cover (24 mm), the identified average values were 27.7, 20.4, and 23.3 mm; whereas, a double increase in concrete thickness to 40 mm resulted in average values that were equal to 43.3, 31.9, and 39.7 mm. Thus, the radar equipment was found to measure concrete thickness with the greatest precision.

In all measurements of covers, the difference from the nominal value was ±6,5 mm, that is, within the limits of allowable execution deviations Δ_{minus} = 10 mm specified by the standard EN 13670:2011 [35].

Tests that were conducted on lightweight concrete specimens confirmed the results obtained from tests on precast lintel beam. The greatest inaccuracies of measurements were observed for rebars with lower diameters (≤10 mm) at close spacing (≤20 mm), and with great diameters (20 mm) at close spacing (20 mm) and great depth (40 mm). For greater diameters (≥16 mm), even close spacing did not affect the detection of the reinforcement, including the number of rebars, their diameter, and the thickness of concrete cover. The radar equipment provided a more precise size of concrete cover when compared to the electrical equipment. Rebar diameters that were calculated from the radargram were overcalculated by ca. 15 in comparison to nominal values. However, considerably better results in measuring diameters were noticed in the case of electromagnetic diameters.

8. Conclusions

The conducted tests demonstrated the occurrence of some limitations of two popular methods for detecting reinforcement location in the structure. The devices were able to correctly detect rebars, even those of small diameters, providing that they were located at an adequate spacing. Devices for scanning reinforcement had some problems with defining the correct quantity and diameter of rebars placed close to each other, at a distance shorter than 2–3 diameters.

The results obtained for covers measured with electromagnetic and radar devices did not significantly differ in terms of an average value. For the real cover, the differences did not exceed the acceptable level of few percent (<5%). There was no clear tendency to state that any of the methods artificially distorted indications, for instance, due to the simplification for measurement method validation. However, it should be emphasized that the minimal covers determined by all non-destructive techniques were bigger than the minimal cover specified with regard to size deviations (accepted by standard EN 13670:2011 [35]) and obtained by direct measurements. Thus, the tests on existing building can generate an error that is greater than the acceptable execution deviation. This can lead to wrong conclusions regarding the accuracy of a building. Measurements that were taken in the existing structure should be analysed with caution because the incorrect interpretation of measurement results for reinforcement location in the cross-section can cause a reduction in safety coefficient for steel $\gamma_{S, red1}$ from 1.15 to 1.1 (acc. to EC-2 [33]) and an unintentional reduction of safety level in particularly significant support zones of reinforced concrete structures and prestressed structures.

The tests indicated that a measurement error was significant, particularly for measuring structures that were reinforced with small diameter bars at a small spacing. This could lead to improper interpretation of test results and, consequently, to wrong calculation analyses for structures. A similar error could be observed in the tests on reinforcement of greater diameter, with rebars at a closer spacing when the concrete thickness was twice the rebar diameter or spacing.

To sum it up, the following recommendations for performing tests on reinforcement location can be considered as the practical application of the obtained results:

a)   while detecting the cover with thickness < 20 mm, errors in determining the number and diameter of rebars can be expected regardless of the used test equipment,

b)   while detecting the reinforcement at a depth > 20 mm, the employed equipment can correctly determine the number of rebars if the spacing is greater than 20 mm.

c)   NDT equipment is not suitable for detecting the reinforcement in support zones of beams and slabs, and in joints where reinforcement is the densest, and

d)   this equipment is perfect for detecting the longitudinal reinforcement in slabs, and stirrups in beams.

Studies aiming at connecting both methods, consisting in connecting scans that were taken with devices operating according to the electromagnetic and radar method seem to be reasonable. Hybrid devices are likely to generate accurate results, especially in terms of measuring the reinforcement diameters.

Author Contributions: Conceptualization, Ł.D. and R.J.; methodology, Ł.D., R.J. and W.M.; formal analysis, Ł.D. and R.J. and W.M.; investigation, Ł.D., W.M. and R.J.; writing—original draft preparation, Ł.D.; writing—review and editing, Ł.D. and R.J.; visualization, W.M. and Ł.D.; supervision, R.J. and Ł.D.

Acknowledgments: The authors would like to express particular thanks to Solbet Sp. z o.o. company, for supply of materials used during the research works. We would also like to thank Hilti, Proceq and Viateco for providing research equipment.

References

1.   Malhorta, V.M.; Carino, N.J. *Handbook on Nondestructive Testing of Concrete*, 2nd ed.; CRC Press LCC, ASTM International: Boca Raton, FL, USA; London, UK; New York, NY, USA; Washington, DC, USA, 2004.
2.   Bungey, J.H.; Millard, S.G.; Grantham, M.G. *Testing of Concrete in Structures*, 4th ed.; Taylor & Francis: London, UK; New York, NY, USA, 2006.

3. Drobiec, Ł.; Jasiński, R.; Piekarczyk, A. *Diagnostic Testing of Reinforced Concrete Structures*; Methodology, Field Tests, Laboratory Tests of Concrete and Steel; Wydawnictwo Naukowe PWN: Warszawa, Poland, 2013. (In Polish)

4. Hoła, J.; Schabowicz, K. State-of-the-art non-destructive methods for diagnostic testing of building structures–anticipated development trends. *Arch. Civ. Mech. Eng.* **2010**, *10*, 5–18. [CrossRef]

5. Drobiec, Ł. *Diagnosis of Industrial Structures*; Materiały Budowlane No 2015/2; Sigma-Not: Warsaw, Poland, 2015; pp. 32–34.

6. Zima, B.; Rucka, M. Detection of debonding in steel bars embedded in concrete using guided wave propagation. *Diagnostyka* **2016**, *3*, 27–34.

7. Hoła, J.; Bień, J.; Schabowicz, K. Non-destructive and semi-destructive diagnostics of concrete structures in assessment of their durability. *Bull. Pol. Acad. Sci.* **2015**, *63*, 87–96. [CrossRef]

8. Ma, X.; Liu, H.; Wang, M.L.; Birken, R. Automatic detection of steel rebar in bridge decks from ground penetrating radar data. *J. Appl. Geophys.* **2018**, *158*, 93–102. [CrossRef]

9. ISO/IEC. *Guide 99:2007 International Vocabulary of Metrology—Basic and General Concepts and Associated Terms*; ISO: Geneva, Switzerland, 2007.

10. Chady, T.; Frankowski, P.K. Electromagnetic evaluation of reinforced concrete structure. In Proceedings of the AIP Conference, Denver, CO, USA, 15–20 July 2013; Volume 32, pp. 1355–1362.

11. Drobiec, Ł.; Górski, M.; Krzywoń, R.; Kowalczyk, R. Comparison of non-destructive electromagnetic methods of reinforcement detection in RC structures. In Proceedings of the Challenges for Civil Construction (CCC 2008), Porto, Portugal, 16–18 April 2008.

12. Ombres, L.; Verre, S. Shear performance of FRCM strengthened RC beams. *ACI Spec. Publ.* **2018**, *324*, 7.1–7.22.

13. Micelli, F.; Cascardi, A.; Marsano, M. Seismic strengthening of a theatre masonry building by using active FRP wires. Brick and Block Masonry—Trends, Innovations and Challenges. In Proceedings of the 16th International Brick and Block Masonry Conference, Padova, Italy, 26–30 June 2016; pp. 753–761.

14. Sivasubramanian, K.; Jaya, K.P.; Neelemegam, M. Covermeter for identifying cover depth and rebar diameter in high strength concrete. *Int. J. Civ. Struct. Eng.* **2013**, *3*, 557–563.

15. Lachowicz, J.; Rucka, M. Application of GPR method in diagnostics of reinforced concrete structures. *Diagnostyka* **2015**, *16*.

16. Lachowicz, J.; Rucka, M. 3-D finite-difference time-domain modelling of ground penetrating radar for identification of rebars in complex reinforced concrete structures. *Arch. Civ. Mech. Eng.* **2018**, *18*, 1228–1240. [CrossRef]

17. Agred, K.; Klysz, G.; Balayssac, J.P. Location of reinforcement and moisture assessment in reinforced concrete with a double receiver GPR antenna. *Constr. Build. Mater.* **2018**, *188*, 1119–1127. [CrossRef]

18. Shaw, M.R.; Millard, S.G.; Molyneaux, T.C.K.; Taylor, M.J.; Bungey, J.H. Location of steel reinforcement in concrete using ground penetrating radar and neural networks. *Ndt E Int.* **2005**, *38*, 203–212. [CrossRef]

19. Chang, C.W.; Lin, C.H.; Lien, H.S. Measurement radius of reinforcing steel bar in concrete using digital image GPR. *Constr. Build. Mater.* **2009**, *23*, 1057–1063. [CrossRef]

20. Shihab, S.; Al-Nuaimy, W. Radius estimation for cylindrical objects detected by ground penetrating radar. *Subsurf. Sens. Technol. Appl.* **2005**, *6*, 151–166. [CrossRef]

21. Ristic, A.V.; Petrocacki, D.; Govedarica, M. A New Method to Simultaneously Estimate the Radius of a Cylindrical Object and the Wave Propagation Velocity from GPR Data. *Comput. Geosci.* **2009**, *35*, 1620–1630. [CrossRef]

22. Idi, B.Y.; Kamarudin, M.N. Utility Mapping with Ground Penetrating Radar: An Innovative Approach. *J. Am. Sci.* **2011**, *7*, 644–649.

23. Zanzi, L.; Arosio, D. Sensitivity and accuracy in rebar diameter measurements from dual-polarized GPR data. *Constr. Build. Mater.* **2013**, *48*, 1293–1301. [CrossRef]

24. Mechbal, Z.; Khamlichi, A. Determination of concrete rebars characteristics by enhanced postprocessing of GPR scan raw data. *Ndt E Int.* **2017**, *89*, 30–39. [CrossRef]

25. Wei, J.S.; Hashim, M.; Marghany, M. New approach for extraction of subsurface cylindrical pipe diameter and material type from ground penetrating radar image. In Proceedings of the 1st Asian Conference on Remote Sensing (ACRS 2010), Hanoi, Vietnam, 1–5 November 2010; Volume 2, pp. 1187–1193.

26. Alhsanat, M.B.; Wan Hussin, W.M.A. A New Algorithm to Estimate the Size of an Underground Utility via Specific Antenna. In Proceedings of the PIERS, Marrakesh, Maroko, 20–23 March 2011; pp. 1868–1870.

27. Wiwatrojanagul, P.; Sahamitmongkol, R.; Tangtermsirikul, S.; Khamsemanan, N. A new method to determine locations of rebars and estimate cover thickness of RC structures using GPR data. *Constr. Build. Mater.* **2017**, *140*, 257–273. [CrossRef]

28. Wiwatrojanagul, P.; Sahamitmongkol, R.; Tangtermsirikul, S. A method to detect lap splice in reinforced concrete using a combination of covermeter and GPR. *Constr. Build. Mater.* **2018**, *173*, 481–494. [CrossRef]

29. Drobiec, Ł.; Jasiński, R.; Mazur, W. Precast lintels made of autoclaved aerated concrete—Test and theoretical analyses. *Cem. Wapno Beton* **2017**, *5*, 339–413.

30. Mazur, W.; Drobiec, Ł.; Jasiński, R. Research of Light Concrete Precast Lintels. *Procedia Eng.* **2016**, *161*, 611–617. [CrossRef]

31. Mazur, W.; Drobiec, Ł.; Jasiński, R. Research and numerical investigation of masonry—AAC precast lintels interaction. *Procedia Eng.* **2017**, *193*, 385–392. [CrossRef]

32. Mazur, W.; Jasiński, R.; Drobiec, Ł. Shear Capacity of the Zone of Supporting of Precast Lintels Made of AAC. *IOP Conf. Ser. Mater. Sci. Eng.* **2019**, *471*, 052070. [CrossRef]

33. EN 1992-1-1. *Eurocode 2: Design of Concrete Structures—Part 1-1: General Rules and Rules for Buildings*; European Committee for Standardization (CEN): Brussels, Belgium, 2004.

34. JCGM 100:2008. *Evaluation of Measurement Data. Guide to the Expression of Uncertainty in Measurement*; Joint Committee for Guides in Metrology: Sevres, France, 2008.

35. EN 13670:2011. *Execution of Concrete Structures*; European Committee for Standardization (CEN): Brussels, Belgium, 2011.

Permissions

All chapters in this book were first published in MDPI; hereby published with permission under the Creative Commons Attribution License or equivalent. Every chapter published in this book has been scrutinized by our experts. Their significance has been extensively debated. The topics covered herein carry significant findings which will fuel the growth of the discipline. They may even be implemented as practical applications or may be referred to as a beginning point for another development.

The contributors of this book come from diverse backgrounds, making this book a truly international effort. This book will bring forth new frontiers with its revolutionizing research information and detailed analysis of the nascent developments around the world.

We would like to thank all the contributing authors for lending their expertise to make the book truly unique. They have played a crucial role in the development of this book. Without their invaluable contributions this book wouldn't have been possible. They have made vital efforts to compile up to date information on the varied aspects of this subject to make this book a valuable addition to the collection of many professionals and students.

This book was conceptualized with the vision of imparting up-to-date information and advanced data in this field. To ensure the same, a matchless editorial board was set up. Every individual on the board went through rigorous rounds of assessment to prove their worth. After which they invested a large part of their time researching and compiling the most relevant data for our readers.

The editorial board has been involved in producing this book since its inception. They have spent rigorous hours researching and exploring the diverse topics which have resulted in the successful publishing of this book. They have passed on their knowledge of decades through this book. To expedite this challenging task, the publisher supported the team at every step. A small team of assistant editors was also appointed to further simplify the editing procedure and attain best results for the readers.

Apart from the editorial board, the designing team has also invested a significant amount of their time in understanding the subject and creating the most relevant covers. They scrutinized every image to scout for the most suitable representation of the subject and create an appropriate cover for the book.

The publishing team has been an ardent support to the editorial, designing and production team. Their endless efforts to recruit the best for this project, has resulted in the accomplishment of this book. They are a veteran in the field of academics and their pool of knowledge is as vast as their experience in printing. Their expertise and guidance has proved useful at every step. Their uncompromising quality standards have made this book an exceptional effort. Their encouragement from time to time has been an inspiration for everyone.

The publisher and the editorial board hope that this book will prove to be a valuable piece of knowledge for researchers, students, practitioners and scholars across the globe.

List of Contributors

Kostiantyn Protchenko, Fares Zayoud, Marek Urbański and Elżbieta Szmigiera
Faculty of Civil Engineering, Warsaw University of Technology, 00-637 Warsaw, Poland

Łukasz Zdanowicz, Szymon Seręga, Marcin Tekieli and Arkadiusz Kwiecień
Faculty of Civil Engineering, Cracow University of Technology, 31-155 Cracow, Poland

Bohdan Stawiski
Faculty of Environmental Engineering and Geodesy, Wrocław University of Environmental and Life Sciences, pl. Grunwaldzki 24, 50-363 Wrocław, Poland

Tomasz Kania
Faculty of Civil Engineering, Wrocław University of Science and Technology, Wybrzeże Wyspiańskiego 27, 50-370 Wrocław, Poland

Tomasz Gorzelańczyk, Michał Pachnicz, Adrian Różański and Krzysztof Schabowicz
Faculty of Civil Engineering, Wrocław University of Science and Technology, Wybrzeże Wyspiańskiego 27, 50-370 Wrocław, Poland

Xiaohu Wang, Yu Peng, Jiyang Wang and Qiang Zeng
College of Civil Engineering and Architecture, Zhejiang University, Hangzhou 310058, China

Erwin Wojtczak and Magdalena Rucka
Department of Mechanics of Materials and Structures, Faculty of Civil and Environmental Engineering, Gdansk University of Technology, Narutowicza 11/12, 80-233 Gdansk, Poland

Dominik Logoń
Faculty of Civil Engineering, Wrocław University of Science and Technology, 50-377 Wrocław, Poland

Karol Grębowski
Department of Technical Fundamentals of Architectural Design, Faculty of Architecture, Gdansk University of Technology, Narutowicza 11/12, 80-233 Gdansk, Poland

Krzysztof Wilde
Department of Mechanics of Materials and Structures, Faculty of Civil and Environmental Engineering, Gdansk University of Technology, Narutowicza 11/12, 80-233 Gdansk, Poland

Tomasz Nowak, Anna Karolak, Maciej Sobótka and Marek Wyjadłowski
Wroclaw University of Science and Technology, Wybrzeze Wyspianskiego 27, 50-370 Wroclaw, Poland

Piotr Mackiewicz and Antoni Szydło
Faculty of Civil Engineering, Wrocław University of Science and Technology, 50-370 Wrocław, Poland

Aleksandra Krampikowska and Grzegorz Świt
Department of Strength of Materials, Concrete and Bridge Structures, Kielce University of Technology, Al. 1000-lecia PP 7, 25-314 Kielce, Poland

Robert Pała and Ihor Dzioba
Faculty of Mechatronics and Mechanical Engineering, Department of Machine Design, Kielce University of Technology, Al. 1000-lecia PP 7, 25-314 Kielce, Poland

Zbigniew Ranachowski, Przemysław Ranachowski and Tomasz Dębowski
Experimental Mechanics Division, Institute of Fundamental Technological Research, Polish Academy of Sciences, Pawińskiego 5B, 02-106 Warszawa, Poland

Łukasz Drobiec, Radosław Jasiński and Wojciech Mazur
Department of Building Structures, Silesian University of Technology; ul. Akademicka 5, 44-100 Gliwice, Poland

Index

Printed in the USA
CPSIA information can be obtained
at www.ICGtesting.com
LVHW082346150324
774517LV00005B/768

9 781639 897575